AF559855

Environment, Biodiversity and Traditional System

Environment, Biodiversity and Traditional System

— *Editors* —

D.R. Khanna
G.S. Solanki
S.K. Pathak

— *Associate Editor* —

Fouzia Ishaq

2014

BIOTECH BOOKS®

ISBN 978-81-7622-307-2

Published by: **BIOTECH BOOKS®**
4762-63/23, Ansari Road, Darya Ganj,
New Delhi - 110 002
Phone: +91-011-23262132
E-mail: biotechbooks@yahoo.co.in

Printed at: **Chawla Offset Printers**
Delhi - 110 052

PRINTED IN INDIA

Preface

Earth and its environment provide livelihood support system for existence of human being and other living organisms. Each one of us relies on the nature for a range of services and resources. The functioning of nature in its own way caters the need of everyone for their survival but the men's greed has change the nature's cycle. Short term benefits and industrialization has become serious culprit of ecological and environmental catastrophe. Advancement in technology, expansion of industries, globalization and urbanization has led to economic and social development at the cost of public health and environment. Environmental degradation, global warming, recession of glaciers, scarcity of potable water, rise in sea level and increasing air and water pollution are some of the consequences of human interference with nature. Human health is jeopardizing in the present situation. It is time to realize that nature has nothing to loose, but it is our survival which is threatened.

Earth is not the property of humans, every organism on the earth has equal right to live, survive and flourish to continue their progeny. Human beings and other life form on the planet are complementary to each other and support for their survival. The range of life forms and complexities within and between them is broadly known as biological diversity. Rosen (1985) gave the abbreviation biodiversity for biological diversity. The predicted number of total species present on earth varies from 5-50 million, but known and described number of the species is between 1.7 and 1.8 millions which is less than 15 % of actual number. Mammals make up one of the smallest groups, just 5490 species. Human civilization and its survival depends directly or indirectly upon biodiversity for their basic needs like food, fodder, fuel, fibre, fertilizer, timber, leather, medicines, and several raw materials. The diversity of life on the earth provides many ecological services that maintain ecosystem health.

Ethnozoology is the study of the past and present interrelationships between human cultures and the animals in their environment. It deals with the animals and their uses through the indigenous knowledge system by the ethnic groups that is inherited in their culture. It means the traditional knowledge and customs of the people based on animals and other living forms. Wild species of animals and plants have been the source for the survival of human. Thus, the survival of wildlife and people are intricately linked. Socio-economic conditions of ethnic group are often dependent on good management of wildlife and other natural resources, and the survival of many animal species on better understanding of use and their management.

Hunting of animals from wilderness for sustenance has been the way of life in many tribal communities or indigenous groups. In recent decades the sustenance has taken over by commercial hunting, the most common cause of animals extinction followed by habitat destruction. Hunting of wild animals for meat on socio-cultural and magical and religious occasion has been a part of traditional lifestyle in many tribal societies in African and Asian countries. Traditional hunting techniques have been documented in different geographical regions in accordance with the cultural entities of ethnic groups.

All the aspects are illustrated in the book in the form of chapters. Each chapter contains valuable information for conservation biologist and wildlife managers.

D.R. Khanna

G.S. Solanki

S.K. Pathak

Contents

Environment, Biodiversity and Traditional System, 2014 *Pages 1–5*
ISBN 978-81-7622-307-2
Editors: D.R. Khanna, G.S. Solanki and S.K. Pathak
Published by: BIOTECH BOOKS, NEW DELHI

Chapter 1

Rapid Analysis of Spatial α, β and γ-diversities of Birds in a Landmosaic in Manipur

Rajkumari Ashalata Devi and Laishangbam Sanjit*

**Assistant Professor of English, Eco-criticism Lab.*
Department of English, Manipur University, Canchipur, Imphal
E-mail: -drslai@rediffmail.com

ABSTRACT

We reported the results of a rapid analysis of spatial α, β and γ diversities of birds in four distinct landscape elements in Bishnupur district and its adjoining areas of Manipur, North-East India. The spatial relations among α (local), β (turnover), and γ (landscape) diversity of birds in a sub-tropical land mosaic were also assessed. We recorded 76 species of birds from four transects (1 km each) placed in the four landscape elements and sampled four times during January, February and March, 2003. α-diversities for each landscape element are (28) forest, (25) suburban, (34) wetland, and (21) agricultural field. The average α-diversity is 27 and γ diversity is 76. β-diversity is (1.4^{-1}) or (0.7). Forest area has (19) unique species whereas suburban has only (9); and wetland area has also (19) unique species whereas agricultural field has only (6) such species. Our study concluded that the studied area is a topographic and habitat heterogeneous land mosaic, to which birds respond well; and bird species turnover between landscape elements contribute a lot to the overall spatial diversity of birds in the studied land mosaic.

Key words: Birds, Subtropical land mosaic, Landscape elements, Spatial α, β and γ-diversities, Unique species.

Introduction

Landscape preservation has been presented as a practical alternative to species-based approaches for the conversation of biological diversity (Noss, 1983). A variable number of distinct landscape elements or habitat types usually combine to form a heterogeneous and dynamic landscape. Although the landscape has certain stability in the number of species it contains, this stability is not found among the elements that make up the landscape. Study of the, β, and γ-components of biodiversity can be useful for measuring and monitoring the effects of human activities on biodiversity (Halffter, 1998). The γ–diversity of a landscape may remain unmodified despite significant changes in α and β–diversities (Halffter, 1998). The three components of diversity were originally proposed for studies of changes along continuous environmental gradients (Whittaker, 1972, 1977; Routledge, 1977; Cody 1975, 1993). But they have been hardly applied to spatially heterogeneous mosaic-like landscapes (Moreno and Halffter, 2001). The protection of landscapes depends on their effective identification and evaluation. But landscapes are changing systems that can be defined at a number of scales (Jongman, *et al.*, 1987; Turner, *et al.*, 1989; Forman, 1995), and they are constantly modified by many human activities.

Conservation biology must centre its attention on those processes that take place in landscapes with distinct grades of human modification (Halffter, 1998). Though some records of mid-winter waterfowl census of the Loktak Lake are available; little is known about the avifauna of Manipur. Moreover, there are hardly any published works or recent records of bird surveys on the status of avifauna of the state. And it has come to our knowledge that a lot of alterations, modifications and development works (human activities) are taking place both within and between the elements of the studied landscape. In this present work, we focus on the spatial dynamic of species diversity in a landscape-mosaic.

Material and Methods

Our study area lies in the southern part of the Manipur basin (24'25"N-24'41"N latitudes and 93'20"E-94'47"E longitudes) in the Eastern Himalayan region (North-east India). Elevation ranges from 765m to 1250m (asl). Our landscape elements are patches of forested areas (4-km.sq.), suburban Moirang (5-km.sq.), agricultural fields (15-km.sq.) and some part (about 15-km.sq.) of the Loktak lake, a Ramsar site. Census of birds were done by walking along a transect line (1-km) dropped in each landscape elements. Surveys were conducted between 6: 30 a.m. and 9: 00 a.m. in the morning and between 3: 00 p.m. and 5: 00 p.m. in the evening. All guidelines established for bird census techniques were followed to reduce observer bias and the effects of environmental variations on the results.

Spatial α–diversity is the total number of species recorded in the total number of replicates per habitat or landscape elements. Spatial β–diversity is calculated by Whittaker's diversity index:

$$\beta = (a1/a2)-1$$

where, a1 is the total number of species and a2 is the mean number of species in the lists.

Spatial γ–diversity is calculated with the formula proposed by Schluter and Ricklefs (1993):

$$\gamma = \alpha 1 \times \beta \times n$$

where α 1 is the average α – diversity in the n habitats that form the region, β is the inverse of average number of habitats occupied by a species, and n is the total number of habitats (or LE's).

Results and Discussion

Unique species are referred to as those species which are exclusively found in each habitat. We found the highest number of unique species occurred in the forest areas and the wetland areas (19 each). This indirectly shows the major contributions of both landscape elements (or habitat types) to the overall landscape mosaic. Wetlands have been widely cited for their disproportionate contribution to biological diversity in a variety of landscapes (La Roe, 1995). And tropical rain forests are one of the richest areas in the world in terms of biodiversity. The remaining two landscape elements showed a moderately high number of unique species: suburban (9) and agricultural field (6). This can be interpreted as they also have some contribution to the overall landscape mosaic. Now we assume that our studied landscape is a heterogeneous land mosaic with lots of disparities of habitats. To confirm our assumption, we will look into the spatially α, β and γ –diversities of our landscape mosaic.

Table 1.1: Total number of bird species, individuals, unique species and area for the different landscape elements of Bishnupur district and adjoining areas of Manipur.

Landscape Element	*No. of Species*	*Individuals Recorded*	*Unique Species*	*Area in Sq. km.*
Forest area	28	106	19	4
Suburban area	25	132	9	5
Wetland area	34	170	19	15
Agricultural field	21	145	6	15
Total study area	76	553	53	39

Table 1.2: Whittaker's measures of spatial β-diversity between pairs of landscape elements in the land mosaic.

Habitats	*Forest*	*Suburban*	*Wetland*	*Agricultural Field*
Forest	*	0.8113	0.8709	0.7959
Suburban	0.8113	*	0.6949	0.6522
Wetland	0.8709	0.6949	*	0.6727
Agricultural field	0.7959	0.6522	0.6727	*

Spatial α-diversities of each landscape elements were 28 for forest, 25 for suburban, 34 for wetland and 21 for agricultural field. The average α diversity was

27, and spatial γ-diversity was 76. Average α-diversity was quite low as compared to spatial γ-diversity. In homogenous landscapes, the number and identity of the species occurring would be very similar to those that are found in the richest element (with the highest γ-value). On the other hand, in heterogeneous landscapes the number of species (76) would be clearly superior to that observed in the richest of its elements (34). This richness at the landscape level would be due to a strong complementarily between distinct landscape elements. Our mean spatial β-diversity was 0.74965 (minimum: 0.6522, maximum: 0.8709) with Whittaker's index. This very high value (0.74965 or 1.4) of spatial β-diversity indicates that our studied landscape is a topographic and habitat heterogeneous area; and birds are sensitive to environmental changes within the spatial scale of the landscape. Our results suggest that the spatial γ-diversity of birds in a palaeotropical landmosaic is largely influenced by species turnover between landscape elements.

Acknowledgement

The Authours would like to thank all the Staff members of English Department, Manipur University, both teaching as well as non-teaching members, for their valuable help in establishing the Ecocrticism Laboratory at The Department. The second authour would like thank Professor (Dr.) Dinesh Bhatt, of the Department of Zoology and Environmental Sciences, Gurukula Kangri University for his valuable guidance during the work.

Rferences

Cody, M.L., 1975. Towards a theory of continental species diversity. Bird distribution over Mediterrannean habitat gradients. In: *Ecology and Evaluation of Communities*, (Eds.) M.L. Cody and J.M. Diamond. Massachussetts–Harvard University Press, Cambridge (The Belknep Press of Cambridge), pp. 214–257.

Cody, M.L., 1993. Bird diversity components within and between habitats in Australia. In: *Species Diversity in Ecological Communities: Historical and Geographical Perspectives*, (Eds.) R.E. Ricklefs and D. Schutler. University of Chicago Press, pp. 147–158.

Formann, R.T., 1995. *Land Mosaics: The Ecology of Landscape and Regions*. Cambridge University Press, Cambridge, U.K., New York, USA.

Halffter, G., 1998. A strategy for measuring landscape biodiversity. *Biology International*. 36: 3–17.

Jongman, R.H.G., Ter Braak, C.J.F. and Van Tongmen, V. (eds), 1987. *Data Analysis in Community and Landscape Ecology*. Pudoc, Wageningen, The Netherland.

Moreno, C.E. and Halffter, G., 2001. On the measure of sampling effort used in species accumulation curves. *J.Appl. Ecol.*, **38**: 487–490.

Noss, R., 1983. A regional landscape approach to maintain diversity. *Bioscience*, 33: 700–706.

Routledge, R.D., 1997. On Whittaker's components of species diversity. *Ecology*, 58: 1120–1127.

Turner, M.G., O'Neill, R.V. Gardner, R.H. and Milne, B.T., 1989. Effects of changing spatial scale on the analysis of landscape pattern. *Landscape Ecology*. 3: 153–162.

Whittaker, R.H., 1972. Evaluation and measurement of species diversity. *Taxon*. 21: 213–251.

Whittaker, R.H., 1977. Evolution of species diversity in land communities.In: Hecht, M.K., Steere, W.C. and Wallace, B. (eds.) *Evolutionary Biology*. Vol.10: pp.1–67. Plenum Press, New York.

Environment, Biodiversity and Traditional System, 2014 Pages 7-18
ISBN 978-81-7622-307-2
Editors: D.R. Khanna, G.S. Solanki and S.K. Pathak
Published by: BIOTECH BOOKS, NEW DELHI

Chapter 2

Diversity and Status of Lycaenids (Lepidoptera : Lycaenidae) in Central Nepal

B. Khanal[1], M.K. Chalise[2] and G.S. Solanki[3]*

[1]*Natural History Museum, Swayambu, Kathamdu, Nepal*
[2]*Central Department of Zoology, Tribhuvan University, Nepal*
[3]*Department of Zoology, Mizoram University, Mizoram, India*

ABSTRACT

Out of the reported eleven families of butterflies from Nepal, members of the family Lycaenidae are widely distributed at different ecological zones along vertical range of the central region. These small blues are very sensitive to their habitat preferences. This study considers the methodology as adapted by the capture and release method in two ecologically diverse but physically inter-linked districts of Bhaktapur and Sindhupalchok of central Nepal that shows interesting features of distribution for butterflies within the elevation of 1300 m to 3000 m. This paper deals with 61 species of Lycaenids from those district studied between 2010 and 2011 considering different seasons. The study also reveals the rare and endangered Lycaenidae such as Esakiozephyrus mandara, Ramelana jangala, and Tajuria yajna. Conservation issues related to Lycaenids have also been presented in this paper.

Keywords: Families, Butterflies, Distributed, Elevation, Districts.

Lepidoptera, an order of butterflies and moths are the second largest order among insects and is made up of approximately 150,000 species so far known (Haribal,

* *Corresponding Author:* E-mail: baya2000@live.com

1992). Of the entire Lepidopteron insects known till date, 17820 species are butterflies and rest are moth species (Shields, 1989).

Lycaenidae which is one of the fifteen families of butterflies constitutes about 41 per cent of the total butterfly species that equals to 6000 species worldwide. Family Lycaenidae has different subfamilies like Polymmatini, Lycaeninae, Theclinae, Miletinae, Lipteninae, Curetinae and Poritiinae. These small butterflies are brightly coloured and some members display metallic gloss. Forelegs of the males are reduced and lack claws.

The geographical complexity with altitudinal changes at the lower flat land to the high mountain areas in this country offer several habitat choices for diverse butterfly species. The total record of 653 species of butterflies in this country is classified into 11 families. Family Lycaenidae represents maximum diversity of 176 species which are classified into 72 Genera (Smith, 1989). Rasuwa district which is located at the interconnecting zone of Nuwakot and Sindhupalchok districts represented the highest diversity record of Lycaenidae which included 36 species within the elevations of 1500 m to 4300 m (Khanal *et al.*, 2012).

Material and Methods

Study Area

Nuwakot District

Climate of Nuwakot district is subtropical to warm temperate. This district covers the area of 1193 sq.km with geographical stand at 84° 58′ to 85° 30 E and 27°48′ to 28° 6′ N.

This district is influenced basically with tropical climate at 550 m to 950 m of elevation where the current study was limited. About 80 per cent of this part is dominated by Sal forest (*Shorea robusta*) with the blend of *Schima wallichi, Dalbergia sisoo, Acacia catechue, Saraka indica, Michelia champaca, Castanopsis indica, Alnus nepalensis, Bombax ceiba, Juglens regia, Pinus roxburghii* and *Euphorbia* species. Faunal species sheltered here include Barking Deer (*Muntiacus muntijak*), Common Leopard (*Panthera pardus*), Sloth Beer (*Melursus ursinus*), Wild Boar (*Sus scrofa*), Leopard Cat (*Felis bengalensis*), Jackal (*Canis aureus*), Langur (*Presbytis entellus*), Indian Hare (*Lepus nigricollis*), Indian Procupine (*Hystrix indica*) and Common Otter (*Lutra lutra*). Among the bird life, Babblers, Parakeets, Pheasants, Quails, Doves, Pigeons, Partridges, Peafowls, and Bulbuls are the notable species (Anonymous, 1971)

Sindhupalchok District

This district covers an area of 2,528 sq.km and stands at 27° 36′ to 28°13′ N to 85°06′ E of the central region. Altitudinal variation follows stepwise pattern at 850 m to 4480m thus offering a wide range of natural habitats for butterflies of the tropical to the Himalayan climates. The distribution of butterflies corresponds to the change in biophysical gradients at different altitudinal pockets with less diversity display at highland areas compared to the lower belts.

The mixed deciduous forest of this district has occurrence of vegetations like *Juniperus, Rhododendron arboretum, Juglens regia, Betula utilis* (higher elevation), *Daphne*

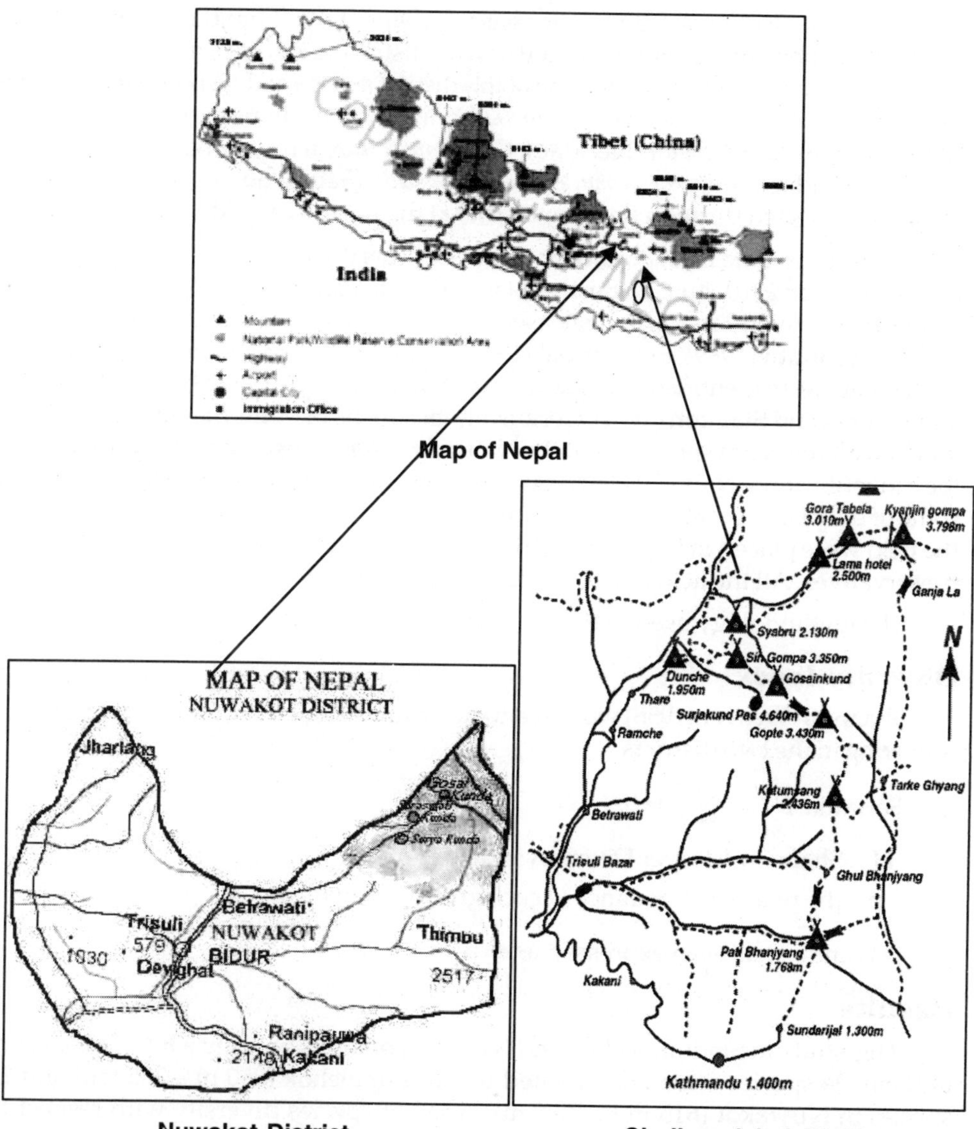

Map of Nepal

Nuwakot District

Sindhupalchok District

bholua, *Cinnamon* and Bamboo species. Prominent fauna sheltered in this district are Barking Deer (*Muntiacus muntijak*), Ghoral (*Naemorhedus goral*), Thar (*Hemitragus jemlahicus*), Musk Deer (*Moschus chrysogater*), Common Leopard (*Panthera pardus*), Clouded Leopard (*Neofelis nebulosa*), Red Panda (*Ailurus fulgens*), Himalayan Beer (*Selenarctos thibetanus*), Leopard Cat (*Prionailurus bengalensis*) and Pikas (*Ochotona* sps) (Pradhan *et al.*, 2009). Avian diversity in this district includes Danfe Pheasant (*Lopophorous impejanus*), Monal Pheasant (*Tragopan satyra*), Finches, Thrushes, Warblers Choughs, Magpies, and Babblers (Anonymous, 1971).

Parnassius hardwickei and *Parnassius epaphus*, both Himalayan natives are distributed above the elevation of 3200 m with distributional gaps at different places of the same range. The temperate zone of this district accommodate popular butterflies like *Precis iphita* (Nymphalidae), *Menelaides helenus* (Papilionidae), *Papilio polyctor* (Papilionidae), *Colias fieldii* (Pieridae), *Childrena childreni* (Nymphalidae), *Callerebia scanda* (Satyridae), *Aulocera swaha, Aulocera saraswati* (Satyridae) *etc.* Both the districts are interconnected through Rasuwa district at the north-west side.

The species data incorporated in this paper was documented at various periods during the year 2010-2011. Butterfly net was used to capture and identify specimens in the field which later were released in nature. Field identification was done consulting Smith (1989) and Haribal (1992). Vegetation components in and around the habitats were identified using keys and photographs provided by Stainton (1988) and Polunin and Stainton (1984). Confusing specimens of butterflies in the field were tallied with reference specimens at the Natural History Museum to bring accuracy in their identification. National Harbarium Section in Lalitpur was consulted to identify some of the plant specimens brought from the field. Collected butterfly specimens in the field were placed in transparent triangular envelope using Naphthalene balls as preservatives. All the field data were recorded accurately.

The methodology used in this research is as following.

Diversity Index

Shannon–Weaver (1958) was used to determine the diversity index of Lycaenids occurring in the two districts.

Formula: $H' = -\sum p_i \ln(p_i)$

H′: Shannon-Weaver Diversity Index

p_i: the relative abundance of each group of organisms

S-W index has been expressed as $e^{H'}$

Results

This study in Sindhupalchok and Nuwakot districts brought a list of 61 species, of them, 34 species were documented in Sindhupalchok (850 m – 3050 m) and 27 species in Nuwakot (615 m to 1000 m). A list of species diversity with respect to altitudes, associated forest and local status is also dealt (Appendix 1).

Nuwakot District

Altitudinal range at 720 m to 850 m of Nuwakot accommodated high diversity including rare species like *Arophala amantes, A. atrax, Everes lacturnus, Catachrysops strabo, Jamindes bochus, Ramelana jangala* (Endangered), *Panchala birmana*, and *Curetis dentata*. Other species occurring below the elevation of 700 m and above 850 m are mostly common in status.

Total Diversity

Altitude:	600-700 m	700-800 m	800-900 m	900-1000 m
Total species:	9	13	11	4

Sindhupalchok District

Species diversity was noted high at 1360 m- 2127 m of elevation where rare species like *Spindasis lohita, Chrysozephyrus birupa, Tajuria illurgis, Tajuria yajna* and *Ramelana jangala* occurred.

The diversity was found declining above 2500 m of elevation where mostly common species were prevalent. Rare species recorded above the elevation of 2150 m to 2575 m included *Pratapa icetas* (2425 m), *Euaspa miliona* (2430 m), *Panchala paraganesa* (2575 m), *Everes huegelii* (2450 m), and *Esakiozephyrus mandara* (2320 m).

Total Diversity

Altitude:	1300-1700 m	1700-2100 m	2100-2500 m	2500-2900 m	2900-3300 m
Total species:	7	18	11	5	2

Diversity at 700- 1000 m in Nuwakot Diversity at different altitudes in Sindhupalchok

Diversity Index of Lycaenids in Sindhupalchok (1300 to 3040 m) = **3.772078145.**

Diversity Index of Lycaenids in Nuwakot (600-1000 m) = **2.685390618.**

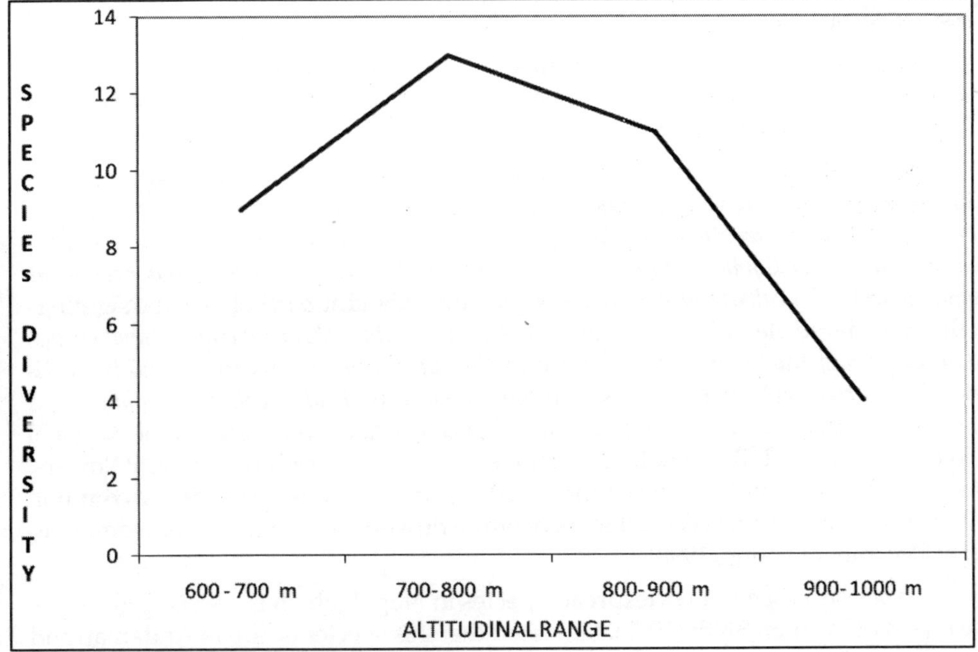

Figure 2.1: Species richness in Nuwakot.

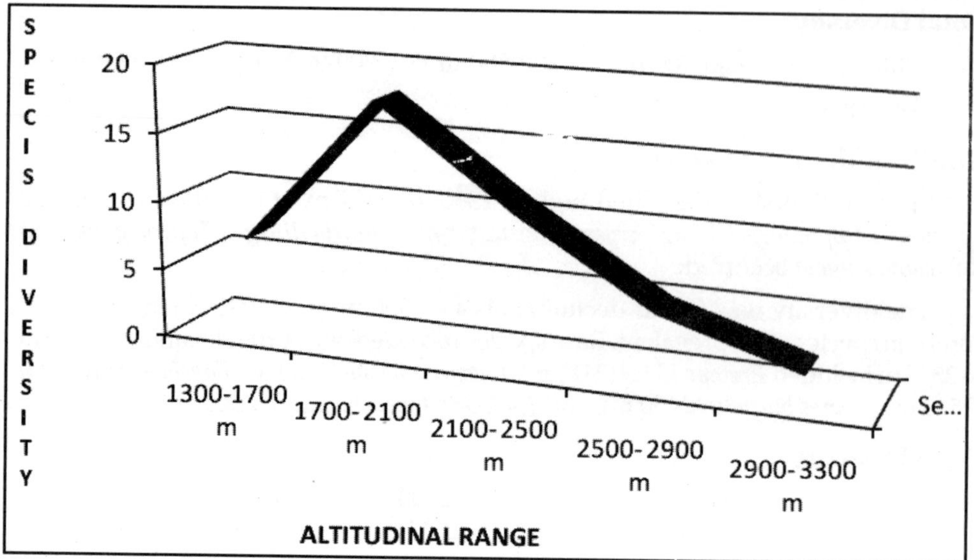

Figure 2.2: Species richness in Sindhupalchok.

Elevation range at 1300–3040 m in Sindhupalchok offered many preferential habitats for diverse Lycaenids. So the Diversity Index of lycaenids in this district was little higher than the Nuwakot which provided less preferential habitats for diverse species

Discussion

Different months considered for this study revealed a sum of 61 species in both the districts. Of the total record, 34 species were documented at Sindhupalchok and 27 species in Nuwakot districts.

Species richness was noted fluctuating at different altitudinal pockets at 1350-2300 m in Sindhupalchok and 600 – 1000 m in Nuwakot districts. *Everes argiades, Spindasis lohita, S. nepalicus, Esakiozephyrus mandara, Arophala atrax, Curetis bulis, Euaspa miliona, Heliophorus tamu, Celastrina argiolus, Lycaena phlaeas, Esakiozephyrus mandara* and *Udara albocerulea* were recorded only in Sindhupalckok with no sightings at the considered elevations of Nuwakot. *Curetis dentata, Arophala* sps., *Rapala nissa, Loxura atymnus, Iraota timeleon, Caleta caleta, Castalius rosimon, Freyeria putli, Zizina otis* and *Heliphorus epicle* were confined to Nuwakot only. *Esakiozephyrus mandara* and *Ramelana jangala,* are very rare species listed under endangered status of the National Red Data Book (NRDB, 1995). Both the sexes of *Ramelana jangala* seldom visit flowers. However, individuals have been found feeding on *Mikania guaco* (DESC.) a common climbing composite with tiny clusters of white flowers blossoming in November to early December (Young, 1998).

Actyolepsis puspa is a widespread species in Nepal which is also supported by the report of Wynter-Blyth (1957). He stated that this species is rare in Andaman and very Rare in Nicobars.

Species richness of Nepalese butterflies varies at different geographical locations. March to October display high diversity in lowland, early May to September in midland and late May to early October in highland regions. Fire played a significant role in determining species composition in fire-afflicted areas and affected flight periods of some species but did not affect species richness. Grazing had a major impact on species composition and it favoured only those Lycaenids whose caterpillars feed on herbs. In case of one of the sites where phenophases of the larval food plant and population trend of a small Lycaenid was documented, the population showed rapid increase at the time when the plants were in suitable phenophase for growth of the caterpillars (Kunte, 1997).

Conservation Issue

Butterflies are bio-indicators of the health of forests and environment. These fragile insects are under serious threats due to urbanization activities leading to habitat modifications. Since conservation of butterflies essentially means the conservation of their habitat, the spots rich in butterfly diversity are high conservation priority areas (Sidhu, 2011).

The current scenario of deforestation and habitat destruction are imposing direct impact on many Lycaenid species. Both the districts have high tourism potentiality so physical constructions for tourist's facilities especially at Sindhupalchok are directly linked to the expense of the original habitats of many rare species. An effective policy specially designed for butterfly conservation is a serious demand which should be based on the local environment condition.

Acknowledgement

Professor Dr. Keshab Shrestha, Chief, Natural History Museum in Kathmandu is acknowledged for providing official facility. Thanks are due to Dr. Nirmala Pradhan, Associate Ptofessor at Natural History Museum and Senior Botanist Dr. M.K. Adhikari, National Herbarium Section for their generous help and cooperation in identifying the herbarium of the field. The first author is thankful to Mr. Laba Dongol and Mr. Madan K. Shrestha for their support during field study programs.

Refrences

Anonymous, 1971. Mechi Dekhi Mahakali, Part 2,. Department of Information, Nepal, Government, Kathmandu, Nepal.

Haribal, M., 1992. *The butterflies of Sikkim Himalaya*. Sikkim Nature Conservation Foundation, Gangtok, Sikkim, India, 91–123.

Khanal, B., Chalise, M. K. and Solanki, G.S., 2012. Diversity of butterflies with respect to Altitudinal rise at various pockets of the Langtang National Park, central Nepal, *International Multidisciplinary Research Journal*, **2**(2): 41–48.

Kunte, J. Krushnamegh, 1997. Seasonal patterns in butterfly abundance and species diversity in four tropical habitats in northern Western Ghats. *J. Biosci.*, **22**(5): 593–603.

NRDB, 1995. *Red Data Book of the Fauna of Nepal*. Biodiversity Profile Project Technical Publication No.4, Department of National Parks and Wildlife Conservation, Nepal Government, Babar Mahal Kathmandu, Nepal.

Polunin, O. and Satinton, A., 1984. *Flowers of the Himalaya*. Oxford University Press. Pp. 1–580.

Pradhan, N., Khanal, B., Siwakoti, M., Bhattarai, S., Thapa, V. and Thapa, G., 2009. *An Assessment of Biodiversity in Panch Pokhari of Sindhupalchok District of Central Nepal Emphasizing Conservation Needs*. Report Submitted to Rufford Small Grant Foundation, http://www.ruffordsmallgrants.org/rsg/projects/nirmala pradhan

Sidhu, A.K., 2011. Changing biodiversity scenario in the Himalayan ecosystem: Mussoorie, Uttarakhand, India, as revealed by the study of blue butterflies (Lycaenidae). *Journal of Threatened Taxa* 3(2): 1559–1563.

Shields, O., 1989. World number of butterflies. *J. Lep. Soc.*, **431**(3): 178–183.

Smith, C., 1989. *Butterflies of Nepal*. Majupuria Publication, Tecpass Press, Bangkok, Thailand

Stainton, A., 1988. *Flowers of the Himalaya: A Supplement*. Oxford University Press, Oxford India Paperbacks, New Delhi.

Wynter-Blyth, M.A., 1957. *Butterflies of the Indian Region*. Bombay Natural History Society, Bombay.

Young, J.J., 1998. A brief note on the life history of *Ramelana jangala* (HORSFIELD) Lepidoptera, Lycaenidae in Hong Kong, PRC. Reprint from *FUTAO* No. 28, Entomological reports of FUTAO–KAI, Hong Kong, PRC.

Appendix

Sl.No.	*Area*	*Genus and Species*	*Altitudein Meter*	*Forest Type*	*Local Status*	*Global Distribution*
1.	NUWAKOT DISTRICT	*Zizeeria maha* Kollar	600-800	*Shorea robusta*	Common	Iran, Tibet, India, Burma, China, Japan.
2.		*Jamides celeno* Cramer	690	*Shorea robusta/ Schima wallichii*	Common	India, Sri Lanka, Thailand, Malaysia, Ciebes
3.		*Arophala pseudocentaurus* Doubleday	695	*Shorea robusta/Schima wallichii*	Rare	Bengal to Sikkim, Assam
4.		*Arophala atrax* Hewitson	660	*Shorea robusta/Dalbergia sissoo*	Rare	Bengal, Mussorie, to Sikkim, Assam, N. Burma
5.		*Arophala single* Deniceville	680	*Shorea robusta/Dalbergia sissoo*	Rare	Mussorie, Kumaon to Sikkim, Bhutan, Assam and North Burma
6.		*Arophala amantes* Hewitson	855	*Shorea robusta/Jugens regia*	Common	Central India (Orissa, Bengal, Mussorie), Sikkim
7.		*Iraota timelion* Stoll	810	*Shorea robusta/roxburghii*	Rare	India, Sri Lanka and China
8.		*Pratapa deva* Moore	910	*Shorea robusta/roxburghii*	Rare	India, Sri Lanka to Malaya and Java
9.		*Panchala birmana* Moore	800	*Shorea robusta/roxburghii*	Common	Burma, Hongkong and Sumatra
10.		*Curetis dentata* Moore	770	*Shorea robusta forest*	Common	South India to China and Japan
11.		*Rapala nissa* Kollar	850	*Pinus roxburghii*	Common	North Burma, Malaya, West and Central China
12.		*Loxura atymnas* Horsfield	700	*Shorea robusta*	Rare	India and Sri Lanka to Malaysia, Philippines
13.		*Spindasis lohita* Horsfield	700	*Shorea robusta*	Common	Himalaya, Sri Lanka, Burma, Thailand, Malaysia
14.		*Everes lacturnus* Godart	750	*Shorea robusta/Pinus roxburghii*	Rare	India, Sri Lanka, Malaya, to Australia, and China
15.		*Caleta caleta* Hewitson	600	*Shorea robusta/Schima wallichii*	Common	SriLanka, Sikkim, Burma, Philippines, Clebes

Contd...

Appendix–*Contd...*

Sl.No.	*Area*	*Genus and Species*	*Altitudein Meter*	*Forest Type*	*Local Status*	*Global Distribution*
16.		*Castalius rosimon* Fabricius	675	*Shorea robusta*	Common	India, Malaysia and Philippines
17.		*Actyolepsis puspa* Horsfield	660-720	*Shorea robusta/Bombax cieba*	Common	India and Sri Lanka to Philippines, New Guinea
18.		*Euchrysops cnejus* Fabricius	650-700	*Shorea robusta/Schima wallichii*	Rare	West China, to India, Malaysia and Australia
19.		*Freyeria putli* Kollar	780	*Shorea/Albizzia*	Common	India, Sri Lanka, Burma
20.		*Zizina otis* Fabricius	800	*Shorea robusta/Pine roxburghii*	Common	Tibet, India, Formosa and Australia
21.		*Jamides alecto* Felder	820	*Shorea robusta/Pinus*	Common	Sri Lanka, Sikkim, Burma
22.		*Catachrysops strabo* Fabricius	830	*Pinus roxburghii*	Common	India, Sri Lanka, Burma
23.		Lampides boeticus Linnaeus	600-900	*Shorea robusta*	Common	Sri Lanka, India, Burma, Andaman and Nicobar
24.		*Jamides bochus* Stoll	900	*Pinus roxburghii*	Common	India, Sri Lanka, Burma, Malaya, Formosa, Central China, Australia
25.		*Prosotas nora* Moore	680	*Shorea robusta*	Common	India, Sri Lanka, Burma, Formosa, Philippines, Java
26.		*Anthene emolus* Godart	770	*Shorea robusta/Albizzia sps.*	Rare	South India, Sikkim, Burma, Malaya, the Solomon Island
27.		*Heliophorus epicle* Godart	650-780	*Shorea robusta/Pinus roxburghii*	Common	West Himalaya (Kumaon), South Burma
28.		*Curetis bulis* Westwood	715	*Dalbergia sisoo/Bombax ceiba*	Common	Burma, Assam and South India
29.		*Curetis dentata* Moore	770	*Dalbergia sisoo/Bombax/Shorea*	Rare	South India to China and Japan
30		*Deudoryx epijarbus* Moore	655			Across India, Sri Lanka, Burma
31.		*Celatoxia marginata* DeNiceville	900	*Pinus/Schima*	Common	West Himalaya to Burma
32.		*Arophala oena* Hewitson	715	*Bombax ceiba/Dalbergia sisoo*	Rare	Mussoorie to Sikkim, Assam, and North Burma to Hainan

Contd...

Appendix–*Contd...*

Sl.No.	Area	Genus and Species	Altitudein Meter	Forest Type	Local Status	Global Distribution
33.		*Spindasis syama* Horsfield	820	*Saraka indica/Euphorbia sps.*	Rare	India, Malaya, Philippines and China
34.	SINDHU-PALCHOK	*Jamides bochus* Stoll	1550-1660	*Schima wallichii*	Rare	India, Sri Lanka, Burma, Malaya, Formosa, Central China, Australia
36.	DISTRICT	*Everes huegeli* Gistel	2540	*Picea smithiana/Rhododendron arboreum/Quercus semicarpifolia*	Rare	Chitral to Kumaon
		Everes argiades Pallas	1720	*Pinus roxburghii/Schima wallichii*	Rare	Europe, Korea to Japan, India and Malaysia
37.		*Zizeeria maha* Kollar	1700–1780	*Pinus roxburghii/Schima wallichiana*	Common	Iran, Tibet, India, Burma, China, Japan
38.		*Celastrina huegeli* Moore	1810	*Quercus semicarpifolia*	Common	Kashmir to Bhutan
39.		*Udara dilecta* Moore	2940	*Rhododendron arboreum/Psuga dumosa/Quercus semicarpifolia*	Common	North Burma to West and Central China
40.		*Celastrina argiolus* Linnaeus	2520-3040	*Quercu semicarpifolisa/Juglans regia/Picea smithiana*	Common	North America, Asia, India, Malaya, Europe, Japan to N. Africa
41.		*Actyolepsis puspa* Horsfield	1890- 2350	*Rhododendron/Alnus nepalensis*	Common	India and Sri Lanka to Philippines
42.		*Celatoxia marginata* DeNiceville	1905 -2250	*Rhododendron/Alnus nepalensis*	Common	West Himalaya to Burma
43.		*Udara albocerulea Moore*	2600	*Rhododendron/Psuga dumosa/ Alnus nepalensis*	Rare	North Burma to west and central China
44.		*Lampides boeticus* Linnaeus	1835	*Schima wallichii*	Common	Sri Lanka, India, Burma, Andaman and Nicobar
45.		Creon cleobis Godart	1860 – 2150	*Schima wallichii*	Rare	India and Malaysia.
46.		*Heliophorus tamu* Kollar	1915	*Quercus semicarpifolia*	Rare	Himalaya from Kumaon to Sikkim, North Burma
47.		Lycaena phlaeas Linnaeus	2300	*Rhododendron arboreun*	Common	North America, Europe, Asia to Japan, North Africa, Abyssinia

Contd...

Appendix–*Contd...*

Sl.No.	*Area*	*Genus and Species*	*Altitudein Meter*	*Forest Type*	*Local Status*	*Global Distribution*
48.		*Esakiozephyrus mandara* Doherty	2320	*Rhododendron arboreum/ Quercus semicarpifolia*	Rare	Kulu to Garhwal and Tibet
49.		*Ramelana jangala* Horsfield	1380	*Schima wallichii/Alnus nepalensis/ Albizzia* sps		Sikkim, Burma, Malaya, Philippines to Hongkong
50.		*Flos areste* Hewitson	1820	*Schima wallichii/ Quercus semicarpifolis*	Common	Sikkim, Bhutan, Assam, North Burma, Malaya
51.		*Chrysozephyrus birupa* Moore	2122	*Pinus roxburghii/Alnus nepalensis/ Rhododendron aeboreum*	Rare	Endemic to west and central Himalaya (Smith, 1989)
52.		*Spindasis lohita* Horsfield	1350	*Schima wallichii*	Rare	Himalaya, Sri Lanka, Burma, Thailand, Malaysia
53.		*Panchala birmana* Moore	2200-2575	*Quercus semicarpifolia/ Rhododendron arboreum*	Rare	Burma, Hongkong and Sumatra
54.		*Ancema ctesia* Hewitson	1400-2000	*Schima wallichii/Alnus nepalensis*	Common	Kumaon to Dawnas, West China
55.		*Prosotas nora* Felder	1454–2122	*Schima wallichii/Pinus roxburghii*	Rare	India, Sri Lanka, Burma, Philippines, Formosa, Java
56.		*Heliophotus androcles* Doubleday and Hewitson	2100	*Pinus roxburghii/Alnus nepalensis/ Rhododendron arboreum*	Common	Kashmir to Kumaon, Sokkom, Assam, West China, Bhutan, Burma
57.		*Pratapa icetas* Hewitson	2425	*Rhododendron arboreum/ Psuga dumosa*	Rare	Himalaya from Kumaon to Sikkim, Assam, West China
58		*Arophala singla* DeNiceville	1800	*Quercus semicarpifolia/ Rhododendron arboreum*	Common	Mussorie, Kumaon to Sikkim, Bhutan, Assam and North Burma
59.		*Tajuria illurgis* Hewitson	2100	*Pinus roxburghii/Alnus nepalensis/ Rhododendron arboreum*	Rare	Mussorie to Sikkim and North Burma
60.		*T. yajna* Doherty	2000	*Rhododendron arboreum/ Alnus nepalensis*	Rare	Himalaya from Mussorie to Sikkim and Burma
61.		*Rapala nissa* Kollar	2060	*Rhododendron arboreum/ Alnus nepalensis*	Common	North Burma, Malaya, West and Central China

Environment, Biodiversity and Traditional System, 2014 *Pages* 19–30
ISBN 978-81-7622-307-2
Editors: D.R. Khanna, G.S. Solanki and S.K. Pathak
Published by: BIOTECH BOOKS, NEW DELHI

Chapter 3

Wildlife and their Behavior in Baghmara Buffer Zone Community Forest, Nepal

B.K. Sharma[1], M.K. Chalise[2] and G.S. Solanki[3]

[1]Ph.D. Scholar, Department of Zoology, Mizoram University, India
E-mail: bhuvan_keshars@hotmail.com
[2]Central Department of Zoology, Tribhuvan University, Kathmandu, Nepal
E-mail: mukesh57@hotmail.com
[3]Department of Zoology, Mizoram University, India
E-mail: gssolanki02@yahoo.co.in

ABSTRACT

Baghmara Buffer Zone Community Forest (BBZCF) is covering 215 ha area and located in the central lowland of Nepal. A study from April 2010 to March 2011 was conducted in the area to find out the behavior of seven species of large wild mammals namely barking deer, spotted deer, wild boar, rhinoceros, rhesus monkey, sambhar and hog deer. Primary data on daily activities of those wildlife were recorded from fixed places as well as from parallel transects (n = 34). Total of 68 kilometer length was covered in all line transact, traversing east to west through all habitat types of BBZCF. Total observation time for the different diurnal behavior was 3638 minutes. Common behavior observed were feeding (browsing and grazing), resting, running, walking, playing, digging and tree climbing. All wildlife species, except hog deer, spent more than 40 per cent of their time on feeding. Digging and tree climbing were specific behavior of wild boar and rhesus monkey respectively. Grasses and forbs were common food plants. More than 70 per cent observed time of rhinoceros were found disturbed from tourism activities.

Keywords: Wildlife behavior, Community Forest, Diurnal activities, Feeding, Food plants.

Introduction

A buffer zone community forest area was designed two decades ago in Baghmara at the adjacent of Chitwan National Park. It is managed forest land created for the benefit of locals and wildlife themselves. The consequences of the long effort of maintenances of habitat provided the opportunity of re-colonization of some endangered flagship species like Asian One Horned Rhinoceros in the forest patch (Sharma and Chalise 2011). This indicates the establishment of a healthy forest habitat maintained by the institutional engagement for the resource management and the involvement of local stakeholder in the process of conservation of the community forest where grazing and hunting is totally controlled, while legal collection of fuel wood and fodder is permitted periodically. Community conservation efforts for twenty years have ameliorated the condition of Baghmara forest. Becoming a pioneer to start wildlife eco-tourism in community managed forest it has been earning more than US$ 50,000 per annum as eco-tourism revenue. The community was less dependent on revenue generated from forest products selling.

In the beginning of community managed forest there were only two species of large wild mammals (Sharma *et al.*, 2011). At present, this forest is becoming habitat for 22 species of mammals of different status (Table 3.1).

Table 3.1: Wild animals recorded from Baghmara community forest Chitwan, 2010-11.

Common Name	*Scientific Name*	*National Status*	*IUCN Status*
Asiatic Elephant	*Elephas maximus*	Protected	Endangered
Barking Deer	*Muntiacus muntjak*		Least concern
Bengal Tiger	*Panthera tigris*	Protected	Endangered
Common Leopard	*Panthera pardus*		Near Threatened
Five-stripe Palm Squirrel	*Funambulus pennantii*		
Golden Jackal	*Canis aureus*		Least Concern
Hanuman Langur	*Semnopithecus entellus*		
Hog Deer	*Axis porcinus porcinus*		Endangered
House Rat	*Mus musculus*		
Indian Crested Porcupine	*Hystrix indica*		
Indian Grey Mongoose	*Herpestes edwardsii*		
Jungle Cat	*Felis chaus*		
Large Indian Civet	*Viverra zibetha*		
One-horned Rhinoceros	*Rhinoceros unicornis*	Protected	Vulnerable
Rhesus Macaque	*Macaca mulatta*		Least concern
Rufous-tailed Hare	*Lepus nigricollis*		
Sambar Deer	*Cervus unicolor*		Vulnerable
Sloth Bear	*Ursus ursinus*		Vulnerable
Small Indian Civet	*Viverricula indica*		
Smooth Coated Otter	*Lutrogale perspicillata*		
Spotted Deer	*Axis axis*		Least concern
Wild Pig	*Sus scrofa*		Least concern

Behavior of an organism is sum of the responses to its internal or external stimuli (Parajuli 2007). Behaviors such as feeding, reproduction are common in all animals and these involve in different occasions differ according to species and their mode. Animal behavior is the means of functional importance from which an animal maintains its relation with the environment. Wildlife behavior is a basic necessity for intelligent conservation and management. Biological data with their behavior are vital for in-situ conservation of wildlife species (Parajuli 2007).

Conservation efforts have brought significant ecological and socio-economic changes in the area. Changes in ecological condition and occurrence of diverse type of wildlife with continued eco-tourism practices may cause alteration of the behavior among wild animals. Such impacts need to be quantified and assessed in future. This study will try to analyze the behavior of large mammals found in the area at the present context that could be baseline data for future use and comparison.

Methods

Baghmara Buffer Zone Community Forest is located in Bachhauli Village Development Committee of Chitwan District, in subtropical region of lowland Nepal (Figure 3.1). It is situated in Buffer Zone area of Chitwan National Park at eastern sector in between 27°34.78′-27°35.53′ northern latitude and 84°28.43′-84°29.40′ eastern longitude elevation of 170 to 200 meter above sea level. The climate is sub-tropical monsoon type with relatively high humidity. Monsoon rain prevails in the study area from late June to September and amount of annual rainfall ranges from 14.04 mm to 602.2 mm. Heavy flooding occurs during the monsoon. The average daily maximum temperature of the area in hot summer days is about 36.8°C. Spring starts from March and is immediately followed by summer and that ends in June. The minimum temperature is about 7.8°C in cool dry winter season which occurs from October to February (Sharma and Chalise 2011).

Observations in fixed places were applied to observe wildlife behavior. Field studies were conducted on daily basis from 1st April 2010 and completed on 31st March 2011. The behaviors of studied wildlife were recorded by scan sampling method for one minute at interval of 30 minutes. Other interesting behaviors of the group

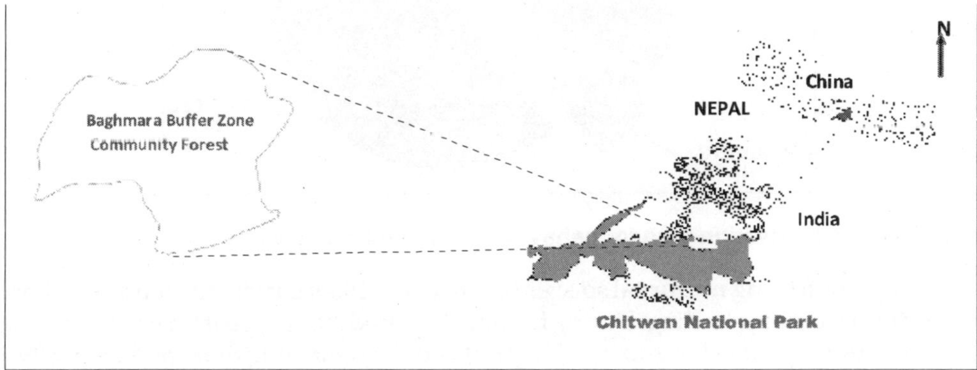

Figure 3.1: Location of Baghmara BZCF in Nepal.

members were recorded whenever they were observed. Five observers collected focal and ad libitum data. Observed diurnal behaviors of studied wildlife were recorded on seven categories as feeding, resting, running, walking, tree climbing, digging and playing. For simplicity grazing and browsing behavior were categorized as feeding. Food habitats were discerned by direct observation during the scan sampling. Field study was conducted 25 days per month (n = 300 days in one year). Daily observation lasted up to 8 hours. Observation was conducted in morning (starts from 6: 00 am), afternoon (starts from 12: 00 noon), and evening (starts from 4: 00 pm). Among 300 observation 123 events (41 per cent) were conducted in the morning, 84 events (28 per cent) were conducted in the day and 93 events (31 per cent) were conducted in the evening.

Results and Discussion

In one year study period, total direct contact time with the wildlife for behavior study was 3638 minutes. The behavior study was conducted only in diurnal time. So, the contact time was also differs according to the species which were more active in the day time or night time.

Spotted deer were more common wildlife observed in the study area. Their numbers were high in comparison to the other wildlife species (Sharma *et al.*, 2011). So, during the observation more contact time (54.7 per cent) was for spotted deer. The more contact time were recorded for diurnally active wildlife like rhesus monkey (10.4 per cent), rhino (8.7 per cent) and hog deer (8.2 per cent). Least contact time were for nocturnal wildlife like wild pig (7.1 per cent), barking deer (5.7 per cent) and sambar (5.1 per cent) (Figure 3.2).

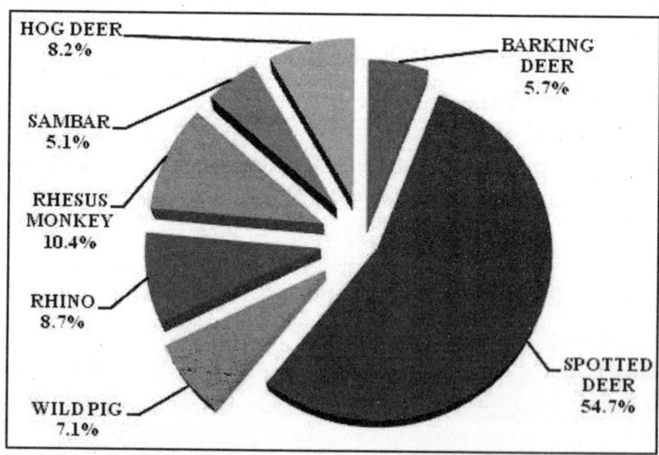

Figure 3.2: Observation time for behavior study of different wildlife.

Although being nocturnal species (Dinnerstein 1980) barking deer used 57.4 per cent diurnal time for feeding. During feeding they used 40.8 per cent time for browsing and 59.2 per cent time for grazing. They utilized the remaining time for walking 39.2 per cent, running 2.9 per cent and resting 0.5 per cent (Figure 3.3). During browsing

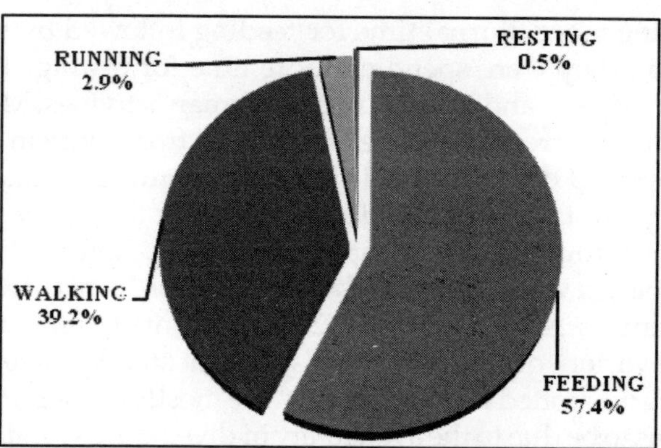

Figure 3.3: Diurnal time spent by barking deer for different behavior.

average feeding schedule was two grabs per minute. While in the grazing average feeding bites was ten grabs per minute. Plant species and their feeding frequency were *Cynodon dactylon* (L.) Pers. (33.3 per cent), *Ageratum conyzoides* L. (16.7 per cent), *Imperata cylindrica* (L.) Beauvois (16.7 per cent), *Digitaria ciliaris* (Retz.) Koeler (16.7 per cent) and *Rungia parviflora* (Retz.) Nees (16.6 per cent). Instead of consuming field crops, as indicated by Pokharel and Chalise (2010) in Kaski, they were more relaying on the forest grasses. This is due to the presence of different patches of grassland inside the study area, which provided more grazing opportunities than the previous study where dense tree canopy did not allowed to grow foraging plant species for this wildlife.

Spotted deer utilized 47.0 per cent of their diurnal time for resting. They spent 42.0 per cent of their time for feeding, 6.0 per cent for walking and 5 per cent for running (Figure 3.4). On the total feeding period they spent 79 per cent time for browsing and 21 per cent time for grazing. During browsing average grabs per minute was 7.7 and that was 14.7 during grazing. Previously it was recorded that spotted

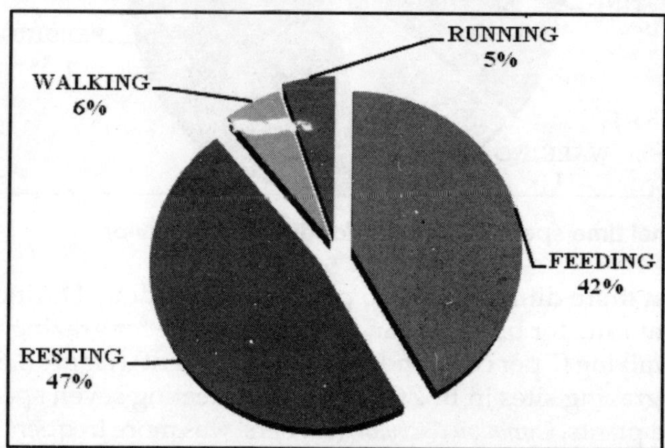

Figure 3.4: Diurnal time spent by spotted deer in different behavior.

deer were spending more diurnal time for feeding followed by walking (Parajuli 2007). In our result they were spending more time for resting. This is due to the difference in the study site and habitat used by human activities. Most of the diurnal time (7 am to 5 pm) of current study site is disturbed from tourism activities, mainly elephant riding. Spotted deer like to hide from the disturbances and will lessen their activities. So, they are taking rest understory during the study period (basically daytime). They were found eating ten species of plants. Among those species more frequent eating species were *Cyanodon dactylon* (29.4 per cent), *Ageratum conyzoides* (20.6 per cent), and *Imperata cylindrica* (20.6 per cent) (Annexure 1). Instead of consuming wide variety of tree and shrubs as indicated by Dinnerstein (1980) in Bardia National Park, spotted deer of BBZCF were mostly consuming forbs and grass species. This seems to be due to the availability of diverse grass and forbs species and presence of few palatable trees and shrubs species in this area.

Wild pigs were spending more diurnal time (47 per cent) for feeding. During feeding they spent 89 per cent time for browsing and 11 per cent time for grazing. They utilized the other time for running 17 per cent, walking 11 per cent, resting 19 per cent and digging 6 per cent (Figure 3.5). Wild pigs are nocturnal opportunistic omnivores and their diet incorporated at least one energy-rich plant (Schley and Roper 2003). Although being nocturnal wildlife species they spent more diurnal time for feeding in BBZCF. This seems to be less palatable food available for them in the forest area which causes them busy in daytime too to find scarce food items in the study area.

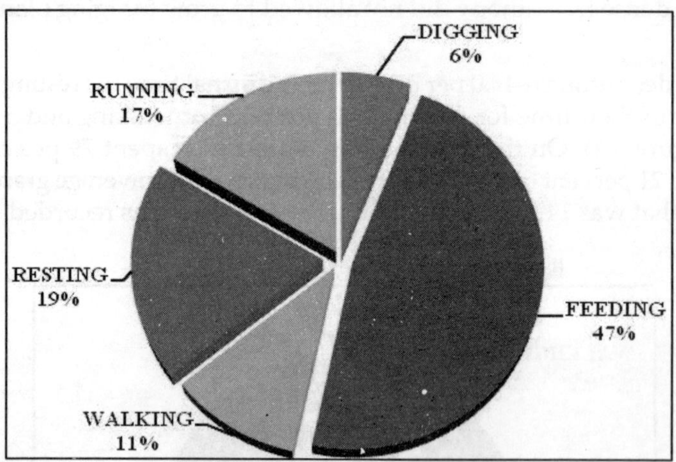

Figure 3.5: Diurnal time spent by wild pig for different behavior.

Rhino spent more diurnal time (75 per cent) for feeding. During feeding they spent 91 per cent time for browsing and 9 per cent time for grazing. They utilized other time for walking (7 per cent) and resting (18 per cent) (Figure 3.6). From direct observation of grazing sites in BBZCF, rhino were eating seven species of plants. Among the food plants, *Cynodon dactylon* (L.) Pers. was more frequently eating grass species (27.3 per cent) followed by *Callicarpa microphylla* Vahl (18.2 per cent) and

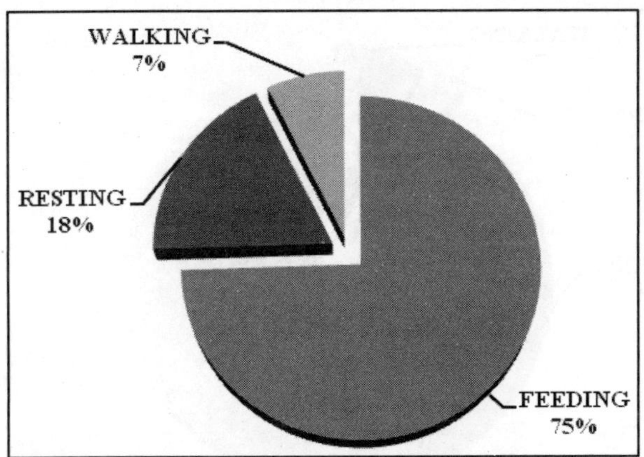

Figure 3.6: Diurnal time spent by rhino in different behavior.

Imperata cylindrica (L.) Beauvois (18.2 per cent) (Annexure 1). We found the same result as mentioned by Hutchins and Kreger (2006) in Chitwan National Park that one-horned rhinoceros used more time for feeding and the foraging species were grasses and aquatic plants.

Rhesus monkey used more diurnal time (42 per cent) for feeding (Figure 3.7). They spent their other time for playing (19 per cent), resting (15 per cent), running (11 per cent), tree climbing (7 per cent) and walking (6 per cent). Previously it was recorded that rhesus monkey spent more diurnal time for resting (Shrestha *et al.*, 2001). In our study they spent less time for resting. These differences were due to the change in the study sites and time of daily monitoring. Previous study was conducted in between 9 am to 5 pm in the temple area where they rely on pilgrim people and needed less effort for food.

Sambar used to spend more diurnal time (55 per cent) for feeding (Fig. 8). During feeding they spent 45 per cent time for grazing and 55 per cent time for browsing.

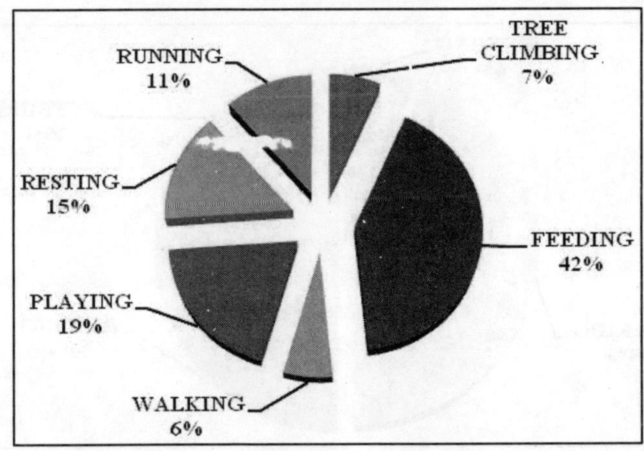

Figure 3.7: Diurnal time spent by rhesus monkey in different behavior.

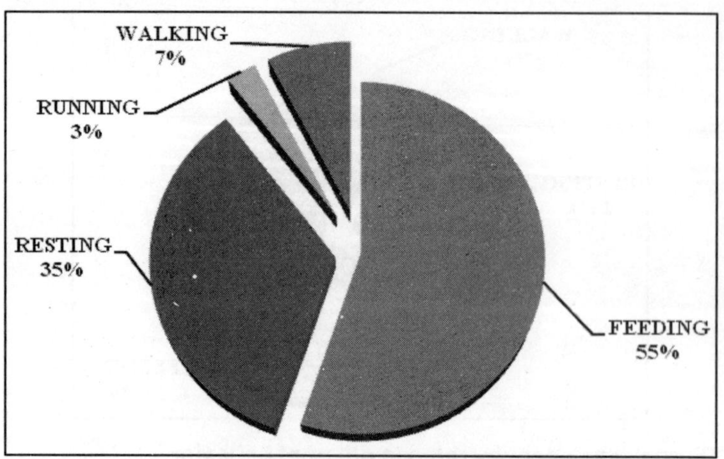

Figure 3.8: Diurnal time used by sambar in different behavior.

They utilized other diurnal time for resting (35 per cent), walking (7 per cent) and running (3 per cent). Sambars are nocturnal in nature but being large bodied animals they used to spend more time for feeding. These results were similar with the previous research (Hutchins and Kreger 2006). Our finding on the habitat used by Sambar was different from the previous studies in Bardia National Park (Dinnerstein 1980). Dinnerstein found the Sal forest was prime habitat for Sambar, but they were well distributed in our study area which was Tropical Evergreen Riverine Forest (Sharma *et al.*, 2011). The cause of this different can be the microclimatic condition and less competition with other large deer in our study area for the food. But, there was competition of sambar with Blue Bull in riverine area of Bardia National Park (Dinnerstein 1980).

Hog deer used more diurnal time (44 per cent) for walking and 26 per cent diurnal time for feeding (Figure 3.9). For feeding they used more time for browsing (93.5 per cent) than grazing (6.5 per cent). They utilized other time for resting (24 per

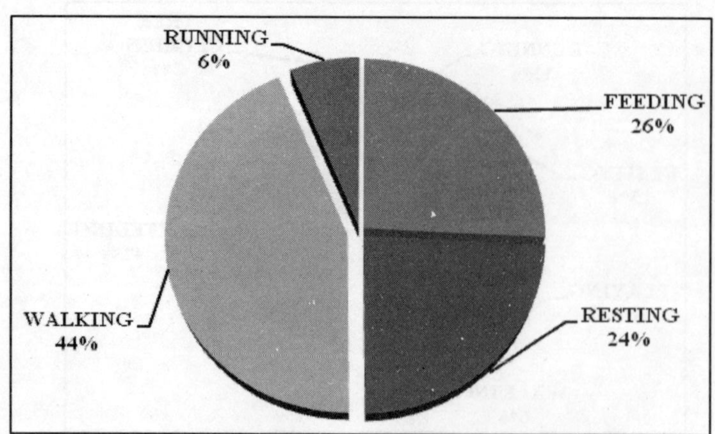

Figure 3.9: Diurnal time spent by hog deer in different behavior.

cent) and running (6 per cent). They are solitary wildlife and their favorable habitats were tall floodplain grassland (Dinnerstein 1980). We found this wildlife in the borders of ox-bow lakes where the floodplain grasses were short. Only small numbers of this wildlife (Sharma *et al.*, 2011) preferred this area due to the presence of similar microclimatic condition of floodplain grassland area.

Among the studied wildlife barking deer, rhino, wild pig, rhesus monkey and sambar deer spending more diurnal time for feeding. Hog deer were spending more diurnal time for walking and spotted deer were using more time for resting. Such patterns are similar to the result of the earlier findings (Hutchins and Kreger 2006) that the feeding time of the wildlife depends on their body size, availability of food and habitat surroundings (Chalise 2000). Although being a nocturnal animal barking deer, sambar, and wild pig (Dinnerstein 1980, Pokharel and Chalise 2010) used less diurnal time for resting (Figure 3.10).

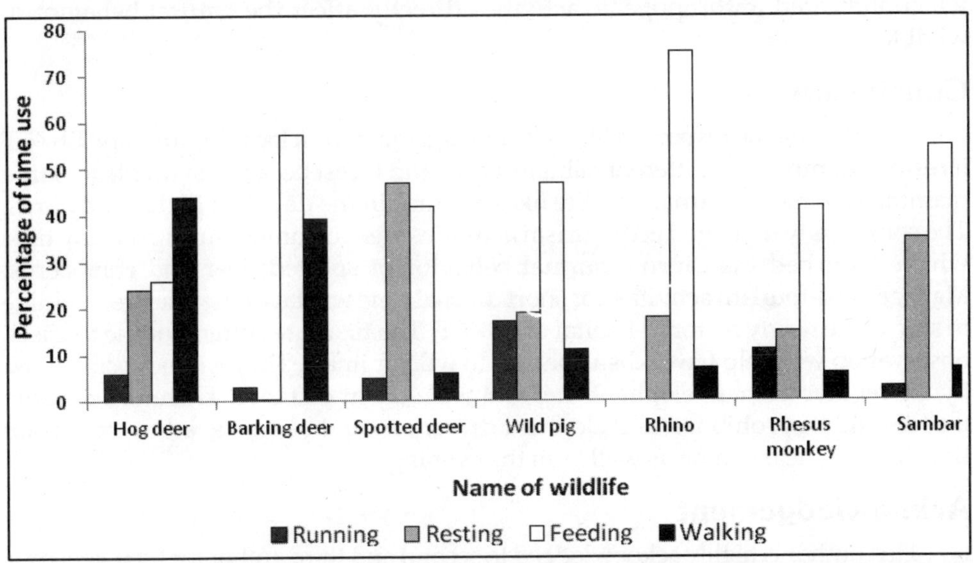

Figure 3.10: Percentage of time spent by different wildlife in different behavior in BBZCF.

Tourism and its Impact on Wildlife Behavior

In total 15.2 per cent of spotted deer's diurnal time was disturbed from the tourism activities. During the study it was observed that they could easily identify the sites from where they were going to be disturbed. So during the preparatory phase of disturbance period they left their ongoing normal behavior (like grazing, resting, playing). Before observing the touristic elephant initial 22.7 per cent time period termed as preparatory phase was used to identify the areas from where they were going to be disturbed. In this phase they tried to escape and hide from the disturb sites. The behavior observed during the preparatory period were looking around by erecting ears, fawns were running, wakeup from resting, and gathering all team members in one group. After these behaviors they tried to hide from the disturbances.

If they were unable to hide and contacted with the touristic elephant, they used to escape inside the dense bushes or forest.

Similarly, 71 per cent of the rhino's diurnal time was disturbed from the tourism activities. Preparatory phase for the disturbances ended within 5 minutes. In preparatory phase they stop their ongoing normal behaviors (like grazing, milking to calf, walking and browsing). In this time they used to watch towards the disturbed sites. After getting contact with the touristic elephants they waited for a maximum of 5 minutes and look whether the elephant to come towards them or not. In this waiting phase scaring behaviors like pepping, snoring and running were recorded. After waiting phase, rhino used to escape from the disturbances. During escaping behaviors like hiding in bushes or forests, calling to calf, and standing inside dense bushes were observed. Our findings were also similar with Yasuè and Dearden (2006) about the impact of the anthropogenic activities in the natural behavior of bird in Thailand, which indicated anthropogenic activities directly affect the natural behavior of wildlife.

Conclusion

After taking the responsibility of its management from local community, BBZCF it restored a mosaic of different habitats types and forest became favorable for large mammals. Some nocturnal wildlife likes to graze in the day time in the study area. The community managed eco-tourism activities were common during the day time which disturbed the diurnal normal behavior of spotted deer and rhinoceros. Managed eco-tourism activities for short diurnal time will favor the emerged wildlife to stay in the newly restored habitat of BBZCF. The fix route system for the tourists' observation will help fewer disturbances to wild animals. They will be accustomed by the regular human/elephant trail and will stay at distance. Therefore random trail should be prohibited and elephant drivers should be restricted to the certain time limit in the morning as well as in the evening.

Acknowledgement

First author is highly acknowledged to second and third author for their generous effort to initiate and guidance of research work. Mizoram University acknowledged for providing opportunity to study, Department of National Parks and Wildlife Conservation Nepal for providing research permission, Baghmara Buffer Zone Community Forest for providing research permit, Nepal Conservation Research and Training Centre for logistic support including assistants and elephant for monitoring. Harkaman Lama Field technician, Manoj Chaudhari secretary of BBZCF were highly acknowledged for the kind support.

Abbreviation Used

BBZCF: Baghmara Buffer Zone Community Forest

References

Chalise, M.K., 2000. Biodiversity of wild food plants in relation to the monkeys feeding ecology. In the *Proceeding of Eighth International Workshop of BIO–REFOR*, Japan, under Biotechnology Applications for Reforestation and Bio-diversity Conservation, p. 289–294.

Dinerstein, E., 1980. An ecological survey of the Royal Karnali–Bardia Wildlife Reserve, Nepal. Part III: Ungulate Population. *Biol. Conserv. 18:* 5–37.

Hutchins M. and Kreger, M.D., 2006. Rhinoceros behaviour: implications for captive management and conservation. *Int. Zoo Yb.,* **40**: 150–173.

Ostner J., Chalise, M.K., Koeing, A., Launhardt, K., Nikolei, J., Podzuweit, D. and Borries, C., 2006. What Hanuman Langur males know about female reproductive status. *American Journal of Primatology*, 68: 701–712.

Parajuli, D. P., 2007. Population status, distribution and general behavior of spotted deer (*Axis axis* Erxleben, 1777. in Bardia National Park, Bardia, Nepal. M Sc dissertation submitted to Central Department of Zoology, Tribhuvan University. 82 Pp.

Pokharel, K. and Chalise, M.K., 2010. Status and Distribution Pattern of Barking Deer (*Muntiacus muntjak* Zimmermann) in Hemja VDC, Kaski. *Nepal Journal of Science and Technology* 11: 223–228.

Schley, L. and Roper, T.J., 2003. Diet of wild boar Sus scrofa in Western Europe, with particular reference to consumption of agricultural crops. *Mammal Review*. **33**: 43–56

Sharma, B.K., Chalise, M.K. and Solanki, G.S., 2011. Vegetation types and wildlife occurrence in Baghmara Buffer Zone Community Forest, Nepal. *International Multidisciplinary Research Journal* (in press).

Sharma, B.K, Chalise, M.K. and Solanki, G.S., 2011. Large wildlife population in Baghmara Buffer Zone Community Forest, Nepal. *Ecoprient* (in press).

Sharma, B.K. and Chalise, M.K., 2011. Assessment of resources in Baghmara Community Forest of central lowland Nepal. *Journal of Natural History Museum Nepal* (in press).

Shrestha, B., Koirala, M. and Chalise, M. K., 2001. Study of agonistic behavior of rhesus monkey in Pashupati area, Nepal. *Journal of Natural History Museum*, TU Kathmandu, 20: 149–158.

Yasuè, M. and Dearden, P., 2006. The potential impact of tourism development on habitat availability and productivity of Malaysian plovers *Charadrius peronii. Journal of Applied Ecology*, **43**: 978–989.

Annexure 1: Name of plant species with eating frequency by chital and rhino during their feeding time.

Name of Plant Species	*Eating Frequency (per cent)*	
	Spotted Deer	*Rhino*
Cynodon dactylon (L.) Pers.	29.4	27.3
Ageratum conyzoides L.	20.6	–
Imperata cylindrica (L.) Beauvois	20.6	18.2
Rungia parviflora (Retz.) Nees	8.8	9.1
Oxalis corniculata L.	5.9	–
Hemarthrua compressa (L. f.) R. Br.	2.9	–
Eleusine indica (L.) Gaertn.	2.9	–
Mikania micrantha Kunth	2.9	–
Bombax ceiba L. (flower)	2.9	–
Artemisia indica Willd.	2.9	–
Callicarpa microphylla Vahl	–	18.2
Digitaria ciliaris (Retz.) Koeler	–	9.1
Circium arvense (L.) Soop	–	9.1
Diplazium esculentum	–	9.1

Environment, Biodiversity and Traditional System, 2014 *Pages* 31–37
ISBN 978-81-7622-307-2
Editors: D.R. Khanna, G.S. Solanki and S.K. Pathak
Published by: BIOTECH BOOKS, NEW DELHI

Chapter 4

Importance of Biodiversity Conservation and its Role in Management of Diverse Forest Resource Production with Special Reference to NEH Region, India

Bandi Gopichand[1] and Bandi Diddi Naidu[2]*

[1]Professor and Head, Department of Forestry, Mizoram University, Aizawl – 796 004, India
[2]MSc. Agricultural Biotechnology (India) and MBA (USA), Washington, DC, USA

ABSTRACT

Biodiversity conservation plays a crucial role in Forest resource management. Environmental management is a pivotal issue and hence is gaining significant momentum in recent years due to the merits it provides not just to the human species but also to the other forms of life on this planet. India is a mega-diversity country of which NEH region is considered to be one of the biodiversity hotspots in the world and is home for a variety of diverse plant and animal species. Forests are of ecological and economical importance to mankind and by serving as a habitat to diverse life forms they contribute to our life support system. Biodiversity forms a basis to a wide range of goods and services that are provided by forests; therefore, biological diversity is directly proportionate to diversified forest resources. But there is continuing loss and degradation of forests in many developing

* *Corresponding Author:* E-mail: gopiagri1@gmail.com

countries which is limiting the progress towards sustainable forest management. There is a need and necessity for biodiversity conservation at regional as well as global level. Our main goal should be to maximize the forest resources to improve economic, social and environmental conditions of people while ensuring biodiversity conservation to meet the needs of future generations. Hence, creating awareness among the people and educating them about the gravity of environmental protection and its conservation is of utmost priority.

Keywords: ***Biodiversity conservation, Forest resources, Ecosystem, Sustainable forest management, Hot spots, Diversity.***

Introduction

Biodiversity conservation plays a key role in providing ecological services and acts as a support system for the entire mankind. It is crucial for survival and sustenance of human life on earth. In simplest of terms, it can be described as protecting and preserving of biological diversity for future generations to cherish nature's wealth and beauty. The US Congress Office of Technology Assessment (1987) in 'Technologies to Maintain Biological Diversity' proposed a definition which is widely quoted– "Biological diversity refers to the variety and variability among living organisms and the ecological complexes in which they occur. Diversity can be defined as the number of different items and their relative frequency. For biological diversity, these items are organized at many levels, ranging from complete ecosystems to the chemical structures that are the molecular basis of heredity. Thus, the term encompasses different ecosystems, species, genes, and their relative abundance."

Forests are very valuable natural resources that are available to the mankind. Forests are important ecologically and economically. Forestry can be defined as "The theory and practice of all that constitutes the creation, conservation and scientific management of forests and utilization of their resources" (Khanna and Chaturvedi, 2008). *Ecological importance:* Forests are niche` for diverse plant, animal and microbial species and are habitat to a large number of earth's species. They enhance the quality of environment by influencing the life support systems and are potential resources that can be renewed if used optimally and managed wisely. Hence their conservation is highly essential for sustainable development. *Economic significance:* Forests are essential source of livelihood and fuel for the poor and marginalized sections of the society. They also provide wood and other products for the human use. India and NEH Region in particular is one of the most diverse zones with rich biological diversity. However, this rich biodiversity is under severe threat owing to habitat destruction, degradation, fragmentation and over exploitation of resources. According to The Food and Agriculture Organization of the United Nations, "To cope with all these challenges and uncertainties a large reservoir of genetic and species diversity will need to be maintained and sustainably used. This diversity will further help maintain and rehabilitate productive ecosystems to supply future generations with abundant food and agriculture."(www.fao.org). In order to maintain and conserve the resources, Sustainable forest management approach should be practiced which involves the use of forests and forests land in a way and at a rate that maintains their productivity, regeneration capacity, vitality, biodiversity and their potential to fulfill, now and in the future, relevant ecological, economic and social functions at local, national and global levels, and that does not cause damage to other ecosystems. (Paavilainen, 1994).

Biodiversity, Forest Resource Management and Conservation

Biodiversity contributes to ecological, economical and aesthetic value and welfare of human species. It consists of–(a) Genetic diversity (b) Species diversity and (c) Ecosystem diversity. The United Nations proclaimed 2010 to be the International Year of Biodiversity and the year offers a unique opportunity to increase understanding of the vital role that biodiversity plays in sustaining life on Earth. (www.fao.org). It is our responsibility to work for safeguarding this irreplaceable natural wealth and reduce biodiversity loss.

Table 4.1: Forest area in Asia and the pacific, 1990-2010.

Subregion	Area (1 000 ha)			Annual change (1 000 ha)		Annual change rate (%)	
	1990	2000	2010	1990–2000	2000–2010	1990–2000	2000–2010
East Asia	209 198	226 815	254 626	1 762	2 781	0.81	1.16
South Asia	78 163	78 098	80 309	-7	221	-0.01	0.28
Southeast Asia	247 260	223 045	214 064	-2 422	-898	-1.03	-0.41
Oceania	198 744	198 381	191 384	-36	-700	-0.02	-0.36
Total Asia-Pacific	733 364	726 339	740 383	-703	1 404	-0.10	0.19
World	4 168 399	4 085 063	4 032 905	-8 334	-5 216	-0.20	-0.13

Note: Countries and areas in the Asia and the Pacific region are grouped into the following sub regions:

East Asia: China, Democratic People's Republic of Korea, Japan, Mongolia, Republic of Korea-

South Asia: Bangladesh, Bhutan, India, Maldives, Nepal, Pakistan, Sri Lanka-

Southeast Asia: Brunei, Cambodia, Indonesia, Lao People's Democratic Republic, Malaysia, Myanmar, Philippines, Singapore, Thailand, Timor-Leste, Viet Nam-

Oceania: American Samoa, Australia, Cook Islands, Federated States of Micronesia, Fiji, French Polynesia, Guam, Kiribati, Marshall Islands, Nauru, New Caledonia, New Zealand, Niue, Norfolk Island, Northern Mariana Islands, Palau, Papua New Guinea, Pitcairn, Samoa, Solomon Islands, Tokelau, Tonga, Tuvalu, Vanuatu, Wallis and Futuna Islands.

Source: FAO, 2010a.

Forest resources having great many uses with their value being immeasurable and unique, need to be preserved. The main functions of forest resources are: Firstly, Protective functions such as soil erosion, floods, droughts, intense radiation, etc., Secondly, productive functions such as source of timber, non-timber and other byproducts, and lastly, accessory functions such as recreation, aesthetic etc.

Forests provide more than 10 per cent of the GDP in many of the poorest countries and is estimated that the forestry sector provides formal employment for 10 million people and informal employment for additional 30 to 50 million people in developing countries. The diversity of forest trees, shrubs and other related resources play a significant role in the daily life of rural communities as sources of wood and non-

Table 4.2: Forest area and area change.

Country / area	Extent of forest 2010			Annual change rate			
	Forest area	% of land area	Area per 1 000 people	1990–2000		2000–2010	
	(1 000 ha)	(%)	(ha)	(1 000 ha)	(%)	(1 000 ha)	(%)
Bangladesh	1 442	11	9	-3	-0.2	-3	-0.2
Bhutan	3 249	69	4 729	11	0.3	11	0.3
India	68 434	23	58	145	0.2	304	0.5
Maldives	1	3	3	0	0	0	0
Nepal	3 636	25	126	-92	-2.1	-26	-0.7
Pakistan	1 687	2	10	-41	-1.8	-43	-2.2
Sri Lanka	1 860	29	93	-27	-1.2	-22	-1.1
Total South Asia	**80 309**	**19**	**51**	**-7**	**0**	**221**	**0.3**

Source: FAO, 2010a.

wood products, as contributors to soil and water conservation, and as repositories of aesthetic, ethical, cultural and religious values. Even the forest animals are a main source of nutrition and income to many people, and have vital roles in forest ecology, such as pollination, seed predation, dispersal and germination, and predation on potential pest species. The United Nations General Assembly designated the year 2011 as 'The International Year of Forests' which builds on momentum already generated in other international arenas, such as those related to climate change and biodiversity, to bring even greater attention to forests worldwide. (www.fao.org)

Table 4.3: Area of planted forests in Asia and the pacific, 1990-2010.

Subregion	Area (1 000 ha)			Annual change (1 000 ha)		Annual change rate (%)	
	1990	2000	2010	1990–2000	2000–2010	1990–2000	2000–2010
East Asia	55 049	67 494	90 232	1 244	2 274	2.06	2.95
South Asia	6 472	7 999	11 019	153	302	2.14	3.25
Southeast Asia	10 059	11 737	14 533	168	280	1.56	2.16
Oceania	2 583	3 323	4 101	74	78	2.55	2.12
Total Asia–Pacific	74 163	90 553	119 884	1 639	2 933	2.02	2.85
World	178 307	214 839	264 084	3 653	4 925	1.88	2.09

Source: FAO, 2010a.

The Food and Agriculture Organization of the United Nations, states that, "The conservation and sustainable use of biodiversity for food and agriculture play a critical role in the fight against hunger by ensuring environmental sustainability while increasing food and agriculture production. It is imperative to do so in a

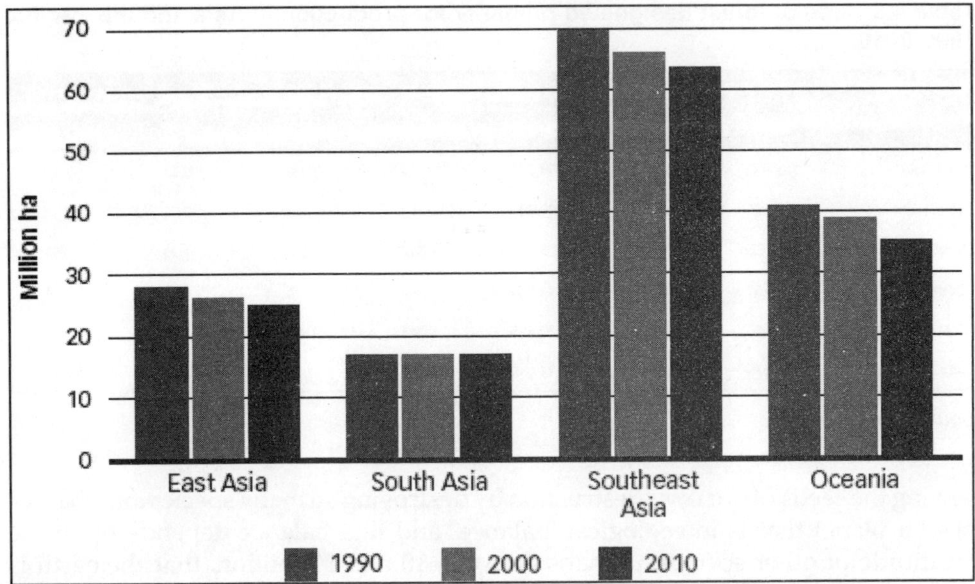

Figure 4.1: Area of primary forest in Asia and the Pacific, 1990-2010 (million ha).
***Source*: FAO, 2010a.**

sustainable way: harvesting resources without compromising the natural capital, including biodiversity and ecosystem services, and capitalizing on biological processes." (www.fao.org).

Conclusion

The contribution of Biodiversity to human well being is immense, incomparable and irreplaceable. In the year 2007, General Assembly of the United Nations approved 'Forest Biological Diversity' as one of the seven thematic elements of the concept of Sustainable Forest Management, together with the Non-Legally Binding Instrument on All Types of Forests. In 1989, Tudge stated that " It has been suggested that we are

Table 4.4: Area of forest designated primarily for conservation of biological diversity in Asia and the Pacific, 1990-2010.

Subregion	Area (1 000 ha)			Annual change (1 000 ha)		Annual change rate (%)	
	1990	2000	2010	1990–2000	2000–2010	1990–2000	2000–2010
East Asia	10 167	10 798	14 889	63	409	0.60	3.26
South Asia	15 037	15 530	22 191	49	666	0.32	3.63
Southeast Asia	32 275	35 475	38 655	320	318	0.95	0.86
Oceania	7 196	8 412	8 234	122	-18	1.57	-0.21
Total Asia–Pacific	64 675	70 215	83 969	554	1 375	0.83	1.80
World	270 413	302 916	366 255	3 250	6 334	1.14	1.92

Source: FAO, 2010a.

Table 4.5: Area of forest designated primarily for production in Asia and the Pacific, 1990-2010.

Subregion	Area (1 000 ha)			Annual change (1 000 ha)		Annual change rate (%)	
	1990	2000	2010	1990–2000	2000–2010	1990–2000	2000–2010
East Asia	126 936	119 592	94 711	-734	-2 488	-0.59	-2.31
South Asia	18 255	18 684	19 713	43	103	0.23	0.54
Southeast Asia	96 554	109 973	104 526	1 342	-545	1.31	-0.51
Oceania	7 241	11 180	11 569	394	39	4.44	0.34
Total Asia–Pacific	248 986	259 429	230 519	1 044	-2 891	0.41	-1.17
World	1 181 576	1 160 325	1 131 210	-2 125	-2 911	-0.18	-0.25

Source: FAO, 2010a.

sowing the seeds of our own destruction by destroying so many species and that we need a planet that is in ecological 'balance' and that balance depends upon the multitude of other species, perhaps between 10 and 30 million, that the earth is thought to contain." According to Briggs and Walters (1997), "Man is endangering species through–the loss, degradation and fragmentation of natural habitats; the over exploitation of natural resources; the catastrophic effects of introduced plants and animals, including pests and diseases; and the effects of environmental pollution, resulting not only in localized loss of biodiversity but also in the wider endangerment of plants and animals."

Habitat destruction leading to the likelihood of extinction of many species is a major concern today and hence there is a need for–understanding mother nature, attitude change, sustainable usage of natural resources and safeguarding and conservation of biodiversity, because the Food and Agriculture Organization of the United Nations States that, "Losing forest diversity means missing opportunities for medicines, food, raw materials and employment opportunities, in one word: welfare." (www.fao.org).

References

Briggs, D and Walters, S.M., 1997. *Plant Variation and Evolution*, 3rd edn. Cambridge University Press, Cambridge, UK.

Freeman, J., 2007. *Science 101 Ecology*. HarperCollins Publishers, New York, USA.

Khanna, L.S and Chaturvedi, A.N., 2008. *Handbook of Forestry*. Khanna Bandhu, Dehra Dun, India.

Paavilainen, E., 1994. The Concept of Sustainable Management in Boreal and Temperate Forests. *IUFRO News*, 23/3: 8–9

Tudge, C., 1989. The rise and fall of *Homo sapiens sapiens*. *Philosophical Transactions of the Royal Society*, London, B, 325, 479–88.

U.S. Congress, Office of Technology Assessment (March, 1987). Technologies to Maintain Biological Diversity (OTA–F–330) Washington, DC, USA: U.S. Government Printing Office.

www.fao.org/biodiversity/en/

www.fas.org/ota/otareports

Environment, Biodiversity and Traditional System, 2014 *Pages* 39–48
ISBN 978-81-7622-307-2
Editors: D.R. Khanna, G.S. Solanki and S.K. Pathak
Published by: BIOTECH BOOKS, NEW DELHI

Chapter 5

Potential Benefits to Biodiversity by Using Marine Renewable Energy

Guruswaamy

Assistant Professor,
Department of Electrical Engineering,
Mizoram University, Mizoram, India
E-mail: guru_revana@yahoo.com

ABSTRACT

The evidence for anthropogenically induced climate change is overwhelming with the production of greenhouse gases from burning fossil fuels being a key driver. In response, many governments have initiated programmes of energy production from renewable sources. The marine environment presents a relatively untapped energy source and offshore installations are likely to produce a significant proportion of future energy production. Wind power is the most advanced, with development of wave and tidal energy conversion devices expected to increase worldwide in the near future.

Concerns over the potential impacts on biodiversity of marine renewable energy installations (MREI) are habitat loss, collision risks, noise and electromagnetic fields. These factors have been posited as having potentially important negative environmental impacts. Conversely, we suggest that if appropriately managed and designed, MREI may increase local biodiversity and potentially benefit the wider marine environment. Installations have the capacity to act as both artificial reefs and fish aggregation devices, which have been used previously to facilitate restoration of damaged ecosystems, and marine-protected areas, which have proven successful in enhancing both biodiversity and fisheries.

MREI have the potential to be both detrimental and beneficial to the environment but the evidence base remains limited. To allow for full biodiversity impacts to be assessed, there exists an urgent need for additional multi and inter-disciplinary research in this area ranging from engineering to policy. Whilst there are a number of factors to be considered, one of the key decisions facing current policy makers is where installations should be sited, and, dependent upon site, whether they should be designed to either minimize negative environmental impacts or as facilitators of ecosystem restoration.

Keywords: ***Artificial reefs, Environmental impact, Fish aggregation devices, Marine-protected areas, Wave power, Wind farm, Wind power.***

Introduction

It is now widely recognized that there must be a paradigm shift in energy production from fossil fuels to alternative energy sources if we are to mitigate the effects of anthropogenically induced climate change. The marine environment represents a virtually untapped source of energy, which could theoretically, meet the total global demand for power; offshore renewables are likely to play a major part in a suite of technologies.

Primary amongst these is wind power, which has rapidly increased in capacity in recent years. Onshore wind farms are often in competition with other land users and cause aesthetic concerns. This, coupled with the better wind conditions in offshore areas, has resulted in an escalation in the development of offshore wind farms. The potential to capture energy from waves has seen increasing interest, with pilot developments in a number of countries. Although the technology is behind that of offshore wind power, it has the potential to provide a significant contribution to renewable energy production in countries with suitable wave conditions.

Given the demand for renewable energy it seems likely that for countries with large offshore wind and wave resources, an increasing proportion of their offshore coastal water will be turned over to marine renewable energy production. Renewable energies are often viewed as environmentally benign, especially when compared with the current vilification of carbon based and nuclear energy supplies. While on a global scale the advantages of renewable energy are not in doubt, the impacts on the local environment must also be carefully considered. The coastal marine environment is currently experiencing unprecedented anthropogenic pressure. Areas including Western Europe, North America, and the Far East which are amongst the sites likely to be targeted for marine renewable energy generation are also the areas experiencing a high degree of environmental stress.

The increased development of marine renewables has been mirrored by an increasing number of studies highlighting the potential impacts on the local environment, and most have tended to concentrate on the potential negative effects. Although a number of large scale offshore wind farms are under development, the marine renewable sector is still in its infancy. As such, we are now ideally placed to assess the environmental impacts of current marine renewable energy installations (MREI) and hence shape the of future marine renewable development, learning from the mistakes made when assessing the impacts of terrestrial wind farms on birds.

Despite having been studied for over 10 years, the impact of terrestrial wind farms on avian populations remains unclear. While a number of studies suggest that wind farms may have detrimental impacts on birds, a recent systematic review demonstrated that the evidence-base remains poor, largely because of methodological weakness and short time scales of previous studies. Particularly worrying are the lack of both replication and baseline comparisons. The systematic review process produces an unbiased and objective synthesis from the available (potentially conflicting) findings in a given field, with the aim to guide decision making. This approach, along with good study design, should be central to assessing the impacts of MREI.

While it is clearly critical to assess, minimize and mitigate any detrimental effects, such effects should be considered within a wider context as a few studies have indicated the potential positive environmental impacts of marine renewables

Possible Negative Impacts on Biodiversity

Some potential negative impacts will be specific to the MREI, but there are also a number of generic threats, and impacts will differ significantly between the construction, operational and decommissioning stages.

Habitat Loss/Degradation

Habitat loss because of MREI will vary depending on the type and size of the installation, the location, whether it is situated in degraded or pristine habitat, and the stage of the life cycle of the installation. Offshore MREI are however generally considered unlikely to result in significant habitat losses, although inappropriate placing has the potential to cause deleterious effects for certain species. For examples sea ducks, whose restricted foraging habitat may also be suitable for MREI, may be displaced from affected areas. Wave energy convertors will have less impact on the seabed than wind farms, as they generally float on the water surface or are suspended within the water column, and are only anchored to the seabed during operation. Impacts will also vary during the lifetime of MREI with the greatest expected impacts during construction and decommissioning as a result of factors such as direct habitat destruction, noise and altered sedimentary process. While MREI may result in some habitat loss, the infrastructure associated with the seabed, particularly from wind turbine piles, may act as artificial reefs, thus increasing the amount of available habitat for some species.

Collision/Entanglement

The collision hazards presented by MREI are divided into two main sections: first, avian collisions with sections of MREI above the water, particularly illuminated areas and wind turbines; and secondly, collision entanglement of marine vertebrates with underwater structures. In this paper we consider the current designs for wind turbines, although new designs to capture wind power at lower altitudes are already under development which may significantly reduce the collision risks for birds. The effects of collisions with wind farms on birds are still a contentious issue, and a number of factors remain unclear. Importantly, it must be noted that the majority of studies so far have been conducted on terrestrial wind farms. The study of the impacts on offshore avian populations is likely to be more challenging and will require the development and application of novel detection strategies. In addition, MREI will carry navigation lights, which have the potential to attract seabirds, increasing the collision risks.

The impacts are highly dependent on the site and the nature and conservation status of both resident and migratory species utilizing the area. Generally, wind farms have a negative impact on local bird abundance. Given that actual levels of mortality caused by collisions with structures appears to be low, reductions in abundance may be attributed to avoidance rather than the direct effect of collisions. Furthermore on investigation to the non-lethal effects of wind farms, such as disturbance or reduction in habitat quality, although initial findings demonstrate that wind farms have minimal impacts and those birds are able to habituate to them.

Virtually nothing is known about the potential for collision between submarine animals and MREI. Wind turbines are unlikely to represent a significant risk, being large and static, whereas there is a vast array of designs for wave and tidal energy collectors which may pose greater hazards. The most advanced of these are only at the pilot stage, so understanding the impacts of large scale installations or arrays is difficult. Using analogies with other anthropogenic marine activities, a number of factors which may find to be important; fixed submerged structures are likely to pose little collision risk, while cables, chain, power lines and components free-moving on the surface or in the water column will pose a much higher risk of collision. A number of proposed devices have rotating turbines, which have the potential to seriously injure or kill organisms. In addition, a variety of marine organisms are attracted to marine light sources which may be present on MREI, and have the potential to increase collision risks.

Noise and Vibration

The impact of anthropogenic underwater noise and vibrations on marine life is a growing concern, with an increasing body of evidence demonstrating its adverse effect over a range of species. MREI will undoubtedly introduce extra noise, which is likely to impact on local marine life, particularly acoustically sensitive species such as marine mammals. Disturbance will be most severe during construction with pile-driving having been observed to directly affect the behavior of seals and cetaceans, possibly through masking vocalization in the latter group. Estimates suggest fish can detect pile-driving noise over large distances, and that the noise may affect intra-specific communication, or cause injury or mortality at close range.

Noise during the operational phase is likely to be less intrusive. Particular attention should be paid to identify the range of frequencies utilized by marine organisms and minimize the production of noise within this frequency range. It is estimated that the operational noise of wind turbines will be audible by harbor porpoises at around 100 m, and by harbor seals Phoca at over 1 km. Fish react to the noise generated by wind farms, and during operation fish avoided wind turbines at a distance of 4 m; however, although noise may mask communication and orientation signals, they experience no destructive hearing effects. While the impacts of wind farms on bird abundances have received much attention, the behavioral mechanisms and clues used by birds to detect and avoid wind farms. It is found that neither turbine noise nor movement affected common eiders and concluded that they used vision to avoid the structures. Indeed, for birds, extra noise from turbines may provide an extra sensory input to aid in avoidance.

Anthropogenic noise is thought to be detrimental to sea turtles, whilst sea otters and manatees are sufficiently restricted to inshore waters so as not to be significantly affected, and marine invertebrates are not thought to be affected. Most work has been carried out on wind farms, yet the effects of noise from other MREI are likely to be highly variable. An important assumption made here is that minimization of noise from MREI is desirable.

Electromagnetic Fields

In addition to the structures used to harness marine energy, submarine electrical cables are needed to transfer power between devices, to transformers and to the mainland, producing a high concentration of cabling at MREI. These cables will produce electromagnetic fields (EMF) which may be detected by a number of marine organisms including electro-sensitive fish. The magnetic component of EMF will be of similar strength to that of the earth in close proximity to the cables, and so have the potential to affect magneto-sensitive species such as bony fish, elasmobranchs, marine mammals and sea turtles. It is also possible that magnetic fields could affect animals using geomagnetic clues as an aid to navigation during migration, although the importance of these clues remains unclear. The evidence for actual effects again remains very poor, and presents an opportunity for future research.

Possible Benefits to Biodiversity

The environment of our coastal waters is being increasingly compromised by anthropogenic activities, although this is not necessarily irreversible and may be restored, to a degree, by establishing marine reserves, for example, which generally show some benefits. There is an increasing body of evidence to suggest that, with appropriate design, placing, and management, MREI may actually have the potential to produce (with some caveats) positive environmental impacts. Of course, it should be noted that placing of MREI in pristine habitats not threatened by human activity would be unlikely to produce any benefits.

Artificial Reefs

The proposed construction of MREI will increase the amount of hard substrate in coastal environments and thus may have a significant impact. Man-made structures placed on the sea bed attract many marine organisms. These 'artificial reefs' are often used to enhance fisheries, for habitat rehabilitation, for coastal protection and to attract eco-tourists. Other anthropogenic structures fixed to the seabed whose primary function is not to act as artificial reefs, such as oil platforms and piers, have also been reported to attract marine organisms, and have been called secondary artificial reefs. The presence of novel structures effectively creates new habitat capable of supporting more epibiota and fish, and have consistently been demonstrated to increase both the density and biomass of fish when compared with surrounding soft bottom areas and even local natural reefs. The species composition of artificial reefs may, however, not be the same as natural reefs, and their presence may also influence the biodiversity of surrounding areas. Artificial reefs may also promote the establishment and spread of non-native species and harmful algal blooms. In addition it remains unclear whether the artificial reefs facilitate recruitment in the local population, or whether the effects are simply a result of concentrating biomass from surrounding areas. If the latter is true then it has been suggested that artificial reefs may perhaps have deleterious effects by increasing both fishing effort and catch rates and by causing exploitation of previously unexploited stock segments, and concentration of currently exploited stock. However, this point becomes less significant if the artificial reefs are out of bounds to fisheries.

It is highly likely that MREI have the potential to act as artificial reefs, and preliminary findings support this. It is found that greater abundances of fish within the vicinity of wind turbines than in surrounding areas, although the species richness and diversity showed little difference. Abundance was greater on the turbine monopiles, although species richness and diversity were lower than in surrounding areas. Preliminary evidence also suggests that the foundations of wave energy converters can act as secondary artificial reefs, with structures becoming rapidly colonized by both epibenthic assemblages and fishes. Additionally the diversity of species increased with time, and was dominated by species resident to the areas.

The extent to which these secondary artificial reefs attract marine life and the nature of the species attracted will largely be shaped by the design of the components of the installation, with structural complexity of exposed surfaces being a key driver of the extent of colonization. Also important is the extent to which they are designed to either attract any species, only attracting local species artificial reefs with similar structural features to local natural reefs will attract local fauna or repel colonization. Given that the desirability of artificial reefs still remains a matter of conjecture, it is important that factors including local species abundance and diversity be considered before deployment, particularly if MREI are sited in pristine sites where increased biodiversity may be undesirable.

Fish Aggregation Devices

While wind turbines are restricted to relatively shallow waters as current designs are attached directly to the seabed, a number of manufacturers are currently developing floating turbines which, along with wave energy devices, will be anchored to the sea bed, but will be free to move on the surface, or within the water column. These floating devices may well have extensive moorings and some underwater engineering that can still function as, perhaps, more of a patch reef, and are in addition likely to act as fish aggregation devices (FAD). Many species of fish aggregate around floating objects and fishermen have known and taken advantage of this phenomenon to increase their catches for centuries. Additionally, offshore mussel farms can attract fish, which are at similar scales to MREI. FADs are now utilized in numerous fisheries globally, particularly in tropical and subtropical waters. Furthermore, as FAD act to concentrate fish stocks rather than increase recruitment, the possibility for resultant overexploitation is clear. The utility of FAD as a tool for conservation and ecological restoration therefore remains questionable, although if fish are being attracted to areas that are free from fisheries pressure, which we envisage being the case for MREI, the benefits appear more tangible.

Marine-Protected Areas

The possibility of collision and gear entanglement means that even without enforcement the immediate vicinity of MREI will not be able to be fished using many gear types. Larger installations with multiple arrays of devices, especially wave energy and tidal stream generating sites, will probably be enclosed within enforced exclusion zones for both safety and protection of the installations and act as de facto marine-protected areas (MPA) to most fisheries. MPA, where all fisheries and other forms of extraction are excluded, are increasingly being called for fisheries management, conservation and ecological restoration tool. As well as protecting and enhancing fish stocks, the implementation of such MPA will also enrich benthic biota, by lifting the pressure from towed bottom fishing gear, which have chronic effects on seabed communities and are likely to affect ecosystem function. Indeed, cited fisheries impact studies as analogues to illustrate the potential negative ecological impacts and consequences of MREI construction. It is noteworthy, however, that construction impacts are temporally discrete, whereas fishing grounds are often repeatedly trawled damaged, leaving little chance of recovery.

The efficacy of MPA, while controversial, is largely dependent on the management objectives. There is increasing empirical evidence to suggest that, under effective management, MPA do work, and represent a relatively cost-effective means of habitat restoration. Despite this, much more targeted research is needed to fully understand the underlying mechanisms at work. MPA were effective in increasing density, biomass, size of organism and diversity within reserves when compared with the surrounding areas. Additionally, even the smallest reserves had a positive impact, with effects increasing directly with the size of the reserve. More recent results strongly suggest, however, that both size and age of the MPA are important determinants of success, at least in the case of commercial fisheries. In addition to size and age, the habitat type and efficacy of the management regime are also key to success.

While MREI acts as MPA it is important to remember that the location, and hence habitat protected by the MPA, may not be particularly valuable in terms of conservation, restoration or fisheries management. The placing of MREI will be influenced by the strength of the energy resource to be harnessed, suitability of the sea-bed, ease of connecting the infrastructure to the main-land, and consideration of other stake-holders in the marine environment. Even so, MPA which have previously been designed and sited with no intent to conserve biodiversity or enhance fisheries have still been shown to protect fish stocks within the MPA.

Interaction of Positive and Negative Impacts

MREI have the potential to produce significant anthropogenic influence on marine ecosystems, and the positive and negative impacts on the marine environment will certainly interact in complex and unpredictable ways. These impacts may also be cumulative, both in time, and with the introduction of increasing number of MREI, hence it is critical that we consider a wider marine ecosystem rather than focusing on the effects of individual installations. Furthermore, for all species of marine organisms, including birds whose flight paths may be affected by MREI, it is critical that we assess whether effects on individuals, at specific sites, are strong enough to produce population level effects.

Given that MREI have the potential to create additional habitat (by creating artificial reefs), attract marine organisms (via artificial reef and FAD effects) and create an area free from fisheries pressure, it seems possible that the overall effects on marine fauna will be positive. There is increasing evidence to suggest that tropical cascades play a pivotal role in ecosystem function, and have recently been invoked as a mechanism by which lower trophic levels are impacted by anthropogenic disturbances at higher levels. Conversely, conditions at lower trophic levels have the capacity to significantly influence the dynamics of higher level organisms. Hence by enhancing conditions at lower trophic levels we suggest that higher trophic levels including predatory fish, marine mammals and birds may also benefit from well-managed MREI. A key consideration here is that we seek to restore and enhance species native to the area, rather than encourage an increase in non-native or invasive species, which may be more competitive in colonizing new habitats created.

The scale over which these effects will propagate is also important and will be strongly affected by fisheries pressure in the vicinity. Recent study has demonstrated that networks of MPA have a rapid positive effect on fish abundance and, as such, the introduction of networks of MPA associated with MREI may provide a powerful tool for restoration ecology. The potential linkage of conservation needs and renewable energy generation requirements should be explored to create such a network. Paradoxically, if potential benefits of certain MREI can be demonstrated, placing of these developments within areas requiring conservation may assist in the protection of wider areas of the marine environment.

Conflicts and Solutions

There are a number of key conflict areas associated with MREI, because of the large number of stakeholders with conflicting interests and objectives. Therefore, this paper suggests that a fruitful line of future research may be modeling a number of management strategies within a multi-objective optimization framework, as have been used in previous environmental management scenarios.

Environmental Concerns

Should marine renewable devices be designed to have minimal negative environmental impacts, or to attract and increase biodiversity? It is critical that policy makers and stakeholders rapidly come to some agreement. The worst possible outcome would be designs that are neither intended to produce a minimal impact nor enhance biodiversity. As it is highlighted here, there is an urgent need for attention of this work, particularly into the potential negatives of MREI, but also the prospect of MREI for habitat enhancement and even environmental recovery. It is essential that currently deployed MREI are studied appropriately to quantify impacts and benefits, the results from which can then be fed back into the design and deployment stages of future installations and used to model the likely impacts of larger scale developments.

Whether designed to be relatively environmentally benign or to enhance biodiversity, it is critical that all stakeholders, including energy companies, engineers, local communities, governmental and non-governmental organizations, fisheries, and academic institutions, are involved at all stages from design, placing, pre-construction monitoring impact assessment construction, operation and decommissioning. Furthermore research should be, as far as is commercially viable, transparent, with results published in scientific journals, to both maintain the scientific credibility, and accessibility to other researchers. Environmental monitoring of wind farms is already a condition of license requirements in many EU counties, and MREI should be designed with this in mind. Finally, it is critical that we consider the longer term future of MREI. If MREI are engineered to enhance biodiversity, particularly in degraded habitats, then we must ensure these enhancements are not lost during decommissioning. Consideration should be given to living new habitat forming infrastructure in situ and building this factor into the consenting process.

Fishing Concerns

The potential loss of access to areas containing and surrounding MREI will be a prime concern to the fishing sector, which should be involved as a key stake-holder from the earliest consultation stages. Outreach and training programmes should also be in place to highlight the potential benefits of MREI, which should also facilitate with the enforcement of fisheries exclusion from MREI. While there will be loss of some fishing areas, this may be countered by spill-over effects, and recruitment subsidies. Indeed it has been suggested that MPA may in fact increase fisheries profitability in comparison to other forms of fisheries management. Additionally it has been suggested there is an opportunity for some forms of aquaculture, such as macroalgae or mussel shellfish farming, within MREI.

Operators

The primary objective of MREI operators will be financial; although many either are or aspire to be environment-friendly, hence additional environmental benefits are desirable. Biodiversity enhancing features of MREI, however, may represent a number of technical problems for operators, such as the potential for bio-fouling to damage infrastructure and components. For larger devices consideration should be given, at the design stage, to the potential for birds using the structures as roosts, or seals using them as haul-out sites.

Conclusions

Undoubtedly MREI will increasingly be part of the seascape of tomorrow. It is clear that decisions made in the near future as to whether installations should be sited, sized and engineered to either minimize negative environmental impacts or maximize biodiversity and fisheries yield in coastal seas, will have impacts on the state of the marine environment in years to come. Since an earlier, some progress has been made in our understanding of how marine renewable energy will affect the marine environment and, as outlined above, there may be positive as well as negative effects. For example, the positive effects of MPA are becoming better understood as is evidence that MREI will act as artificial reefs. It is also becoming evident that wind farms may not have such a detrimental effect on avian populations as earlier studies suggested. Despite this, the evidence-base for the impacts, both positive and negative, of marine renewables remains poor and there exists an urgent need for additional multi and inter-disciplinary biodiversity orientated research ranging from engineering to policy. Given the diverse number of stakeholders interested in the coastal seas, all such initiatives must take an inclusive approach for best effect. Given the already seriously degraded nature of our coastal seas the paper suggests that, if research and development programmes are targeted at identifying and promoting environmental benefits, marine renewable energy has the capacity to enhance biodiversity in degraded marine habitats, thus, representing an excellent example of 'win–win ecology'.

References

Abbasi, S.A. and Abbasi, N., 2000. The likely adverse environmental impacts of renewable energy sources. *Applied Energy*, 65: 121–144.

Balmford, A., Gravestock, P., Hockley, N., McClean, C.J. and Roberts, C.M., 2004. The worldwide costs of marine protected areas. *Proceedings of the National Academy of Sciences*, 101: 9694–9697.

Bohnsack, J.A., Harper, D.E., McClellan, D.B. and Hulsbeck, M., 1994. Effects of reef size on colonisation and assemblage structure of fishes at artificial reefs off south-eastern Florida, USA. *Bulletin of Marine Science*, 55: 796– 823.

Drewitt, A.L. and Langston, R.H.W., 2006. Assessing the impacts of wind farms on birds. *Ibid*, 148: 29–42.

Gill, A.B., 2005. Offshore renewable energy: ecological implications of generating electricity in the coastal zone. *Journal of Applied Ecology*, 42: 605–615

Kennedy, M.C., Ford, E.D., Singleton, P., Finney, M. and Agee, J.K., 2008. Informed multi-objective decision-making in environmental management using Pareto optimality. *Journal of Applied Ecology*, 45: 181–192.

Luschi, P., Benhamou, S., Girard, C., Ciccione, S., Roos, D., Sudre, J. and Benvenuti, S., 2007. Marine turtles use geomagnetic cues during open-sea homing. *Current Biology*, 17: 126–133.

Stewart, G.B., Pullin, A.S. and Coles, C.F., 2007. Poor evidence-base for assessment of wind farmimpacts on birds. *Environmental Conservation*, 34: 1–11.

Environment, Biodiversity and Traditional System, 2014 **Pages 49–57**
ISBN 978-81-7622-307-2
Editors: D.R. Khanna, G.S. Solanki and S.K. Pathak
Published by: BIOTECH BOOKS, NEW DELHI

Chapter 6

Plants Used in Hair Care by the Ethnic Communities in Thoubal District of Manipur

Mohd. Habibullah Khan[1] and P.S. Yadava[2]

[1]Directorate of Environment, Government of Manipur, Porompat, Imphal East – 795 010
E-mail: habibullah_mohd@yahoo.co.in
[2]Centre of Advanced Study in Life Sciences, Manipur University, Imphal – 795 003, Manipur
E-mail: pratap_yadava@ymail.com

ABSTRACT

Since time immemorial the ethnic communities have been utilizing various plant species for hair care in Thoubal District of Manipur. 42 plant species were recorded during ethnobotanical field trip in different part of the district which were used for hair care, such as hair fall, hair growth, baldness, lice problems, dandruff, hair tonic, and as hair wash by the ethnic communities. The detail information on their habit, habitat and uses in haircare have been reported in the present study.

Keywords: Chinghi, Hair care, Loi, Meitei, Meitei-pangal, Thoubal.

Introduction

Thoubal district of Manipur is situated between 23°45' N and 24°45' N Latitudes and 93°45' E and 94°15' E Longitudes occupying the south eastern part of the Manipur valley offering immense scope for etnobotanical studies, since it is inhabited by numerous ethnic communities, aboriginal tribes and castes *viz.* Meitei, Meitei-pangal(Muslim), Chiru, Gangte, Hmar, Kabui, Kom, Lamgang, Maring, Paite, Simte, Tangkhul, Vaiphei, Zou, Loi, Yaithibi etc. which have developed their own indigenous culture, customs and tradition for using various plants and animals from their surrounding for their food, medicine, material culture, religious rites, myths etc. by empirical reasoning, gained from their age long experiences which are well reflected in their folklore.

In the recent years due to the influence of modern civilization the people have given up their age old culture, traditions and beliefs *etc.* and also many of the valuable plants are disappearing along with the so called development. Thus the need to record and document the valuable information and knowledge available on the use of ethnomedicinal plants became very important as the wealth of information which is an unwritten *materia medica* may be wiped out with the death of the few medicinal practitioners and the older generation. The paper reports 42 plants used especially for hair care by the different ethnic communities of the Thoubal District of Manipur in North East India.

Materials and Methodology

Ethnomedicinal information of hair care practiced by the different ethnic communities of the district was collected through field visits in those areas that are inhabited by the different ethnic communities. The procedures for getting information differ from person to person. As for the present study, methods adopted by Jain and Goel (1995) were followed with slight modifications under local context.

The collected plant material that were used by the ethnic communities were identified by comparing with related published floras of the region and for authentic identification, flora and monographs have been consulted especially Flora of British India Vol.1-7 (Hooker, 1872-1897); Flora of Assam, Vol.1-4 (Kanjilal *et al.*, 1934-40) and Vol. 5 (Bor, 1940); Flora of India, Vol. 1-3, (Sharma *et al.*, 1993a,b,c), Flora of Manipur, Vol. I. (Chauhan *et al.*, 2000) and finally herbarium of Botanical Survey of India, Shillong were consulted for confirmation of the specimens. After identification, the correct nomenclature were written on the herbarium sheet and kept in the herbarium centre of Manipur University, Manipur.

Observation

After conducting regular exploration trips in all the seasons during the year 2005-2008 in most part of the district, ethnomedicinal information of 42 plant species used for hair care by the ethnic communities were collected and are enumerated below in alphabetical order. For each plant species listed, the following information is reported: Latin name as per the requirements of the International Code for Botanical Nomenclature, followed by Family names in brackets in bold letters immediately after the species name; in case of genera with more than one species the first alphabet of the genus has been given in the subsequent species under the same genus. The most local vernacular names are given in Manipuri as most of the people of different dialects living in the district serve Manipuri language (Man.) Similarly Hindi as (Hin.), English as (Eng.) and names in other dialects is used with the name of the tribe within brackets if available followed by Specimen number after the abbreviation HK;

a short note on the habit and locality of the plant followed by ethnomedicinal uses including the form of preparation and name of the ingredients used are mentioned where ever it is possible.

Enumeration

Ageratum conyzoides **L.; (Asteraceae); 'Khongjainapi' (Man.); 'Goat weed' (Eng.) HK 1190**

It is an annual herb commonly grows along dried up streams, low moist situation and swampy areas on roadside and homestead compounds.

The Meitei and the Meitei-pangal communities uses the plant as an important ingredient in the preparation of local hair care lotion '**chinghi**' for luxurious hair growth and the plant is also boiled with *Commelina bengalensis* in rice soaked water which is sometimes used as hair wash by the women folk of Meitei community.

Albizia stipulata **(Roxb.) Boivin. (Mimosaceae) 'Khok' (Man); 'Siris' (Hin.); 'Siris tree' (Eng.) HK 1231**

It is a tall unarmed evergreen tree commonly planted as an avenue tree.

The boiled extract of the bark is used to wash the hair in dandruff by the Meitei and the Meitei-pangal communities.

Aloe barbadensis **Mill.; (Liliaceae) 'Ghrita kumari' (Man.); 'Ghee kanvar'(Hin.); 'Indian aloe'(Eng.) HK 1080**

It is a perennial plant commonly grows in homestead compound and also planted as a pot plant in the garden.

The crushed leaf extract mixed with sesame oil is applied on the head against falling hair by the Meitei community.

Ananas comosus **(L.) Merr.; (Bromeliaceae) 'Kihom'(Man); 'Ananas'(Hin.); 'Pineapple'(Eng.) HK 1223**

It is a herb having numerous, elongated and finely toothed rosulate leaves which is commonly cultivated in the hill slopes.

The crushed extract of the fruit is applied on the scalp in dandruff by the Loi community.

Anisomeles indica **(L.) O. Kuntze; (Lamiaceae) 'Thoiding angouba' (Man.) HK 1506**

It is a tall erect branching suffruticose herb occasionally cultivated.

The leaves of the plant is used as an ingredient in the preparation of local hair care lotion along with rice soaked water extract '**chinghi**' by the women folk of every community.

Areca catechu **L.; (Arecaceae) 'Kwa' (Man.); 'Supari' (Hin.); 'Betel nut' (Eng.) HK 1244**

It is a tree having a straight, slender and unbranched stem rarely cultivated.

The nut is rubbed on a stone with water and the extract obtained is applied on the head in hair fall by the Meitei-pangal community.

Artabotrys hexapetalus **(L. f.) Bhandari; (Annonaceae) 'Chinichampa'(Man.) HK 1066**

It is a climbing shrub occassionally planted in the gardens.

The leaf and the flower is used as an ingredient in the preparation of local hair care lotion '**chinghi**' along with rice soaked water extract by the Meitei community.

Artemisia maritima **L.; (Asteraceae) 'Ching laibakngou'(Man.); Kirmala'(Hin.) HK 1077**

It is a decidious shrub which grows in the foothill, roadsides and other wasteland.

The leaves are sometimes used as an ingredient in the preparation of traditional local hair care lotion '**chinghi**' along with rice soaked water extract by the Meitei community.

A. nilagirica **(C.B.Clarke) Pamp; 'Laibakngou'(Man.) HK 1272**

It is a tall aromatic shrub commonly grows in the wastelands, foot hills, roadsides.

The leaves are used as an ingredient in the preparation of traditional hair care lotion '**chinghi**' along with rice soaked water extract by the Meitei community.

Blumea densiflora **DC.; (Asteraceae) 'Karpoor macha'(Man.) HK 1261**

It is an annual herb which grows wild in waste places.

The plant is used as an ingredient in the preparation of local hair care lotion '**chinghi**' along with rice soaked water extract that is used to get rid from lice and for good hair texture by both the Meitei and the Meitei-pangal communities.

Calamintha umbrosa **Benth.; (Lamiaceae) 'Vicks pambi'(Man.) HK 1576**

It is a small hairy herb rarely planted in homestead compound.

The tender shoot is also used as an ingredient in the preparation of local hair care lotion '**chinghi**' along with rice soaked water extract by the Meitei community.

Callistemon linearis **DC.; (Myrtaceae) 'Brushlei'(Man.) HK 1025**

It is a small evergreen tree cultivated as an ornamental in homestead compound.

The plant is used as an ingredient in the preparation of a local hair care lotion '**chinghi**' along with rice soaked water extract by Meitei and Meitei-pangal communities.

Citrus aurantium **L. (Rutaceae) 'Champra'(Man.); 'Nimbu'(Hin.); 'Lemon'(Eng.) HK 1058**

It is a small tree commonly planted in domestic compounds.

Fresh juice of the fruit mixed with sesame oil is applied in dandruff by Meitei and Meitei-pangal communities. The leaves and rind is used as an ingredient in the preparation of local hair care lotion '**chinghi**' along with rice soaked water extract by the Meitei community.

***C. maxima* (Burm.) Merr.; 'Nobab'(Man.); 'Maahanibu'(Hin.); 'Paradise apple' (Eng.) HK 1404**

It is a small profusely branching tree commonly planted in homestead compound.

The spongy fruit rind is used as an ingredient in the preparation of a local hair care lotion '**chinghi**' along with rice soaked water extract by almost all the communities.

***Coffea arabica* L.; (Rubiaceae) 'Coffee' (Man.) HK 1045**

It is a glabrous shrub or small tree rarely cultivated in some homestead compound.

The extract obtained by boiling the seed in rice soaked water is also used to wash the hair in dandruff by the Meitei and the Meitei-pangal communities.

***Commelina benghalensis* L.; (Commelinaceae) 'Wandengkhoibi'(Man.); 'Kanchara'(Hin.) HK 1589**

It is a glabrous and diffused small herb rooting at lower nodes commonly grows in moist area and also sometimes planted as an ornamental pot plant.

The plant is boiled along with *Ageratum conyzoides* in rice soaked water and the extract is used as hair wash by the women folk of Meitei and Meitei-pangal communities.

***Dicrocephala latifolia* (Pers.) DC.; (Asteraceae) 'Lallukok'(Man.) HK 1301**

It is an annual herb commonly grows wild in moist waste places and other wastelands.

The leaf is used as an ingredient in the preparation of local hair care lotion '**chinghi**' along with rice soaked water extract by the Meitei and the Meitei-pangal communities.

***Hibiscus cannabinus* L.;(Malvaceae) 'Sougree'(Man.); 'Ambari'(Hin.); 'Deccan hemp'(Eng.) HK 1448**

It is a shrub having prickly stem very commonly cultivated in homestead compound.

The leaf is used as an ingredient in the preparation of local hair care lotion '**chinghi**' along with rice soaked water extract by the Meitei-pangal community.

***H. rosa sinensis* L.; 'Juba kusum angangba'(Man.); 'Jasut'(Hin.); 'Hibiscus'(Eng.) HK 1149**

It is an arborescent shrub commonly planted as hedge plant and also as an ornamental in the garden.

The leaf is used as an ingredient in the preparation of local hair care lotion '**chinghi**' along with rice soaked water extract which is mainly used to soften and to make the hair oily by the Meitei community.

***H. subdariffa* L.; 'Silo-sougree'(Man.) HK 1465**

It is an annual erect glabrous herb commonly cultivated in the kitchen garden as vegetable.

Boiled extract of the leaf and flower is used as an ingredient in the preparation of a local hair care lotion '**chinghi**' along with rice soaked water extract by the Meitei community.

***Homskioldia sanguinea* Retz.; (Verbenaceae) 'Kharam leithong'(Man.) HK 1194**

It is a large straggling shrub occasionally grows wild at the outskirt of forest areas.

The leaf is used as an ingredient in the preparation of local hair care lotion '**chinghi**' along with rice soaked water extract by the Meitei-pangal community.

***Imperata cylindrica* (L.) P. Beauv.; (Poaceae) 'Imom'(Man.); 'Ulu'(Hin.); 'Thatch grass'(Eng.) HK 1142**

It is a perennial grass having solid culms that grows wild in marshy areas.

The leaf of the plant is used as an ingredient in the preparation of local hair care lotion along with rice soaked water extract '**chinghi**' by the Meitei community.

***Kaempferia galanga* L.; (Zingiberaceae) 'Yai thamnamanbi'(Man.) HK 1614**

It is a herb having tuberous rootstocks occasionally cultivated as well as grows wild in floating masses like the phoomdis in the wetlands.

The boiled extract of the rhizome is applied on the head in baldness by the Meitei community.

***Kydia calycinia* Roxb.; (Malvaceae) 'Khabi koknal'(Man.) HK 1176**

It is a tree occasionally grows wild in the hills.

The boiled extract of the bark is used to wash hair in dandruff, headache, hairfall and for good hair texture by the Meitei community.

***Lawsonia intermis* L.; (Lythraceae) 'Mehendi'(Man. and Hin.); 'Henna'(Eng.) HK 1358**

It is a glabrous and much branched shrub commonly planted as hedge plant or as an ornamental in the garden.

The smashed fresh leaf is applied as poultice on the head in dandruff and as hair dye by the Meitei-pangal community.

***Magnolia hodgsonii* (Hook.f. and Thoms) Keng.; (Magnoliaceae) 'U-thambal'(Man.) HK 1563**

It is a small tree occasionally grows in the foothill and sometimes planted as an ornamental.

The flower and the leaf is used as an ingredient in the preparation of local hair care lotion along with rice soaked water extract '**chinghi**' by Meitei and Loi communities.

Michelia champaca L.; (Magnoliaceae) 'Leihao'(Man.); 'Champa'(Hin.); 'Champak'(Eng.) HK 1309

It is a tall handsome evergreen tree commonly planted as an ornamental and also grows wild in the hills.

The crushed extract of the leaf and flower is applied on the head for eradicating lice by almost all the communities.

M. montana Blume; 'Leihao leisang'(Man.) HK 1310

It is a tall glabrous tree occasionally cultivated in homestead compound as an ornamental.

The uses are more or less similar to *M. champaca*.

Millettia pachycarpa Benth.; (Fabaceae) 'Ngamuyai'(Man.) HK 1390

It is a large climber common in the hills.

The crushed root extract of the plant is applied on the scalp as hair wash for removing lice and controlling dandruff by the Meitei community.

Mimusops elengi L.; (Sapotaceae) 'Bokul'(Man. and Hin.) HK 1025

It is a medium sized to large evergreen tree occasionally planted in the homestead compound, road sides.

The crushed extract of the leaf is applied on the scalp in falling hairs by the Meitei-pangal community.

Mussaenda roxburghii Hook.f.; (Rubiaceae) 'Hanuba lei'(Man.) HK 1109

It is a large erect shrub commonly grows wild along foothill and in outskirt of forests.

The leaves are used as an ingredient in the preparation of local hair care lotion '**chinghi**' along with rice soaked water extract which is preferably used for proper hair growth by the Meitei and the Meitei-pangal communities.

Nyctanthes arbor-tristis L.; (Oleaceae) 'Singarei'(Man.); 'Harsinghar'(Hin.); 'Night jasmine'(Eng.) HK 1450

It is a large shrub or small tree occasionally cultivated in homestead compound.

The crushed root is applied as pulp on the scalp in dandruff by the Loi community and Maring tribes.

Oryza sativa L.; (Poaceae) 'Phou'(Man.); 'Paddy'(Eng.) HK 1430

It is an annual grass very commonly cultivated.

The rice water after washing the dehusked grain is used to wash the hair for softening hair and it is also the medium for preparation of indigenous hair care lotion '**chinghi**' with other ingredients which is used by almost all the communities.

Plectranthus ternifolius **D.Don.; (Lamiaceae) 'Khoiju'(Man.) HK 1177**

It is a robust and densely tomentose undershrub rarely grows wild in the hills.

The leaves and inflorescences are used as an ingredient in the preparation of local hair care lotion '**chinghi**' along with rice soaked water extract by almost all the communities.

Pogostemon parviflorus **Benth.; (Lamiaceae) 'Sangrei' (Man.) HK 1445**

It is a suffruticose herb which grows along foothill and sometimes planted as an aromatic plant in homestead compound.

The leaves and tender shoot are used as an ingredient in the preparation of local hair care lotion '**chinghi**' along with rice soaked water extract by almost all the communities.

Rhus javanica **L.; (Anacardiaceae) 'Heimang'(Man.); 'Kakra singi'(Hin.); 'Wild varnish tree'(Eng.) HK 1119**

It is a deciduous shrub or small tree planted as well as grows wild in the foothill.

The water soaked extract of the seed is applied on the hair in early greying by the Meitei and the Meitei-pangal communities.

Rumex maritimus **L.; (Polygonaceae) 'Torong-khongchak'(Man.); 'Jangli palak'(Hin.); 'Golden doc'(Eng.) HK 1498**

It is an annual herb which grows wild in moist waste places.

The extract obtained by boiling the leaf in rice soaked water is applied on the scalp in dandruff by the Meitei community.

Salix tetrasperma **Roxb.; (Salicaceae) 'Uyum'(Man.) HK 1535**

It is a medium sized tree which grows wild along river banks and low lying moist areas.

The boiled extract of the leaf is applied on the head in dandruff by the Meitei-pangal community.

Spondias pinnata **(L.f.) Kurz.; (Anacardiaceae) 'Heining'(Man.); 'Amara'(Hin.); 'Wild mango'(Eng.) HK 1121**

It is a glabrous tree planted in homestead compound.

The extract obtained by boiling the bark of the plant in rice soaked water is applied on the scalp in infection and dandruff by almost all the communities.

Zanthoxylum acanthopodium **DC.; (Rutaceae); 'Muthrubi tingkhang panbi'(Man.); 'Darmar'(Hin.) HK 1335**

It is an erect aromatic shrub cultivated and sometimes grows wild in the hills.

The crushed seed made into pulp is applied on the scalp in immature greying and falling of hairs by some of the Loi community.

Z. armatum DC.; 'Muthrubi'(Man.) HK 1667

It is an aromatic shrub cultivated and also grows wild near foothill.

Uses are more or less similar to Z. *acanthopodium*

Z. rhetsa (Roxb.) DC.; 'Ngang' (Man.); 'Badrang'(Hin.) HK 1668

It is a tree occasionally grows wild in the hills and sometimes cultivated.

The smashed seed is applied as pulp on the scalp in baldness by the Meitei and the Meitei-pangal communities.

Discussion

42 plant species belonging to 23 families used in hair care, maximum of five plant species were from each of the two families i.e, Asteraceae and Rutaceae, whereas four plant was from Malvaceae; three from Magnoliaceae; two from each of the three families *i.e.* Rubiaceae, Poaceae and Anacardiaceae and one from each of the 15 families *i.e.* Mimosaceae, Liliaceae, Bromeliaceae, Arecaceae, Annonaceae, Myrtaceae, Commelinaceae, Verbenaceae, Zingiberaceae, Lythracaeae, Fabaceae, Sapotaceae, Oleaceae, Polygonaceae and Salicacaeae.

The study reveals that the traditional knowledge about hair care were used by the forefathers and inherited through oral folklore and there is urgent need to promote the traditional knowledge for hair care which is ecofriendly instead of modern hair care lotions.

References

Bor, N.L., 1940. *Flora of Assam*. Vol. 5 (Gramineae). Government of Assam.

Chauhan, A.S., Wadhwa, B.M., Singh, D.K. Singh, K.P., Chakraborty, P., Sanpru, R. and Dam, D.P., 2000. *Flora of Manipur, Vol. 1. Botanical Survey of India*, Kolkata, Government of India, pp 1–600.

Hooker, J.D., 1872–1897. *The Flora of British India*. Vol. 1–7. London.

Jain, S.K. and Goel, A.K., 1995. *A Manual of Ethnobotany*, (Ed.) S.K. Jain. Scientific Publishers, Jodhpur, pp. 142–153.

Kanjilal, U.N., Kanjilal, P.C., Das, A. and Bor, N.L., 1934–1940. *Flora of Assam*, Vols. 1–4. Government of Assam, Shillong.

Sharma, B.D., Balakrishnan, N.P., Rao, R.R. and Hajra, P.K., 1993a. *Flora of India, Vol. I: Ranunculaceae–Barclayaceae*. Botanical Survey of India, Calcutta, Deep Printers, New Delhi.

Sharma, B.D., Balakrishnan, N.P. and with assistance from Sanjappa, M. (Eds.), 1993b. *Flora of India*, Vol. II. Botanical Survey of India, Calcutta, Deep Printers, New Delhi.

Sharma, B.D. and Sanjappa, M. with assistance from Balakrishnan, N.P. (Eds.), 1993c. *Flora of India*, Vol. III. Botanical Survey of India, Calcutta, Deep Printers, New Delhi.

Environment, Biodiversity and Traditional System, 2014 ***Pages* 59–68**
ISBN 978-81-7622-307-2
***Editors:* D.R. Khanna, G.S. Solanki and S.K. Pathak**
***Published by:* BIOTECH BOOKS, NEW DELHI**

Chapter 7

Joint Forest Management in Arabari, W.B., India

*Aveek Samanta and Samidh Baran Jana**

Faculty of Botany, Mahishadal Raj College,
Mahishadal, Medinipur (E) – 721 628, W.B.

ABSTRACT

To bring a reasonable success in forest management and sustainable use of the forest land, 1988 Forest Policy imposed immense importance on community relationship and environmental factors. It advised to forest departments to bring a collaborating management system with the forest fringe communities, which is later resulted into an effective forest management and popularly known as Joint Forest Management (JFM).

The first pioneer of JFM was Dr. Ajit Banerjee, who trialed this in Arabari Forest Range, in West Bengal. This paper is based on the background and the present scenario of Arabari– the source site of a breakthrough forest management system.

Introduction

For any country, forest is it's a very important resource from many aspects. Forest not only gives immense financial support but it is a rich source of biodiversity where flora and fauna are naturally nourished and nurtured.

In India, this treasure is of 19.27 percent (Forest Survey of India-2010-2011) of the whole land area which has a greater contribution to the total annual income of the country. In the year of 2004 the estimated income only from the rural forest was 220 million dollars (World Bank Report, 2006). But due to poverty and lack of awareness this open treasure is looted and degraded day to day by the local

* *Corresponding Author:* E-mail: samidhbaran@gmail.com

communities. Various approaches initiated to conserve this valuable treasure, but without the help and spontaneous co-operation from the side of local communities the conservation of forest is nearly impossible. To overcome the limitations and successful implementation of the conservation process, government established a joint management arrangement between the local communities and the forest department. It is estimated that about 15 per cent of the country's population is the natural inhabitants close to forests and they are important users of forests produces everywhere. Besides this, in some parts of the country forest is traditionally conserved by them without imposing any Govt. Law.

Given the backdrop, the bureaucrats as well as Government have effectively initiated the Joint Forest Management (JFM) Program. So, JFM shows a new horizon to create responsive and positive relationship between local inhabitants and forest departments which has been widely accepted as a promising approach to forest management (Campbell, 1992; Poffenberger and Singh, 1989).

But all the success of JFM have not come through a easy way. West Bengal and Haryana– the two states played very important and initiative roles in genesis of the JFM. In West Bengal it was successfully experimented in 1970 at the Arabari Forest Range, whereas in Haryana it was Sukhomajri village in 1975 (K. Balooni and M. Inoue, 2009).

This paper focuses on the background of the JFM started and the present scenario of Arabari–the mother site of JFM.

Background

Arabari Forest Range is situated in the district of Paschim Medinipur (formerly Medinipur) under the Medinipur Forest Division. It is very close to the district head quarter Midnapur and about 25 km. distant from Midnapur town. It is constituted of five Beat Offices, namely –*Arabari, Socio, Mirga, Dakshinsole and Gopalbandh*. National Highway-60 (NH-60) is passing through Arabari and connects another forest –rich district headquarter—Bankura.

Arabari Forest Range was a very productive forest mainly for timber of *Shorea robusta (Sal)* and *Acacia moniliformis (Akashmani/Sonajhuri)*, commercial tobacco leaves *(kendu)*, wood for paper pulp. But in the late 1960s to early 1970s the forest cover of that area was gradually degrading. At last it was observed by 1972 that about 1256 Hectares forest area was totally degraded and the remaining were tending the same path as well (Socio-Economic Project, Forest Research (South) wing, Arabari, 1987).

The loss of such a drastic change in forest cover created various ecological problems including soil erosion. On the other hand, the Forest Range was no longer able to produce commercial timber and other forest products. Soil erosion associated with the loss of tree cover had an adverse effect on local Agriculture also (F. J. Rossi, 2007). These were the main facts, for which the Forest Department had to take immediate control of the area.

Then Dr. Ajit Banerjee who was working as Divisional Forest Officer (DFO) for silviculture, took a constructive and positive step to meet the conservation challenge. It was observed that the degradation of forest had been resulted due to the over exploitation such as grazing, felling of immature trees by the local communities, encroaching, etc. Initially through meeting and the contact programs with 618 families comprising 3907 people, he tried to recover the degraded 1256 hectares area (Socio-Economic Project, Forest Research (S) Wing, Arabari, 1987). He understood that local villagers need the economic help to stop the illegal felling. A pilot project well known as the Arabari Socio-Economic Project has been taken by the Forest Department. The villagers in the project understood that by giving their assistance to the Forest Department, they would be given employment opportunities with the forest department, the right to collect firewood at an ecologically sustainable rate from the protected forest, and usufruct rights to the forest (S. B. Roy, 1993).

The Arabari Socio-Economic Project covered the eleven villages of Keshpur Police Station *viz. Satshol, Sankrui, Safadiha, Majhigarh, Mahishadubi, Jorakeudi, Erashol, Sakhishol, Urami, Ghuchishol and Chandmura* (Socio-Economic Project, Forest Research (S) Wing, Arabari, 1987). Most of the forest land as 55 per cent was managed for the re-growth of coppiced *Sal (Shorea Robusta)*. Major portion of the remaining land was cultivated by plantation of *Acacia spp., Eucalyptus spp.,* Cashew nut etc. In addition to the wages for day to day unskilled labourus work, the people of the said project were promised to benefited by 25 per cent of the final timber harvest(F.J. Rossi, 2007). Villagers under the project were offered to collect the Non Timber Forest Produces (NTFPs) also.

The Arabbari experiment thus initiated a decentralized and participatory forest management program to rehabilitate a portion of the state's forest. Saxena states– "this arrangement made it clear to the people concerned that they had a right to enjoy the enhanced benefits of forests, but this right was accompanied by their duty to nurture and protect the forests" (1997).

In this context, the Arabari experiment as Socio-Economic Project created a drastic qualitative change in the degraded forests of Arabari. Addition to this grand success, the concept gave a whole new dimension to the traditional system of Forest management. But at the same time, this may be reminded that there was no legal provision in the forest policies and law which could be impose-on the local people to involve them in the forest management. But the success story of the project attracted great attention by mid-1980s.

By this time, Dr. Banerjee had left the forest department and joined the World Bank (A. Joshi, 1998/99). In 1987, after first 15 years Arabari Project started, the Forest Department officially agreed to share 25 percent of the profit from timber sales after deducting project costs. The 25 percent of the profit has been distributed among the villagers of the project area. (S. B. Roy, 1993).

Another exclusive initiative has been accepted at the same time in Purulia. There another Divisional Forest Officer (DFO) Mr. Palit started a co-operative forest management with the local people, to prevent degradation of forest cover (Palit, 1989). The DFO influenced the villagers to form committees which is well known as Forest Protection Committees (FPCs). Though the DFO there did not offer any profits from forest product sales or any employment to the people, yet in response to his arguments the local people began protecting the surrounding forests (A. Joshi, 1998/99).

During this period, other agencies like World Bank and the Ford Foundation came frontline to study and support the Joint Forest Management System in West Bengal. Two NGOs– the Indian Institute of Bio-Social Research and Development (IBRAD) and the Ramakrishna Mission (RKM) were supported by the Ford Foundation to start the study and carry out the necessary research on JFM. At last, in 1989 a workshop has been held collectively by the forest departments, NGOs and the Ford Foundation. After the grand success of the workshop, the West Bengal Govt. issued the first state wide policy on 12July, 1989. This policy supported the FPC program. By this order Govt. also influenced the FPC formation to protect the statewise forests and offered 25 per cent share of net income from final harvesting (S. B. Roy, 1993).

Government Orders and JFM Committees

According to the Govt Orders 1989, the Ban-O-Bhumi Sanskar sthayee Samiti had the power to select the all FPC members. The beat officer was the secretary of the committee. He also was the convener of the committee meetings. By the GO, DFO had the power to dissolve the FPC, if the committee became adverse to the order or to the Indian Forest Act of 1927.

The 1989's Government Order had two main limitations, *viz.* (1) Govt. instructed to include the members in Committee from economically backward class, keeping in mind that only economically backward class people were interested to participate in JFM. And (2) Limited the role of FPC for only protection of forest. To overcome the limitations Govt. amended the order in 1990. By these amendments, every family living in the vicinity of the forest had got chance to became committee members, including the middle-and-upper income classes which was the most of the population. Besides this, the FPCs had been involved in "various activities" such as constitution of local management plans (micro-planning) by these amendments.

Another important amendment to the 1989 Govt. order was that the Forest Department or local political party– none of these had power to exclude or include anybody from becoming member of FPC.

In the recent year, 2008 Government issued a resolution which states that JFM committees shall be constituted in eight districts, *viz.* Purba Medinipur, Paschim Medinipur, Bankura, Purulia, North 24 parganas, South 24 Parganas, Burdwan and Birbhum. By this resolution there will be Joint Forest Management Committee including the Executive Committee. The Executive Committee would be recommended by "Bon-O-Bhumi Sanskar Sthayee Samiti" and require proper approval from the DFO. The committees may seek help from the Gram Panchayet(s) for smooth and proper functioning. The Executive Committee is constituted of five members, among them the beat officer or his nominee in the rank of Head Forest Guard/Forest Guard/ Ban Majdur/Ban Shramik would be designated as secretary.

Present Scenario

In 1988, the recorded forest land of West Bengal was 13.38 per cent and total forest cover of the state was 14.32 per cent. Whereas the Recorded forest land has not been increased upto the year 2006 by the Govt. Report but the percentage of forest cover has been increased year to year *e.g.* in 1997 up to 15.16 per cent and in 2006 up to 15.68 per cent. Thus recorded forest area of the state is 11,879 square km. of total geographical area 88,752 square km. (State report, WBFD)

In case of Midnapur, according to prior report the total forest area of Purba and Paschim Medinipur districts was 1709 sq.km. which is now for only Medinipur Division is 447.8554 sq km (Table 7.2).

Table 7.1: Details of FPCs in Medinipur Forest Division*

Range	No. of FPC	Male Member	Female Member	Total Member	Members belongs to SC/ST Community
Arabari	29	8126	531	8657	2828
Chandrakona	26	7364	111	7475	2261
Godapeasal	52	4639	310	4949	2385
Nayabasat	63	7236	287	7523	3421
Bhadutala	34	4036	266	4302	2237
Chandra	50	3424	355	3779	1772
Lalgarh	50	4727	219	4946	2306
Medinipur	16	3370	241	3611	1403
Pirakata	42	5116	481	5597	704
Division Total	**362**	**48038**	**2801**	**50839**	**19317**

**Source*: Forest Department of West Bengal.

Table 7.2: Area protected By FPCs (ha.) in the Medinipur Division*

Range	Sal	Plantation	Blank	Total	Per cent of Total Area
Arabari	2,419.50	2,246.25	1,185.34	5,851.09	74.63
Chandrakona	1,857.40	1,748.36	1,239.65	4,845.41	82.15
Godapeasal	2,120.08	1,512.21	127.28	3,759.57	81.17
Nayabasat	4,774.85	1,319.07	533.85	6,627.77	87.75
Bhadutala	2,523.64	1,247.00	260.67	4,031.31	84.87
Chandra	3,966.49	736.82	512.31	5,215.62	88.42
Lalgarh	4,900.64	1,347.22	184.00	6,431.86	88.88
Medinipur	733.04	1,297.85	309.03	2,339.92	96.10
Pirakata	4,001.50	982.10	699.39	5,682.99	95.43
Division Total	**27,297.14**	**12,436.88**	**5,051.52**	**44,785.54**	**86.16**

**Source*: Forest Department of West Bengal.

The amount of minor forest produces such as – Honey, Sal Seeds, Kendu leaves and Citronella grass collected from forest area play an important role in the state economy. Among these produces, a large amount of Sal Seeds and Kendu leaves are collected from Arabari forest area.

According to state reports, as on 31.03.2007, the total no. of FPCs of Medinipur Division was 474 in which the no. of male member is 38254 and no. of female members is 2449. In present state, though the nos. of FPCs slightly decreased (*i.e.*362), but the total no. of members has been increased from 40,703 to 50,839. This statement suggests that the involvement of the local communities into JFM in the division has been successfully amplified. In Arabari Range, 29FPCs are working in collaboration with the Forest Department to protect the forest. In 29 FPCs, 8126 male members and 531 female members are working effectively. 2828 people from the Schedule Caste and Schedule Tribe category are also being the valuable part of the JFM of Arabari and get benefits from the forest department (Table 7.1).

Table 7.3: Details of benefit sharing With FPCs in Medinipur Forest Division*

Felling Year	*Harvested Area (Ha.)*	*Total FPC Share (Rs.)*	*No. of FPC*	*No. of Beneficiaries*	*Share per Beneficiary (Rs.)*
1997-1998	1,220.50	10,285,661.76	112	16691	616.24
1998-99	480.50	5,197,864.00	45	6973	745.43
1999-2000	777.15	7,374,366.00	69	11404	646.65
2000-2001	905.30	11,136,028.77	82	15004	742.20
2001-2002	1,173.25	12,307,864.00	97	16808	732.26
2002-2003	1,329.00	12,540,287.00	105	17168	730.45
2003-2004	1,306.50	13,978,457.00	105	17571	795.54
2004-2005	1,231.00	14,702,607.00	102	18838	780.48
2005-2006	1,144.50	19,092,580.00	95	15303	1,247.64
2006-2007	1,176.50	20,263,897.00	112	17952	1,128.78
2007-2008	1,121.60	15,340,886.00	96	19804	774.63
2008-2009	1,016.50	16,290,221.00	75	11776	1,383.34
2009-2010	642.75	10,92,69,648.00	47	9525	11471.87
2010-2011	720	1,73,97,693.00	52	11779	1477

**Source*: Forest Department of West Bengal.

Keeping in view the regeneration and amplification of forest, Department takes regular afforestation and plantation program, through various schemes. The plantation achievement of the state through the "Project on consolidation of JFM" was 3496.52 hectares area in 2007-2008 in which the contribution of Medinipur Division was of 107 hectares (WBFD). In that financial year, that division successfully worked for various schemes of plantation, such as plantation of quick growing spp., Forestry treatment, plantation for fodder or fuel wood, Artificial Regeneration etc. By these schemes the Division achieved total 731 hectares area of plantation in which Arabari has an important stake. In recent years, it has also met the expectations as the total regeneration area of the division was 650.75 Ha. in 2009-2010 and 727 Ha. in 2010-2011 (Table 7.7).

Table 7.4: FPC share from sale proceeds of Cashew Nut Medinipur Forest Division*

Felling Year	*Total FPC Share (Rs.)*	*No. of Beneficiaries*	*Share Per Beneficiaries (Rs.)*
2001-2002	84,327.00	1469	57.40
2002-2003	87,265.00	1534	56.90
2003-2004	94,769.00	1706	55.55
2004-2005	424,210.00	2754	154.00
2005-2006	716,546.00	2884	248.45
2006-2007	624,793.00	2778	224.90
2007-2008	449,938.00	2892	155.58
2008-2009	436,575.00	2892	150.95
2009-2010	312,277.00	2556	122.17
2010-2011	309,225.00	2556	120.98

**Source*: Forest Department of West Bengal.

Table 7.5: Details of benefit sharing with FPCs in Arabari Forest Range*

Felling Year	*Area Felled (Ha.)*	*No. of Beneficiaries*	*Total FPC Share (Rs.)*
1999-2000	20	980	80,404.00
2001-2002	30		
2002-2003	25	1048	Total of 3 consecutive
2003-2004	40		years: 16,17,147.00
2004-2005	25	1048	339,331.00
2005-2006	33	1048	619,388.00
2006-2007	20	1048	231,565.00
2007-2008	30	1048	490,619.00
2008-2009	50	1048	1,107,258.00
2009-2010	50	1048	636,205.00
2010-2011	30	1048	1,011,872.00

**Source*: Forest Department of West Bengal.

Table 7.6: Works done by IAP scheme, in 2010-2011 by forest departments in Arabari*

Works Done	*Number/Amount*
Digging Deeptubewell	2
Building of Community Hall	1
Building Ring Well	2
Construction of Irrigation Channel	375 meter
Prepared Cylinder Deeptubewell	5

**Source*: Forest Department of West Bengal.

Table 7.7: Achievements in forest regeneration*

Division	*Year*	*Coppiced Sal (Ha.)*	*Plantation (Ha.)*	*Total (Ha.)*	*Grand Total (Ha.)*
Midnapur		276.25	218.5	494.75	650.75
Forest Corporation South	2009-2010	22	—	22	
Forest Corporation West		90	44	134	
Midnapur		274	253	527	727
Forest Corporation South	2010-2011	15	40	55	
Forest Corporation West		85	60	145	

**Source*: Forest Department of West Bengal.

Every year 25 per cent share of total income which is gained by mainly departmental felling and selling of cashew nut, is distributed among the JFM members by Forest Department. The total income as well as total FPC share amount has been increased up to Rs. 10,92,69,648.00 in 2009-2010 financial year, which is the highest amount of share from Departmental felling (Table 7.3). For the latter years, the amount has slightly dropped. In case of share from sale of cashew nut, the share amount has been increased up to Rs.7,16,546.00 in the financial year 2005-2006 (Table 7.4).

Particularly in Arabari, the JFM members got highest amount of share from departmental felling in the financial year 2002-2003. In very recent year 2010-2011, the share amount was also as good as Rs. 10,11,872. Total no of beneficiaries privileged by this amount is 1048 (Table 7.5). From financial year 2002-2003 to 2006-2007, total share amount of Rs.1,77,750.00 from sale of cashew nut has been distributed among total 1000 of beneficiaries.

To give the total economic back up to the local people, Forest Department assured all semi-skilled and unskilled work to them, round the year, which is executed by the forest departments. More the department constructed deeptubewell, community hall, ring well, Irrigation Channel etc. in villages through various schemes, to ease their daily life (Table 7.6).Integrated Action Plan (IAP) plays a very crucial role for these works. This plan has been specially imposed by the central government for the places which are seriously affected by Left Extremists. By these type of activities, the relationship between local communities and forest department is strengthened more and more and people become very spontaneous to run the JFM.

Conclusion

As the success rate of the Participating Management of forest collaborating with the local communities is very very high, all the 28 states along with Andaman and Nicobar Islands implemented the JFM by July 2005 (J. P. Basu, 2010). We may point out the target lines which may be easily touched by joint strategy from versatile issues such as forest management, protection, ecology, traditional forest relations, etc.

- ☆ JFM paves the way for rural development as well as sustainable use of the forest.

- ☆ Protects the forest from over grazing and encroachment.
- ☆ Plays a vital role in regeneration and National Afforestation Program.
- ☆ Helps to conserve the biodiversity.

In this account, Arabari forest range successfully achieved all of these in very early years than the other forest parts of the country. Yet, the JFM is a decades old management strategy and need to be revised to promote the conservation more effectively in the present perspectives. The ministry of environment and forests has very important role to bring the essence.

Acknowledgements

The authors are very much grateful to Mr. A.K. Samanta (I.F.S.), Medinipur Division and Mr. M. S. Mukherjee (W.B.F.S.), Medinipur Division for providing valuable information and help. Necessary help from the Arabari Range Office and the local people are also thankfully acknowledged.

References

Balooni, K. and Inoue, M., 2009. Joint forest ,anagement in India: The management change process. *IIMB Management Review* 21 (1), Bangalore.

Banerjee, A., Ghosh, S. and Springate-Baginski, O., 2010. The creation of West Bengal's forest underclass: An historical institutional analysis of forest deprivations. Discussion Paper Series, No. 51, *IPPG*, JDPM, University of Manchester.

Basu, J.P., 2010. Joint forest management, deforestation and local people participation: A case study in West Bengal, India. *18th Commonwealth Forestry Conference*, Edinburgh.

Campbell, J., 1992. Joint forest management in India. *Social Change*, pp. 22.

Forest Research (South) Wing., 1987. Socio-Economic Project, Arabari, Paschim Medinipur.

Joshi, A., 1998/99. Progressive Bureaucracy: An Oxymoron? The Case of Joint Forest Management In India, RDFN paper 24a, London.

Kant, S., and Nautiyal, J.C., 1994. Forest Ecology and Management: Sustainable Joint Forest Management Through Bargaining: A bilateral monopoly gaming approach, p 251–264.

Palit, S., 1989. Present Status of Forest Protection Committees: Forest regeneration through community protection working group meeting on forest committees, Calcutta.

Poffenberger, M. and Singh, S., 1989. Community management for India's forests– emerging experiences. The Ford Foundation, New delhi.

Rossi, F.J., 2007. Socio–Economic Impacts of Community Forest Management In Rural India (A Dissertation of Ph D), University of Florida.

Roy, S., Dutta, M. and Bandyopadhyay, S., 2001. India: Environmental Management Capacity Building Technical Assistance Project, EERC working Paper Series: CPR–8, MOEF, IGIDR, World Bank.

Roy, S.B., 1993. Legal Frameworks for forest management in Asia: Case studies of community/state relations: Forest Protection Commitees In West Bengal, India.

World Bank., 2006. India Unlocking Opportunities for Forest– Department People in India: Main Report: Vol. I, Feb–6, Agriculture and Rural Development Sector Unit, South Asia Region, Report No.–34481–IN.

Environment, Biodiversity and Traditional System, 2014 Pages 69–74
ISBN 978-81-7622-307-2
Editors: D.R. Khanna, G.S. Solanki and S.K. Pathak
Published by: BIOTECH BOOKS, NEW DELHI

Chapter 8

Traditional Knowledge of Medicinal Plants Used by the Hill Tribes of Senapati District, Manipur, North-East India

A. Kikim[1], P.S. Yadava[2] and M.R. Khan[1]

[1]Department of Botany, Lilong Haoreibi College, Lilong
E-mail: drriyajuddin_khan@rediffmail.com
[2]Centre of Advanced Study in Life Sciences,
Manipur University, Imphal – 795 003
E-mail: yadavaps1@gmail.com

ABSTRACT

The present paper deals with 20 important medicinal plants used traditionally by the hill tribes which are inhabited in Senapati District of Manipur. There are 33 tribes speaking different dialects in Manipur state. Out of which some of the tribes such as Thadou (Kuki), Vaiphei, Hmar, Chiru, Kom, Rongmei, Thangal and Tangkhul etc. were the main tribes which are dominating in Senapati District. The tribes have a good knowledge about the utilization of wild plants as their traditional medicine. The uses of 20 wild plants have been enumerated in alphabetical order on their botanical names, family, local name and also their therapeutic uses.

Keywords: Traditional knowledge, Tribal community, Wild plants.

Introduction

India is one of the richest floristic regions of the world and is well known for its ancient heritage regarding medicinal plants and plant drugs since time of the Rigveda.

India has immense medicinal plant diversity and is situated in the tropical zone and of 12 mega biodiversity countries having a vast variety of flora and fauna. The plants as medicine are used in different systems of medicine, as in Ayurveda, Allopathy, Unani and even in other systems.

The North-eastern region of India comprising seven states harbouring more than 130 major tribal communities out of a total 427 tribal communities recommend in India (Dutta and Dutta, 2005 and Ramakrishnan, 1992). Manipur is one of the special significant for the conservation of biodiversity because it is recognized as one of the rich endemism, ethnomedicinal and also traditional culture. North-eastern region lies in the range of North eastern Himalaya and this, region is included as one of the nine newly identified biodiversity hot spots (Roach, 2005). The hills of Manipur are inhabited by the Naga, Kuki and Mizo tribes while the valley is populated by mostly the Meitei and Meitei Pangal communities. The region constitutes the part of Indo-Burma *"Hot spots of biodiversity"* (Meyers *et al.*, 2000) rich in diversity of flora.

Senapati district of Manipur in the North-East India occupies the North-Western part of Manipur state. The district is inhabited by several ethnic communities. Majority of the tribal communities are living in bordering areas of Senapati district. But some of them are intermixed with other communities. The different tribal communities are maintaining their own tradition and culture since long back. In Senapati district, tribal and other communities are use different plants, traditionally for curing the ailments in their day to day life. The basic objective of the present study was to document the wild plants based on traditional medicinal practices among the rural and tribal community of Senapati district, Manipur, India. An attempt has been made to conserve and document this vanishing knowledge of the medicinal properties of the plants used by the different tribes/communities of Senapati district, Manipur. The knowledge of herbal medicine is an important source of information for scientific community, research workers and medicinal practitioners.

Study Area

Senapati district of Manipur, North East India located between 24°452-24°512N latitudes 93°482 and 93°552E longitudes occupies the bigger portion of the North-Western sites of Manipur Valley which is inhabited by a large number of ethenic communities living in close vicinity of forests *i.e.* Vaiphei, Thadou, Hmar, Chiru, Rongmei, Liangmei, Thangal, Kom, Tangkhul, Meitei and Nepali etc. The people of tribal community which is residing in the Senapati district explored the utility of wild plants as their resources for livelihood as well as for the prescription of healing activities. In the investigation, different wild plants used as traditional medicine by the ethnic communities.

Methodology

Frequent field trips were made during April 2008 – May 2009 in order to survey the inhabited areas of tribal communities in Senapati district, Manipur to collect plants gathered with knowledgeable persons on traditional medicine. Further, the collected plant samples were identified with their local names with the help of local knowledgeable persons, herbalists/physicians, village chiefs, who have the knowledge of the utilization of plants as herbal medicine. The botanical names were identified from BSI, Shillong, Calcutta, National Botanical Institute (NBRI) Lucknow and also consulted from Deb's (1961) flora of Manipur both monocot and dicot, Kanjilal *et al.* (1934) Flora of Assam, Flora of British India (Hooker, 1872-1897); Chauhan *et al.* (2000). The identified plant material were deposited in the herbarium of Centre of Manipur University, Department of Life Sciences.

Results and Discussion

The plants were enumerated alphabetically the botanical names along with their family, local names in Vaiphei (V) and Manipuri (M) languages, Parts used and uses of medicinal plants are given in Table 8.1.

A total of 20 plant species belonging to 19 genera and 14 families were collected. Among the families of wild medicinal plants, Asteraceae and Verbenaceae represented (3 each) exhibited maximum medicinal plats followed by Agaricaceae and Euphorbiaceae (2 each). The remaining 10 families *viz.*, Bigoniaceae, Juglandaceae, Saururaceae, Rutaceae, Anacardiaceae, Smilaceae, Caesalpinaceae, Magnoliaceae, Acanthaceae and Gesneraceae were represented by single medicinal plant each.

The present study revealed that the ethnic group of tribal communities, mostly residing in Senapati district, Manipur, used different plants for different ailments/ diseases. These are Asthma (2), Diabetes (3), Bronchitis (1), Dysentery (3), Cut and Wounds (3), Anaemia (1), Jaundice (1), Cancer (2), Eczema (1), Rheumatism (3), Malaria (1), Gonorrhea (1), Peptic ulcer (2), Cold and Cough (1), Hair lotion (1) Dyspepsia (2), Digestion (2) Snake bite (1) Scorpion sting (1) Blood purification (1) Mouth ulcer (1), Itching (2), Leprosy (1) Stone case (1) Blood pressure (1) Dizziness etc.

One of the important species *i.e. Clerodendrum colebrookianum* Walp. is commonly used by the people of Manipur for reducing high blood pressure by the people of Manipur and not only this which is used as simple boil vegetable curry by tribal community. Juice of *Rhus semialata* Murray seeds are taken orally by the women of Vaiphei tribe for healthier ovary.

An interesting observation has been made regarding the food habits and medicinal uses of wild plants which is more nutritious, it contains vitamin B-complex besides vitamin K, C and E. (Sinha, 1996) and highly marketable also used by tribal communities for their income generations *i.e. Agaricus campestris* L. and *Schyzophyllum commune* Fr. cooked as vegetable item and also used as medicinal purposes. Cooked fruiting body are eaten in diabetes and constipation by the people of rural and tribal community. Decoction of fruiting bodies is prescribed in tonsillitis, dizziness and asthma (Singh *et al.*, 2003).

Conclusion

The present study revealed that the local healthcare tradition particularly that of the tribal communities and their practitioners in this district was self reliant nature were the traditional healing knowledge has been inherited and transmitted by the words of mouth from generation to generation. From the above information it was clear that there were a number of medicinal plants whose medicinal practices and

Table 8.1: Medicinal plants used by different tribes of Senapati District, Manipur.

Sl.No.	Botanical Name/Family	Local Name	Parts Used	Uses
1.	*Agaricus campestris* L. (Agraricaceae)	Chengum (M); Papah (V)	Fruiting body	An excellent source of vitamin B complex besides vitamin K, C and E (Sinha 1996).
2.	*Artemisia vulgaris* L. (Asteraceae)	Laibakngou (M); Gamsai(V)	Tender shoot	Tender shoots are eaten as cooked vegetable after repeated washing. Boiled extract of the leaves used diabetics treatment (Khan and Yadava 2010).
3.	*Bauhinia purpurea* L. (Caesalpinaceae)	Chingthrao (M)	Leaves/barks/ flower	Decoction of leaves or bark is given in leprosy, asthma and stone in gall bladder. Flowers are cooling. Paste of flower is given for cooling treatment.
4.	*Clerodendrum colebrookianum* Walp. (Verbenaceae)	Anphui (V); Kuthap angouba (M)	Leaves	Boiled leaves are taken to get relief from high blood pressure.
5.	*Clerodendrum serratum* (L.) Moon (Verbenaceae)	Moirang khanum (M); Peitanumkhao (V)	Leave/root	Leave extract is given in dysentery, cold and cough. Root is used in rheumatism and dyspepsia.
6.	*Emblica officinalis* Gaertn. (Euphorbiaceae)	Sohlhu (V); Heikru (M)	Fruits/Leaves/ Barks	Fruits are eaten raw and useful in diabetes, anaemia, jaundice and it is the main part of Ayurvedic preparation like Chavanprash etc.
7.	*Eupatorium odoratum* L. (Asteraceae)	Kambirei (M)	Leaves	Boiled extract of the leaf is applied in scabies, eczema, boils and other skin infections. Smashed leaves are applied as cut and wounds and ash of the leaf mixed with honey and water is prescribed orally in asthma and bronchitis (Khan and Yadava 2006).
8.	*Houttuynia cordata* Thunb. (Saururaceae)	Toningkok (M); Ithanglou (V)	Whole plant	Fresh leaf are eaten for promoting blood purification. Decoction leaf and root taken for dysentry and indigestion.
9.	*Juglans regia* L. (Juglandaceae)	Makha (V); Heijuga (M)	Fruit (Nut)/ Leaves	Leaf paste and outer rind of fruits are used for eczema, syphilis and rheumatism. Seed kernel after removing brown husk is used in nervous disorders and improving memory (Sinha, 1996).
10.	*Lantana camera* L. (Verbenaceae)	Nongbanlei (M)	Leaves	Leaf decoction applied for rheumatism and taken for malaria. And dried leaves also are used for pest control in store item like potato.

Contd...

Table 8.1–*Contd...*

Sl.No.	*Botanical Name/Family*	*Local Name*	*Parts Used*	*Uses*
11.	*Macaranga denticulata* Blume. (Euphorbiaceae)	Lakoi (M); Nakhal (V)	Leaves and latex	Gum antiseptic. Plant extract is used for cleaning worms. Stem juice is useful against fresh cuts and wounds.
12.	*Magnolia grandiflora* L. (Magnoliaceae)	Uthambal (M)	Leaf/Bark	Boil extract of leaf is used for diabetics. Bark is stimulant, diaphoretic and tonic; used also for malaria and rheumatism.
13.	*Mikania micrantha* Kunth. (Asteraceae)	Oori-hingchabi (M)	Whole plant	Plant is a rich source of vitamin A, B, C. Leave extract is applied to itches and as poultice to wounds. The plant is used as antidote to snake bite and scorpion sting.
14.	*Oroxylum indicum* Vent. (Bigoniaceae)	Shamba (M); Paklong (V)	Fruit (Bean)	Young fleshy bean are eaten as cooked vegetable item, for the prevention of leukemia bitter in taste. Root bark is tonic and astringent used in diarrhoea and dysentry.
15.	*Phlogacanthus thyrisiflorus* Nee . (Acanthaceae)	Nongmangkha Angouba (M)	Whole plant	Inflorescence is antidote to pox and prevent skin diseases like scabies and sores etc. Fruits and leaves are roasted and taken in fever, cough and cold (Sinha, 1996). Boiled decoction of shoot and leafs are taken in cough, fever, peptic ulcer, intestinal disorders and muscular sprains (Singh *et al.,* 2003).
16.	*Rhus semialata* Murray (Anacardiaceae)	Heimang (M); Hongma (V)	Leaves and Fruits	Fruit powder mixed with egg is prescribed in urinary complaints due to stone. Water soak of dry seeds are used in dyspepsia, peptic ulcers and unhealthy uterus.
17.	*Rhynchotechum ellipticum* A.DC. (Gesneraceae)	Yembum (M); Cheklap (V)	Leaves	Young leaves used as vegetable either raw or cooked for peptic ulcer and constipation.
18.	*Schyzophyllum commune* Fr. (Agaricaceae)	Kanglayen (M); Pahshih (V)	Fruiting body	Decoction of fruiting bodies is prescribed in tonsillitis, dizziness, asthma and diabetics (Sinha, 1996, Singh *et al.,* 2003).
19.	*Smilex lancifolia* Roxb. (Smilaceae)	Kwa-manbi (M)	Root nodule	Boiled extract of the nodule is given in jaundice and diabetics (Khan and Yadava 2010).
20.	*Zanthoxylum acanthopodium* DC. (Rutaceae)	Mukthrubi (M)	Leaves/Fruits	Leaves and fruits are used as a spices for cooking meat items and also serve as mouth refresher and antigastric.

were are not known to scientific community. There is a need to gather this valuable information on these plants especially from rural and tribal communities. Therefore there is an urgent need to make systematic investigations of the plants to establish the pharmacological activities as well as spur conservation of such unique habitat and its resources.

References

Chauhan, A.S., Wadhwa, B.M., Singh, D.K. Singj, K.P., Chakraborty, P., Sanpru, R. and Dam, D.P., 2000. *Flora of Manipur, Vol. 1.* Botanical Survey of India, Kolkata, Government of India, pp. 1–600.

Deb, D.B., 1961. Dicotyledonous plants of Manipur. *Bulletin of Botanical Survey of India,* 3: 253–350.

Deb, D.B., 1961. Monocotyledonous plants of Manipur. *Bulletin of Botanical Survey of India,* 3: 115–138.

Dutta, B.K. and Dutta, P.K., 2005. Potential of ethnobotanical studies in North East India: An overview. *Indian Journal of Traditional Knowledge,* 4(1): 7–14.

Hooker, J.D., 1872–1897. *The Flora of British India.* 1–7 (London).

Kanjilal, U.N., Kanjilal, P.C., Das, A. and Bor, N.L., 1934–1940. *Flora of Assam.* 1–4, Government of Assam, Shillong.

Myers, N., Mittermeier, R.A., Mittermeier, C.G., da Fonseca, G.A.B. and Kent, J., 2000. Biodiversity hotspots for conservation priorities. *Nature.* 403: 853–858.

Ramakrishnan, P.S., 1992. Tropical forests, exploitation, conservation and management. *Impact of Science in Society.* 42: 149–162.

Roach, J., 2005. Conservationists name, nine new "Biodiversity Hotspots" *National Geographic News.* Feb. 2, 2005.

Singh, H.D., Singh, R.S. and Sandhu, J.S., 2003. Herbal medicine of Manipur. *Daya Publishing House,* Delhi, India.

Sinha, S.C., 1996. *Medicinal Plants of Manipur,* pp.238.

Environment, Biodiversity and Traditional System, 2014 Pages 75–85
ISBN 978-81-7622-307-2
Editors: D.R. Khanna, G.S. Solanki and S.K. Pathak
Published by: BIOTECH BOOKS, NEW DELHI

Chapter 9

Ecology of Pika (*Ochotona* spp.) in Gosainkunda Area, Langtang National Park, Nepal

Narayan P. Koju,[1] Mukesh K. Chalise[2] and Randall C. Kyes[3]

[1]Khwopa College, Bhaktapur
E-mail: npkoju_2003@yahoo.com
[2]Central Department of Zoology, TU
E-mail: mukesh57@hotmail.com
[3]University of Washington, USA
E-mail: rkyes@u.washington.edu

ABSTRACT

The study was carried out on the population status and behavior of pika in rainy season in Gosainkunda, Langtang National Park, Nepal during July and August 2011. Pika inhabits in talus field in alpine area. Local people call pika "Bhragomjin" and respect it as the symbol of Buddhist monk. Tharepati (3651masl), Phedi (3825masl), Gosainkunda (4428masl), and Laurebina (3903masl) were selected for population survey and behavior recording. Quadrates of 50m×50m were randomly plotted in each site except in Phedi. Only two species of pika were recorded in those areas of LNP and were *Ochotona marcotis* and *Ochotona roylei*. Total number of 124 pikas was head counted in which 75 were *O. marcotis* and 49 were *O. roylei*, The ratio of age group adult, infant and juvenile in *O. macrotis* is 29: 26: 20 and *O. roylei* 16: 23: 10 population with density 21/hectare and14/hectare respectively. Eleven pikas were observed for 455 minutes to record their behaviours from 5: 30 am to 19: 30 pm. Feeding, foraging, musing, grooming, chasing each other (playing) were the behavior recorded. They were active in dawn and dusk but were absent in rain. They spent their 26 per cent time in feeding, 24 per cent musing, 16 per cent foraging, and 11 per cent grooming. Pikas were observed digging soft rocks, soil and mosses on the rock before discharging pellets. Pika in Gosainkunda was recorded collecting plastics to their burrow. Sometime pikas enter the store room of local resident and steal vegetables. Ten thousands of pilgrims visit Gosainkunda, the only Ramsar site in Langtang National Park each year in rainy season. So the pollution is increasing year by year and was perceived threat in the long term survival and conservation of pika in the region.

Keywords: Behavior, Gosainkunda, Musing, Ochotona, Pika, Pollution.

Introduction

Pikas are small mammals related to rabbit and hare that inhabit in rocky talus field in alpine area. They seem to be relatively primitive animals, having passed their evolutionary zenith. The only living genus, Ochotona, includes few species adapted to sub-alpine and alpine habitats of Asia and North America. Ochotona reached their greatest diversity and widest distribution in Miocene when they spread throughout Europe, Asia, Africa and North America (Bhattacharyya *et al.*, 2009). The presences of four incisor teeth differentiate them from rodents. There are 30 species of Pikas (*Ochotona* spp.) distributed throughout the world (Hoffmann and Smith, 2005). Five species of Pikas have been reported hitherto from Nepal namely *Ochotona curzoniae, O. nubrica, O. thibetana, O. macrotis, O. roylie* (IUCN, 2010). Distribution of *O. himalayana* is uncertain. There is no evidence to support presence of *O. himalayana* from Nepal. Hence, this species was kept as possibly occurring Pika species of Nepal (Thapa *et al.*, 2010). We have very less information about pika. The Himalayan pika resides from 2500m to 5000m (Tak and Lamba, 1985). They are reported from Sagarmatha National Park, Langtang National Park, Rara National Park, Annapurna Conservation Area and Makalu Barun conservation area. Only large ear pika (*Ochotona macrotis*) and Royle's pika (*Ochotona roylei*) are found in Langtang National Park. They both have similar structure and habitat. They are reddish brown fur with pale band over the nape, the winter coat is similar, but may show traces of rufous coloration. Its body length range from 15–20 cm and the diameter of head is 7 cm, its weights 100–150gm (Smith *et al.*, 1990). Although, pika shows external activities at night they are frequently active during hours of dawn and dusk (Kawamichi, 1971).

Study Area

Langtang National Park is located in the central Himalayas of Nepal between 85 15′ E to 86 0′E and 28 20′ North. It is in road distance of 132.2 KM form Kathmandu the capital city of Nepal. The park covers 1710 sq km. 46 mammalian species are recorded from Langtang National Park. The shortest highway that connects India and China through Nepal, Pasang Lhamu Highway passes through this National Park. Langtang Lirung (7245 m) is highest point in the park. Langtang National Park was officially established in March 1976. Its land area includes three districts, Rasuwa, Nuwakot, and Sindhupalchowk (DUHE 1977). The study was carried out on the population status and behavior of pika in rainy season in Gosainkunda, Langtang National Park, Nepal during July and August 2011. Regular observation for

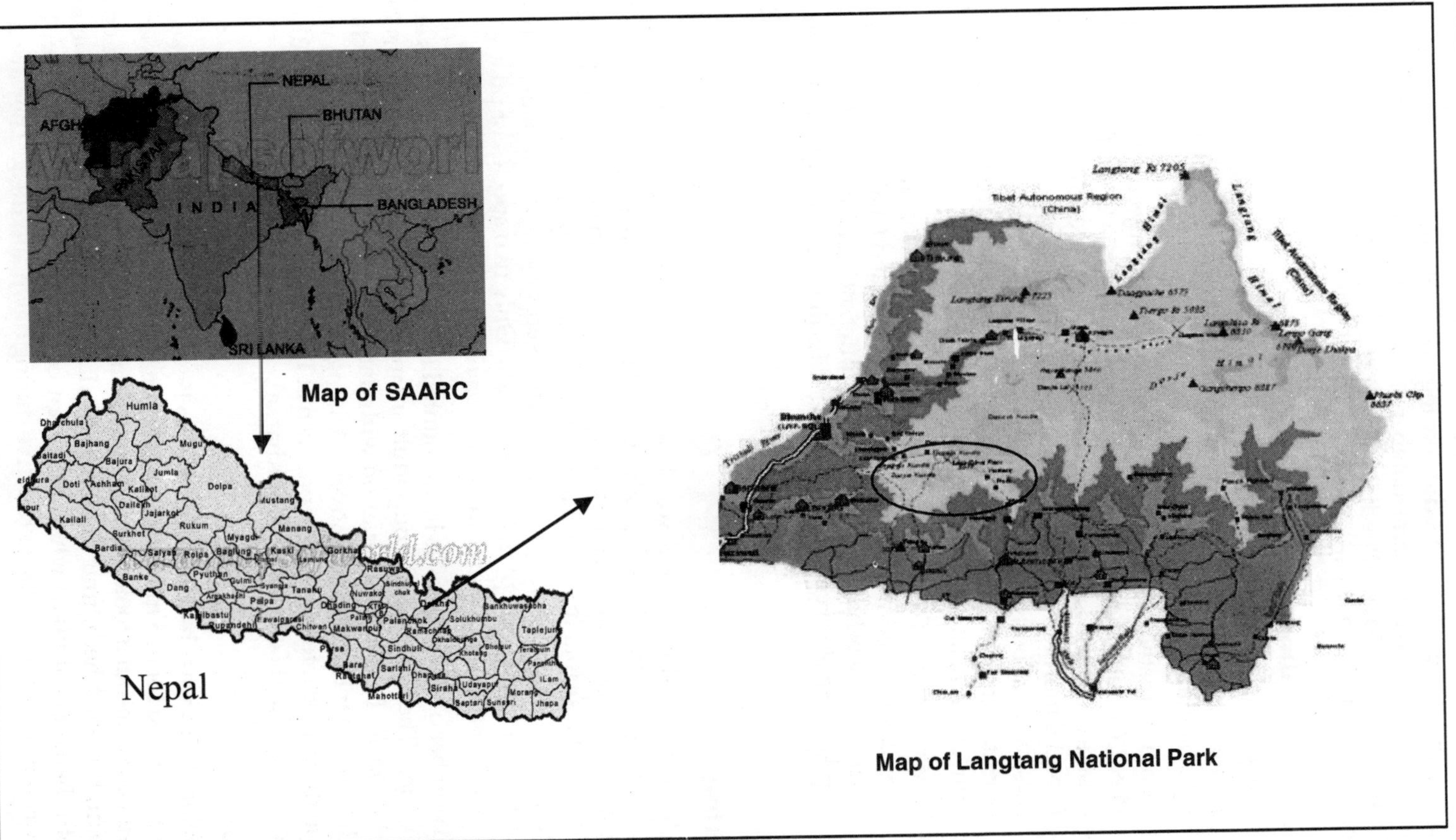

Figure 9.1: Map of study area in Langtang National Park.

population status, behavior and feeding ecology was determined for Gosainkunda area. The observation was concentrated into four spots of Gosainkunda *viz*., Tharepati (N28 °00' 47.6" E085 °29' 31.4"), Phedi (N 28° 03' 29.9" E 085°27' 29.8"), Gosainkunda (N28° 05' 38.9" E085°24' 22.6") and Laurebina (N28 °05' 47.1" E085 °23' 36.1").

Material and Methods

a) Questionnaire/interview: Interview with pre-set questionnaire were performed with local people, local herder and staying tourist to explore the pika habitat.

b) Quadrate method: 50m×50m quadrate was randomly plotted in Tharepati, Phedi, Gosainkunda and Laurebina along with Barna Goth. The population of pika (direct observation) was counted thoroughly within the quadrate. Especial morphological features in individual pika like scratch or tears in ears, wart in body, colour and size were recorded carefully to prohibit repetition in population count. Photographs of all observed pika were taken from different angle for identification.

c) Scan sampling: To calculate population distribution and behavior scan sampling method was applied (Altmann, 1974 and Chalise, 2003). During observation of behavior, plant species (different parts of plants) consumed by pika were observed and collected for further scientific identification.

Result

Population

The total population of pika was head counted in four different places of Gosainkunda route. Total fourteen quadrates of 50mx50m were randomly plotted in those places. They were four in Tharepati, two in Phedi (as this is very steep slope with many waterfalls), four in Gosainkunda and four in Laurebina. All the quadrates were plotted above 3500m in talus area. Quadrates were placed at least 50m far from local residence and hotel except in Gosain Kunda. Only two species of pika were recorded in those areas of LNP and were *Ochotona marcotis* and *Ochotona roylei*.

Total number of 124 pikas was head counted in which 75 were *O. marcotis* and 49 were *O. roylei*, The ratio of age group adult, infant and juvenile in *O. macrotis* is 29: 26: 20 and *O. roylei* 16: 23: 10 population with density 21 individuals per hectare and14 individuals per hectare respectively. Among 14 quadrates, there no pika was observed in three quadrates; they were one each in Tharepati, Gosainkunda (South of *the lake, north facing)* and *Laurebina* (Odar Kharka).

Behavior

Behavior of pika was observed from 5: 30 to 19: 30 for ten days. Behaviors Eleven pikas were observed for 455 minutes that is seven hours and thirty five minutes to record their behaviors but it took 103 hours and 35 minutes field work. They were one in Tharepati, three in phedi, four in Gosainkunda, two in Laurebiana and one in Barna Goth. Behavior of pikas which were not continuously active less than 15 minutes was not recorded. Pikas were active during dawn and dusk. They are not

Figure 9.1: ***O. marcotis.***

Figure 9.2: ***O. roylei.***

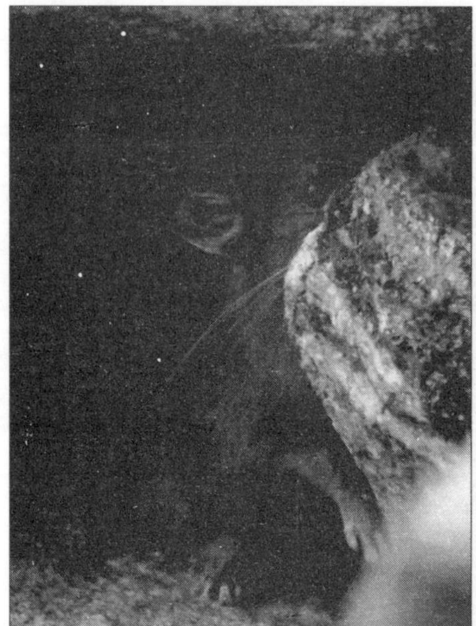

Figure 9.3: Infant *O. roylei.*

Figure 9.4: Juvenile *O. roylei.*

Figure 9.5: Infant *O. marcotis.*

Figure 9.6: Juvenile *O. marcotis.*

active outside and even not seen in raining. Only very few infants were seen at the day time. Camera trap was not used during this field observation, so the behaviors of late evening and nocturnal behavior of pika was not explored. They are active and fast moving. They use numbers of burrows while foraging and feeding. Behavior of pika was recorded in the interval of one minute.

General behaviors as **Musing**: Motionless or just resting on the big stone for short or long time, **Feeding**: Eating grass, flower or other parts of plant, **Foraging**: the

Table 9.1: Total population of pika in different quadrates.

Quadrate No.	Population Pika Observed			Coordinate	Elevation	Location
	O. macotis	*O. roylei*	*Total*			
Q1	3(2+1+0)	2(0+2+0)	5	N28°00'47.6" E085°29'31.4"	3636m	Tharepati
Q2	1(1+0+0)	2(0+2+0)	3	N28°00'48.4" E085°29'32.6"	3605m	
Q3	2(1+0+1)	5(2+1+2)	7	N28°02'05.2" E085°28'53.0"	3524m	
Q4	0	0	0	N28°03'06.1" E085°28'02.7"	3634m	
Q5	9(4–3+2)	8(5+3+0)	17	N28°03'29.9"E 085°27'29.8"	3764m	Phedi
Q6	5(3–0+2)	4(1+2+1)	9	N28°03'31.2"E 085°27'53.8"	3764m	
Q7	19(4+9+6)	14(4+8+2)	33	N28°05'38.9" E085°24'22.6"	4423m	Gosainkunda
Q8	6(3–2+1)	1(0+1+0)	7	N28°05'02.2"E085°24' 37.6"	4436m	
Q9	9(4–0+5)	5(2+1+2)	14	N28°06'04.7" E085°24'42.8"	4404m	
Q10	0	0	0	N28°06'54.7" E085°22'42.8"	4324m	
Q11	13(4+6+3)	5(2+1+2)	18	N28°05'23.1" E085°22'56.9"	3951m	Laurebina
Q12	5(1–4+0)	3(0+2+1)	8	N28°05'47.1"E085°23' 36.1"	3916m	
Q13	Only old pellets were observed			N28°05'12.3"E 085°23'25.2"	3951m	Odhar lekh
Q14	3(2+1+0)	**0**	**3**	N28°05'48.9" E085 °23' 30.5"	3879m	Pulagpati lekh
Total	75	49	124			

movement for searching and selecting food, **Groom**: the process of cleaning body and fur, **Calling**: Making sound by pika, **Galloping**: Jumping from one rock to other rock or in the ground, **Chasing**: playing like behavior recorded among infants at evening time, **Inside burrow**: Pika enters burrow and use number of burrows for short time while foraging and feeding were recorded.

In average they spent maximum time in feeding (26 per cent) and least time in chasing (1 per cent). the chasing behavior was observed only in five pikas at the evening time. Calling behavior is of two types: long call 'Chirr....r' (producing while musing) and short call 'chi chi' (produce during chasing). Musing behavior was observed in all pikas. They performed short musing (less than one minute) and long musing (more than five minutes in a time). They entered burrow for few seconds to less than five minutes while feeding and foraging. Pika groom they body and fur by using hind limbs, sometime mouth and fore limb were used in grooming.

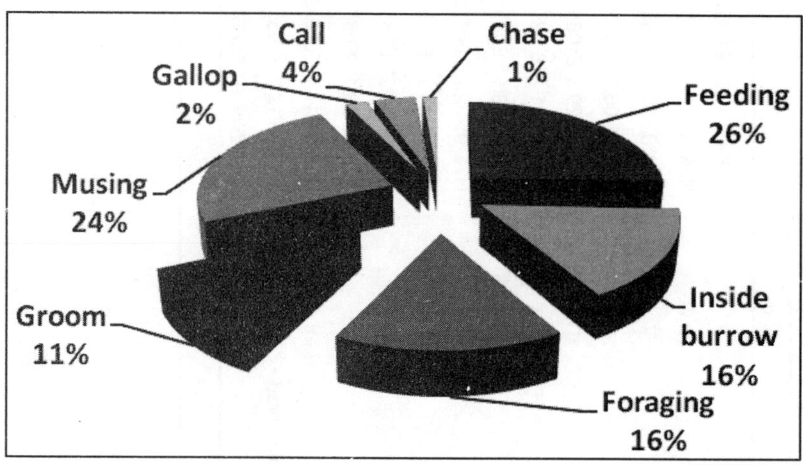

Figure 9.7: Time budget of pika in different places.

Some infant pikas were observed discharging pellets outside the burrow. This behavior was recorded both in the morning trip and evening trip. They dig on soft rock or moss on the rock for few minutes by fore limbs before discharging pellets. Only single pika was observed transporting green plants to its burrow in rainy season.

Threats

Locally Pika is known as 'Bhragomjen' which means rabbit at talus. Pika is believed as the symbol of Buddist monk by local people. Pika steals cabbage, potato and other vegetable from the store of local people especially in winter season. In Laurebina pika starts stealing rice and flour from few years ago; this is not their normal food. Sometime children kill pika for fun (with reference to local respondent). A pika (*O. roylei*) was observed transporting piece of polymer sack in Gosainkunda. Pika's habitat in Gosainkunda was much polluted so this behavior may resemble to adaptation of pika with pollution. It tore the sac and run to about five meters with mouthful of polymer then entered the burrow. It returned back from burrow after 40 seconds without polymer to the same place. It continued this behavior for three times but it was not seen on following days. Plastics, papers, clothes pieces, bottles and other waste were abundant in Gosainkunda. Gosainkunda is known as holy place for both Hindu and Buddishist. Ten thousands of pilgrims visit Gosainkunda, the only Ramsar site in Langtang National Park each year in rainy season. Those pilgrims bring plastics and other pollutant. According to Hindu ritual they drop their cloths in the lake. There are five hotels in Gosainkunda, two in Phedi, six in Laurebina and three in Tharepati, annually thousands of tourists visit this area. Pikas forage and feed together with cattle, goats and horse. In this study period 17 horses and 312 sheep were recorded in Gosainkunda. So the pollution is increasing year by year and was perceived threat in the long term survival and conservation of pika in the region.

Discussion

Pika were not found below 3165 m above sea level in any location In Langtang National Park (LNP) during this study period but Gurung and Singh 1996 reported their presence ranging from 2,300 m to 6,400 m; kawamichi (1971) reported their occurrence from 2650m in Gosaikunda region; Fedosenko, 1974, Formozov, 1982 found at 2800 m to 5000 m and Tak and Lamba, 1985 found the Himalayan pika resides from 2500m to 5000m in Sagarmatha region.

Smith *et al.* (1990) and Smith and Xie (2008) explain *O. macrotis* will occupy higher elevations when sympatric with *O. roylei* from the research in Nepal and the mountainous regions of several provinces of China (Xizang, Sichuan, Xinjiang, Yunnan, Qinghai, and Gansu) respectively but it was not observed in this study site.

Bhattacharya *et al.* (2009) recorded mean density of royle's pika 15.3 individual per hectare in Uttarkhanda, Western India. Similarly Smith *et al.* (1990) found the density of royle's pika 12.5per hectare individuals in Nepal. The population observed in this study is less than Indian population density and more than Smith's report. The population of pika reported by Khatiwada (2004), 350 individuals per hectare in Langtang is very high figure in comparison to other researches.

Kawamichi, 1968 reported Pika in Gosainkunda do not make call. He heard only low cries in winter season but in this study of rainy season two types of calling was recorded and six among eleven pikas were observed making call.

Conclusion

Only two species of pika are found in Gosainkunda route of Langtang National Park they are *Ochotona marcotis* and *Ochotona roylei*. Pikas are active in dawn and dusk but absent in rain. Pollution is increasing in this region which is major threat to pika.

Acknowledgements

We are grateful to Prof. Andrew T. Smith, President, IUCN/SSC Lagomorph Specialist Group, Arizona State University, Prof. Dr. Taeko Kawamichi, Kyoto University, Pror Dr. Erik Beever, US geological survey and Prof. Chris Ray, Colorado University for providing us literature and valuable suggestion.we wish to extend our thanks to Hira K Dongol, Indianapolis USA and Mr. Rajendra Chawal of Bhaktapur, who help us in data collection and field work.

References

Altmann, J., 1974. Observational study of behavior: sampling methods. *Behaviour*, 49: 227–265.

Bhattacharyya, S., Adhikari, B. S. and Rawat, G. S., 2009. Abundance of royale's pika (Ochotona roylei) along an altitudinal gradient in Uttarakhand, Western Himalaya. *Hystrix It J. Mamm.* 111–119.

Chalise, M. K., 2003. Assamese Monkeys (*Macaca assamensis*) in Nepal. Primate Conservation. No. 19: 99–107. The Jour. Of IUCN/SSC Primate Specialist Group, Conservation International, USA.

DUHE., 1977. Durham University Himalayan expedition, Langtang National Park Management plan HMG/UNDP/FAO Project Nep/72/2002, Kathmandu, Nepal.

Fedosenko, A.K., 1974. Some morphological characteristics of Ochotona. *Zool.* Zh 53: 485–486 pp.

Formozov, N.A., 1981. Behavioural Adaptations of Pikas in Rocky Biotopics. In: *Ecology, Population, Structure and Communication Processes in Mammals,* (Ed.), N.P. Naumov. Nauka, Moscow–Leningrad, 245–263 pp.

Gurung, K.K. and Singh, R., 1996. *Field Guide to the Mammals of the Indian Subcontinent.* Academic Press, San diego, California, USA

Hoffmann, R.S and Smith A.T., 2005. Oder lagomorpha. In: *Mammals Species of the World,* (Eds.) D.E. Wilson and D.M. Reeder. Johns Hopkins University Press Baltimore, USA, pp. 185–211.

IUCN, 2010. *IUCN Red List of Threaten Species.* version 2010.4.www.iucnredlist.org. downloaded August, 14, 2011

Kawamichi, 1971.Studies on the genus *Ochotona* Link (Mammalia, Lagomorpha, Ochotonidae), V. *Jour. Fac. Sci. Hokkaido Univ. 8€r. VI, Zool.* 18 (1), 1971.

Kawamichi, T., 1968. Winter Behaviour of the Himalayan Pika (Ochotona roylei) *J. Fac. Sci. Hkkaido University,* ser. Vl. Zool., 16: 582–554 pp.

Kawamichi, T., 1971. Daily Activities and Social Pattern of Two Himalayan Pikas, Ochotona marcrotis and Ochotona roylei, observed at Mt. Everest *J. Fac. Sci., Hokkaido University,* Japan, Vl, Zool 17: 587–

Khatiwada, J.R., 2004. The status of snow leopard and its conflict perception in Langtang National Park. *M.Sc. Thesis,* Tribhuvan University

Smith, A.T. and Xie, y., 2008. *The Mammals of China.* Princeton University Press, Princeton, New Jerssey, USA

Smith, A.T., Formozov, N., Hoffmann, R.S., Changlin, Z. and Erbajena, M.A., 1990. *The Pika Accounts of Genera and Species, Rabbits, Hares and Pikas. Status Survey and Conservation Action Plan.* Chapter 3, IUCN, Gland, Switzerland.

Tak P.C. and Lamba B.S., 1985. Nanda Devi National Park: A contribution to its mammalogy. *Indian Journal of Forestry*, 8: 219–230.

Thapa, A., Dahal, B.V., Koju, N.P. and Thapa, S., 2011. *A Review on Pikas of Nepal. Small Mammals Conservation and Research Foundation*, New Baneshwor, Kathmandu, Nepal, 22pp.

Environment, Biodiversity and Traditional System, 2014 Pages 87–106
ISBN 978-81-7622-307-2
Editors: D.R. Khanna, G.S. Solanki and S.K. Pathak
Published by: BIOTECH BOOKS, NEW DELHI

Chapter 10

Indigenous Knowledge Systems and Ethno-Medicinal Practices among the Tribal Communities in India: A Study of Koyas* in Andhra Pradesh

Ram Babu Mallavarpu[1], *Bikku Rathod*[2] *and Koteswara Rao Mannam*[3]

[1] *UGC Post Doctoral Fellow, Centre for Regional Studies (CRS)*
E-mail: rambabuphduh@gmail.com
[2] *Ph.D . Research Scholar in Department of Anthropology,*
E-mail: bikkurathod@mail.com
[3] *Alumni of Department of History,*
School of Social Sciences, University of Hyderabad,
Hyderabad – 500 046, A.P

ABSTRACT

As per the etiology of Koya community certain diseases are caused by malevolent supernatural forces. They try to cure them by pleasing the supernatural powers through prayers and sacrifices. They also use various locally available medicinal roots and herbs for treating several diseases. Village medicine-man locally known as vejji/vejju who have a

* Koyas are one of the traditional tribal communities, who are settled all along the Godavari River in the East, West Godavari and Khammam Districts of Andhra Pradesh (Rao, 1981).

good knowledge about the herbal plants usually treats the patients. Many elderly persons of the village are also aware about the importance and use of such herbal medicine. The present study analyzes the beliefs, practices and knowledge systems used by the Koya community in the maintenance of their health in India with special reference to Bhadrachalam agency (scheduled) area of Andhra Pradesh (AP) based on the anthropological study.

Keywords: ***Indigenous knowledge systems, Ethno-medicine and Tribal communities.***

Introduction

Ethno-botanical studies related to tribal medicine or ethno-medicine have identified some 45,000 plant species which are available to 550 Indian tribal communities belonging to 160 linguistic groups inhabiting in varied geographical and climatically zones. Tribal people have their own traditional knowledge and wisdom, and their ethno-medicinal systems. The tribal peoples are custodian of unique traditional knowledge systems and their ambient flora and fauna. Scuttles (1962) explained about the Indian ethno-botanical emporia in his words: India with her many living groups of people, having diversified ethnic culture, history of rituals and performances, who are more or less isolated from modern world, and are closely associated with their ambient vegetation is the emporia of ethno-botanical research". The information about the perceptions, pharmacology attitude towards disease diagnoses etc., of the age-old tribal medicine are lying unknown as they are not rendered in writing.

The people who belong to the modern societies are not aware of the rich native knowledge systems. So far, studies in tribal medicine have enabled to identify some 1,600 drug-yielding plants. Thus it has become imperative to collect information and document the same to study them scientifically. This study attempts to bring to light as some aspects of the Koyas' knowledge system on therapeutic uses of plant species.

The Koyas of Boddugudem, like their counterparts in the agency area, have their own institutional mechanisms to (1) perceive health problems (2) protect and promote health (3) prevent and treat illnesses and (4) provide care to the sick. Their institutions are less formalized, simple and more mixed with their day to day family life, religion, magic and ritual.

Among the Koyas, health care activities are embedded in the traditional beliefs, customs, folklore, taboos, prescriptions, etc. These are enacted through family networks, health roles, healers, shamans, priests, medicine-men, magicians etc. Indigenous health practices among the Koyas refer to the measures relating to preventive healthcare, diseases and its curatives which are being used by them as medical therapy. Traditional systems of medicine of the Koyas are mainly based on the relationship of variables like kinds of diagnosis, curing techniques and other preventive acts within their society.

Illness is a common and widespread phenomenon prevailing among the Koyas is not only a personal affair, but also it arouses a wide variety of feelings in the sick. This makes them to search for treatment as an immediate problem. The choices and alternatives for curing diseases are considered among the Koyas within their socio-cultural frame work, *i.e.*, the existing knowledge and experience of the Koyas in relation to identification and curing of diseases. Past experiences of the people with similar diseases dictate the initial definition, diagnosis, and the initial treatment. A lack of response from the initial treatment necessitates a search for more information on the illness from the patient's friends, relatives and other neighbors who discuss and offer alternative and useful suggestions regarding treatment. Subsequent failures lead people to an intensive discussion of choices and a search for expert treatment from the specialists. The choices of therapy among the Koyas are determined by the following factors.

1. Treatment of certain natural diseases and other ailments by using home remedies.
2. Treatment of certain diseases by using the traditional systems of medicine such as herbal medicine, sorcery and witchcraft.
3. Treatment of certain other diseases by seeking available health care facilities at the Governmental and Non-Governmental Institutions.

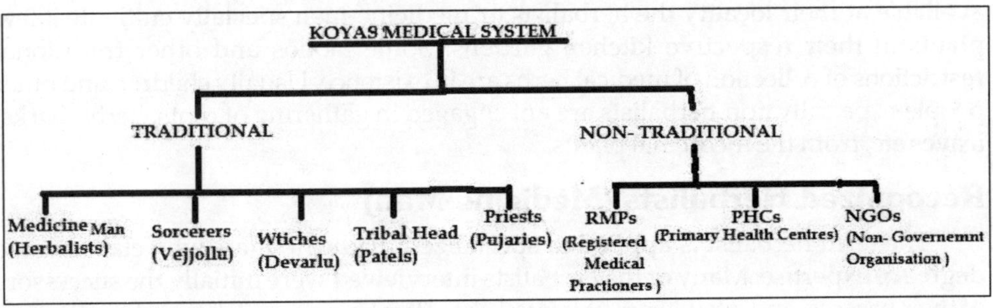

Figure 10.1: Categorization of the Koya's medical system.

Home Remedies

The Koyas of Boddugudem normally keep some medicines in their houses for the treatment of common illnesses like colds, fever, abdominal pains, headaches etc. Home remedies preserved at home by the Koyas include a variety of roots, herbs, drugs or tablets brought from the shops located at nearby towns *i.e.*, Edugurallapalli, Chintoor, and Bhadrachalam. In the initial stage, the Koyas use home remedies to cure certain illnesses and other minor ailments. If they fail to cure the illnesses, then they consult to traditional herbalist available at their locality or neighboring villages.

Traditional Systems of Medicine

The Koyas of Boddugudem are well verses with the use of herbal medicine, sorcery and witchcraft in curing their common diseases and other ailments. Herbalist or medicine-man is considered as a healer. The work of the herbalist or medicine man involves of the task of restoring bodily health to any person afflicted with some illness. This consists of treatment of the sick in such a way that all the parts of his body regain the form and function, which he had before illness. This restoration of health is a much desired goal of medicine-man or herbalist. The task would be completed when he succeeds in curing the sick person and restoring him/her to a useful place in society. Generally herbalists among the Koyas use several roots, herbs, leaves and barks of various plants, and other foods having medicinal values. The knowledge of plant uses particularly about medicinal plants and uses of plants in the treatment among the Koyas are specialized and it is limited only to a few members in the community who are recognized as herbalists or medicine-men. These persons are generally the most responsible members in the society. Each medicine man or herbalist treats illness by a series of trials with different medicines. In general, herbalist or medicine-man treats all kinds of illness but some are specialized in specific illnesses. Some herbalists or medicine-men treat only children or women or only aged. Some other herbalists or medicine-men have inherited knowledge of special remedies and treat only specific illnesses.

Most of the roots, barks and leaves are obtained from various medicinal plants available in the surrounding forests by the herbalists or medicine-men. A few herbs, roots and barks which are not available in surrounding forests of the village are collected from the forests far from their village. Many useful medicinal plants are available at their locality the herbalists or medicine-men specially cultivate a few plants in their respective kitchen gardens. Some taboos and other traditional restrictions of collection of medical herbs are in existence. Usually children and other people-especially non-herbalists are not engaged in gathering of roots, herbs, barks, leaves etc, from the medicinal plants.

Recognized Herbalists (Medicine-Man)

The term herbalist is applied to specialized medicine-man, who claims some degree of expertise. Many of the herbalists interviewed were initially the successors of their parents, and a few were patients, who, after being cured, learned the therapy from their herbalists after paying a certain sum of money. A few other herbalists claim to have attained through dreams or visions. Herbalists are believed to have powers to cure certain illnesses.

Vuike Veeraiah aged around 40 years was a well known herbalist to the Koya of Boddugudem and also in neighboring villages. According to him, he became herbalist after several dreams in which the spirits taught him to use various medicinal plants to cure certain illness.

Veeraiah worships the goddesses Mutyalamma regularly and maintains contacts with the local *Vejjollu* (sorcerers) and *Devarlu* (witches). He sends various patients particularly with pain and severe illness to the *Vejjollu* or *Devarlu* for further treatment. He is familiar with almost all herbs, roots, barks and their uses. He explained that certain medicinal plants are available during rainy, winter, and summer seasons. He also explained his difficulty in finding several roots as they are not identifiable in the hot summer. He said that all the medicinal plants needed to cure various diseases are not available in the forests of his village. He brings certain of the roots used in curing snake bites from the forests of Gundala village (when he goes for collecting *Pilla-pogaaku* from the non-tribal fields).

He gives treatment to his patients either in early hours of the morning or at night. He does not give treatment in the midday. Generally he would not charge any fee in the form of money or kind. But the patients most often, offer their traditional liquor to him. Sometime they may also pay in cash or kind as a gratitude to his services. He cultivates certain medicinal plants in his kitchen garden itself for regular use. Most of his views related to the cause of diseases are attributed to *krimi,* evil spirits, and souls of diseased and other supernatural causes. Generally maintains secrecy in giving medicines. He will not reveal the name of the particular root, herb, or bark used in treating a particular illness. For this purpose he mixes several roots, herbs, barks etc., into a paste and makes some small rounded pills. It is very difficult for the patients to identify the roots, herbs, barks, and leaves used in curing a particular health problem. He brings some of the roots, barks; herbs form the other herbalists in the village or other villages. At times he also visits the hill slopes of the Konda Reddis[1] and brings some roots, barks, etc., from the Konda Reddi herbalists or medicine-man. He does not attend marriage functions and death ceremonies to keep himself away from pollution since; he is a medicine-man.

Unrecognised Herbalists

There are several people who know certain methods of treatment of various minor ailments, and other illnesses. They are not recognized as full-fledged herbalists. According to the villagers, several of these new practitioners have no power to heal diseases and no experience in curing illnesses. Generally, these unrecognized herbalists are of younger age when compared with the recognized herbalists. Vuike Kannaiah (nephew of Vuike Veraiah) aged around 20 years, also learned the art of healing. Earlier he assisted his uncle Veeraiah the popular herbalists in the village. Though Kannaiah learned treating of various ailments and diseases, people do not recognize him as full-fledged practitioner. He expressed that he needs to acquire powers to cure certain of the diseases attributed to the supernatural causes. In his opinion herbalist or medicine-man must have possessed experience and supernatural powers to get the social recognition from the people. Most of the people say that, in the village a particular person can only cure certain disease or illness, why because a particular person hand works in the curing only certain illness. It is god gift to them.

Role of Patels, Pujaries, Sorceres and with-Doctors in Curing Diseases

Patels, Pujaries, Sorcerers, witch doctors form a class among the Koya society who play an important role in maintaining health by curing various diseases.

---

1. a)The Konda Reddis are one of the most primitive tribes of Andhra Pradesh. They are mainly found in the picturesque Papi hill ranges (part of Eastern Ghats) on either side of the Godavari gorge, in West Godavari and East Godavari, besides the hilly tracts of Khammam district Andhra Pradesh. In the past, the Konda Reddis' has mainly depended on shifting cultivation. According to Haeimendorf (1945) they are still at the Neolithic stage with pre-agricultural techniques. After India's independence, the National forest policy has restricted shifting cultivation to certain gradients of hills. At present, a majority of Konda Reddis are settled cultivators and agricultural labourers. The absence of road communication resulted in the isolation of the area from plain for a long time.

   b) Haimendorf (1945)The Reddis of the Bison Hills.

***Patels* (Headman)** are the heads of the Koya community. In Bodugugudem, they direct the people to follow the rules and regulations *i.e.*, taboo/incest/restriction regarding food, sex etc. People generally inform about their illnesses to the *Patels*. *Patels* also give some herbs, roots etc., to people, affected with some illness or minor ailments.

***Pujaries* (priests)** are the specialized priests. People most often consult their Pujaries for minor ailments, *i.e.* snake bites, scorpion bites and epidemic diseases like Ammatalli (chickenpox). *Pujaries* generally treat by conducting some special rituals with the traditional mantras (chanting) and administering herbal medicines.

People of the Boddugudem village approach the ***Vejjollu* (sorcerers)** either male or female to get cured from certain illnesses attributed to supernatural causes. These sorcerers conduct some magico-religious rituals. The affected persons have to provide *pasupu* (turmeric), *kunkuma* (vermillion), *kobbari-kayalu* (coconuts), *nimma-kayulu*,(lemons) etc., used in manipulation of demonic powers attributed to the illness of a victim. They use the bones of animals and human beings to manipulate the supernatural power. They chant some mantras and at times they may bite that part of the affected body with the disease. Then, they spit out some pieces of bones, which were kept secretly earlier in their mouths. They establish the confidence in the minds of victims that the elements which caused the disease, *i.e.* the supernatural spirits were sent out through the tiny bone pieces removed from the disease effected part of the body. He also asks the victims to bring hen, goat, sheep, eggs etc., for offering to various goddesses, spirits which caused the illness. Sorcerers are also available at the neighboring villages of Boddugudem. It is observed that the practice of sorcery is common and it is used for the well being of the society in most cases and in a few cases this is also used for punishing the enemies.

The People of the Boddugudem approach to the ***Devarlu* (witch-doctors)** available at their neighboring villages. Most of the Koyas visit Reddypalem near Bhadrachalam for solving their health problems like; common diseases, infertility, ailments and other epidemic diseases. While going to witch-doctors, people have to carry *pasapu* (turmeric powder), coconuts, *karpuram*, etc. *Devarlu* may be either men or women. During time of curing diseases, they sit on the skin of *jinka*/ledi, (deer). They perform witchcraft in the night time only. They go to trans by chanting some mantras which cannot be understand by others. In that process, they reveal the causes of the illness of the affected persons by showing some gestures and signs. The assistants of the witch-doctors interpret the meanings and the illness to the affected persons as diagnosed by them. It is believed that the witch-doctors have supernatural powers through which they establish contacts with their gods/goddess.

The affected people follow the instructions given by the witch-doctors. They offer their traditional liquor bottles or animals such as hen, eggs, sheep or goat and also pay some money. The witch-doctors give the ritualized turmeric powder to patients to keep *bottus* on their foreheads daily. In most of the cases the practice of witchcrafts are used to remove the evil effects and in rear circumstances it is also used of revenge or punishing the enemies.

Case Study I–Madakam Pantulu (aged around 30 years) was a Patel to one group of the people who are supporters of the CPI (M). He treats people in the village who were affected with cold, headache, fever, fractures, and cures in the traditional manner by using several herbs, roots and barks. According to him majority of the people was suffering from malaria, and jaundice. He also said that the people of the village are not able to visit the PHCs as they were located far from their village. He appreciated the works of the Loyola Integrated Tribal Development Society (LITDS)–a Non-Governmental Organization (NGO) which was providing health care to the people of the entire village. He wants to request the Integrated Tribal Development Agency (ITDA) officials to construct PHC in the village itself for the benefit of his community. He expressed his view that the traditional practices of medicine was powerful than any other systems of medicine. He wants to use their traditional medicinal systems for some common diseases and for treatment of ailments, and the Government (*i.e.* allopathic) medicines for other epidemic diseases (which cannot be treated in the traditional medicinal systems).

Case Study II–Kaniti Jogaiah (aged around 30 years) belongs to a family of the Pujaris of the Koyas. He was the leader of another group of the Koyas (the Roddu Gumpu) in the village, and maintains affiliation with the Telugu Desam Party (TDP). People affected with minor illnesses and other ailments such as scorpion bite (*telu-katu*) take traditional treatments from Jogaiah as he was considered to be an expert in curing these bites. He cures these by chanting mantras and giving medicines of herbs and leaves.

Case Study III–Kaniti Usaiah (aged around 70 years) was a *Pujari* to the Koyas of Boddugudem. As the ritual head of the village, he has a command over certain diseases attributed to their Kula-Devata (Goddess Mutyalamma). People affected with epidemic diseases like chickenpox (ammatalli) generally go to him for treatment, since the cause of these diseases are attributed to the anger of the goddess Mutyalamma. For the treatment of this disease, he chants some mantras to appease the goddess Mutyalamma and gives a ritualized bath to the affected, using the leaves of the *vepa* (*neem*) tree after applying the turmeric powder to the body. According to Koyas, he was an expert in curing common diseases, ailments, and other animal bites by administering medicines of herbs, leaves, and mantras.

Available Medicinal Plants, and the Treatment of Diseases/ Ailments

In general, medicine-man maintains secrecy in using various medicinal plants for curing several diseases. There is a strong belief that the medicines will not work properly, if they disclose the names of plants. Only healer knows some prescriptions and he/she enjoys the respect/status. The healer receives in form of cash or kind towards their services. Plant remedies are mixed into paste or in the form of pills. Most of the medicinal plants are also used for therapeutic purposes of cattle, other animals and birds.

The Koyas have certain beliefs and taboos about medicinal plants. The Koyas of Boddugudem have identified about more than 100 medicinal plants for of curing various illnesses and other ailments (Table 10.1). Plants having milky fluids (latex) are considered as medicinal plants, *e.g. Buggajilledu*. It was observed that the Koyas mostly depend on green plants for medicinal purpose rather than the dried plants. Roots/tubers of plants available in the earth/soil are also used as medicinal remedies. Sacred plants like *rela, ippa, vepa* are also used for curing certain diseases as these plants are known to have magico-religious powers. It is also believed by the Koyas that violation of any taboo makes the medicine ineffective. The Koyas gave names to various medicinal plants with the known values. Names of the Koya medicinal plants are frequently accompanied by two or three words which belong to the original of Dravidian dialect. More than one local name, attributed to one plant has been creating confusion to find the actual plant. Prefixes and suffixes of plant names indicate the habit of the plant, color of flowers, shape of leaves, size of fruits and seeds etc.

Table 10.1: Plants and their specific parts commonly used in ethno-medicine

Sl.No.	*Varnakulam Names (Local Names)*	*Latin Name*	*Usable Parts*
1.	*Adavidonda chettu*	Solena Amplexicaulis	Leaves
2.	*Adivibenda chettu*	Hibiscus lampa	Roots
3.	*Adavitulasi vhettu*		Leaves
4.	*Adivinimma chettu*	Pleispermum alatum	Fruits
5.	*Amrutapalem chettu*		Leaves
6.	*Anadulodu*		Barks
7.	*Arechettu*	Bomhinia racemosa	Barks
8.	*Ballakayala Chettu*	Fluggea virosa	Leaves and Roots
9.	*Balka Chettu*	ridella Montana	Leaves
10.	*Billedu Chettu/Bildu Chettu*	Chloxylan Swietenia	Roots and Leaves,
11.	*Boddi Chettu*	Fluggea Virosa	Leaves
12.	*Bugga Jilledu Chettu*	Celtis orientalis	Milk and Bark
13.	*Chedu Kakara Chettu*	Momordica Charantia	Leaves
14.	*Chedu Anapa Chettu*	Cucubita maxima	Leaves
15.	*Chenchela Chettu*		Leaves
16.	*Chendra Chettu*		Roots
17.	*Chetla pale*		Leaves
18.	*Chinnapale Chettu*		
19.	*Chinta Chettu*	Tamarindus Indica	Leaves and fruits
20.	*Chikkudu Chettu*	Beans	
21.	*Chinna Raite Chettu*		
22.	*Chinna Thumma Chettu*		
23.	*Chepata Chettu*		
24.	*Chinna Virugudu Chettu*		
25.	*Chinnaku Chettu*		

Contd...

Table 10.1–*Contd...*

Sl.No.	Varnakulam Names (Local Names)	Latin Name	Usable Parts
26.	*Danimma Chettu*	Punika Grantatum	
27.	*Dula/Durada Chettu*	Mucuna prurita	
28.	*Eedi Chettu*	Heptaleurun Stellatum	
29.	*Elagadam Chettu*		
30.	*Ganuga/Kanuka Chettu*	Pongamic pinnata	
31.	*Giritonda Chettu*	Antidesmaghaesembilla	
32.	*Golurgu Chettu*	Glycosmis raborea	
33.	*Guppidi Chettu*		
34.	*Guntakalakarru Chettu*	Acanthospermen-Ehispidum	
35.	*Jitty Chettu*	Demia Extensa	
36.	*Jilledu Chettu*	Calotropis-procera	
37.	*Jinny Chettu*		
38.	*Jonna Chettu*	Sorgh-um Halepnse	
39.	*Juttupaku Chettu*		
40.	*Kantuta Chettu*	Andropogon pertus-us	Flowers
41.	*Kancha Kakara Chettu*	Momordica Charantia	Barks
42.	*Kanneru/Ganneru Chettu*	Nerium Indicum	Barks
43.	*Karam Chettu*		Leaves
44.	*Karakka Chettu*	Terminalia Chebula	Roots
45.	*Krishna Tulasi*		Roots and Barks
46.	*Kirusugadala Chettu*		Roots
47.	*Kotar Chettu*		Roots and Leaves
48.	*Korinda Chettu*	Scutia Indica	Barks
49.	*Koyagiri Chettu*		Roots
50.	*Loutonda Pale Chettu*		Leaves
51.	*Maredu Chettu*	Aegle Mamelos	Barks
52.	*Maudava*	Hibiscus Rosa sinensis	Barks
53.	*Maddi Chettu*	Terminalia Alata	Barks
54.	*Mamidi Chettu*	Mangifera Indica	Barks
55.	*Maddu Vedama Chettu*		Roots
56.	*Madi Chettu*	MorindaTomentosa	Roots
57.	*Magasiri gaddala Chettu*		Barks
58.	*Medi Chettu*	Ficus virgate	Leaves and Roots
59.	*Mekala Jilledu Chettu*		Leaves and Barks and Roots
60.	*Moori Chettu*	Buchannania latifolla	Leaves

Contd...

Table 10.1–*Contd...*

Sl.No.	*Varnakulam Names (Local Names)*	*Latin Name*	*Usable Parts*
61.	*Mulaga Chettu*	Moringa pterygo sperma	Leaves
62.	*Mulligera Chettu Jas*	minum Auriculatum	Roots
63.	*Mempela Chettu*		Leaves
64.	*Mitcha Chettu*		Leaves
65.	*Nallavishal Chettu*		Roots
66.	*Nagali dumpa Chettu*	Dalbergia Lanceolaria	Leaves
67.	*Nagamalli Chettu Barks*	Jasminum Arborescens	Leaves
68.	*Nalla tummala Chettu*	Acacia Nilotica	Barks
69.	*Nalla Vsiri*	Clitoria Ternatea	Leaves and Barks
70.	*MelaVsiri*		Roots
71.	*Nela Vemula Chettu-*	Andrographis pinculata	Roots
72.	*Mella Chettu*		Leaves
73.	*Nela Tangedu*	Cassia Kleinii	Leaves and Barks
74.	*Neela Tentam*		Roots
75.	*Nelatadi Chettu*	Myrobalanus Bellerica	Fruits
76.	*Nelamtudu Chettu*		Leaves
77.	*Nernima Chettu*		Fruits
78.	*Nimma Chettu*	Citrus Aurantifolia	Fruits and Leaves
79.	*Padisam Chettu*		Leaves
80.	*Pedda Puli Chettu*		Roots and Leaves
81.	*Pedelli Chettu*	Gardenia Fragrans	Barks
82.	*Poteha Chettu*		Roots
83.	*Potutadi Chettu*		Sticks
84.	*Ratnapurusha Chettu*	Hybanthus Enneaspermus	Leaves
85.	*Redepaku Chettu*		Bark
86.	*Rela Chettu*	Cassia Fistula	Barks
87.	*Rimpa Chettu*		Leaves
88.	*Samdi Chettu*		Leaves
89.	*Sitru mullens Chettu*		Leaves
90.	*Some Chettu*	Panicum Colonum	Roots
91.	*Tangedi Chettu*	Cassia Auriculata	Roots
92.	*Tamara Chettu*	Nelumbo Nucifera	Leaves
93.	*Tamalaku Chettu*	Piper Betle	Leaves
94.	*Tagiresu Chettu*		Leaves
95.	*Tadi Chettu*	Myrobalanus Bellerica	Leaves

Contd...

Table 10.1–*Contd...*

Sl.No.	*Varnakulam Names (Local Names)*	*Latin Name*	*Usable Parts*
96.	*Telia Jilledu*		Barks
97.	*Telia Teega Chettu*		Roots and Leaves
98.	*Telia Teega Chettu*	Grewia Orientalls	Roots
99.	*Telia Naga Musti Chettu*		Leaves
100.	*Telia usiri Chettu*	Sauropus Quadirangularis	Roots
101.	*Telia Uishal Chettu and flowers*		Barks
102.	*Ummetta Chettu*	Datura Metel	Leaves
103.	*Urimutyala Chettu*		Leaves and Barks
104.	*Varni Chettu*		Leaves
105.	*Vemi Pelle Chettu*		Leaves

The Koyas medicine-man/herbalists generally observe certain timings in each any every season for collecting leaves, herbs, roots tubers, barks, latex, flannels, seeds and fruits from the respective medicine plants.

Table 10.2: Specific timings for collection of plants parts by Koya's medicine-man.

Plant Parts	*Time*	*Season*
Leaves	After sun-rise in the morningIn between 7A.M to 9A.M	Rainy and spring Seasons
Roots, Tubers, Flowers, Barks and Latex	Day time in between 7 A. M to 4.P.M	Winter, Summer and Autumn Seasons
Seeds and Fruits	Day time between 7 A. M to 4.P.M	During the available seasons

Medicines of Insects, Animals and other Mineral Origin

The Koyas of Boddugudem also use several organs and products like *Bellam-(Jaoggary), Tene-*(honey*), Pidgurallu, Telalu, Karpooram, ontullipay,*(garlic) *Nimmakayalu-(*lemans*), Vedipalu (*hot milk*) Amudamu-*(castroil*), Tallipalu-*(mother's milk*) Varipottu (*Rice hask*) panchadara-*sugar, *Memidisona-*mango plant *(gum), Godavarikachipa.*

Mollusca available in the Godavari river; *Gaddipoosa* (cudrania javanensis), *miryalu-*(pepper), *gudlu* (eggs), *sara* (traditional liquor), *pasupu* (turmeric), *gede-venna* (ghee), *chedu-chepalu*(fish), *kobbari-nune* (coconut oil), *yerralu* (earth warms), *vellulli* (garlic), *yerra-mirchi* (red chili), *gorojanam* (fat), *pandi-mamsam* (pig meat), *bangaru-purugu* (an insect) manishi- mutram (urine of man), *elugoddu-seru, alli-chemalu* (ants), *eluggoddu-mamsam* (meat of bear), *konda-mutchu-mamsam* (meat of langur), *gede-peda* (buffalos' dung), *eluggoddu-charmam* (skin of bear), *boddinkalu* (cockroaches), etc., which have had therapeutic values.

Plant Conservation Systems of the Koyas

The Koyas of Boddugudem have a wide range of animistic beliefs associated with groves, plant species and forest worship. Certain plants having medico- religious values are not damaged or destroyed by the Koyas under any circumstances. Worshipping of certain plants is common during festivals. Useful plants near or inside the tribal villages are protected for in many cases. They never take off the whole plants or all fruits for use, but leave some re-productive parts for the growth of the plant. Some vegetables like *chikkudu* (beans) is not consumed at the ripening stage of the fruits. Observance of these types of restrictions plays an important role in the conservation of genetic resources. Generally the Koyas avoid peeling the bark from any single plant more than once in a year. Koya medicine-man plucks the medicinal herbs at day time before noon and collects the bark only from one side of the trees. Koya herbalist never cuts the whole tree for collecting the medicinal parts. Collection of gums, resins, latex, sap, etc., is not done from younger plants. Certain fruits or vegetables are not consumed by the Koyas before the performance of specific rituals.

Pharmacology of Herbal Medicines and their Prescriptions

Preparation of drugs and medicines among the Koyas is based on age-old experience. Most of the medicines are prepared either as single drug (s) made of a mixture of different plant parts. A combination of plants, animal organs, rocks and, minerals etc., in their medicines is common. Some of the Koya medicines include infusions, decoctions, mixtures, pastes, powders, pills, plasters, etc. In general, the dose prescribed by the Koya herbalists for each medication is the adult dose. It is proportionately reduced according to the age of the patient. Using the plant or plant parts into a paste or the extract or squeezing the liquid and fermentation are common practices among the Koyas. For rheumatic, swellings and boils, either boiled roots, barks, leaves or the paste of the plant part is applied. Dry plant parts are usually made into powder. In certain cases barks or roots of medicinal plants are chewed and sucked.

Generally Koyas take the medicine either with fresh cold drinking water or with their traditional liquor, or with honey as advised by the medicine-man. Several Koya medicines are prepared in combination with black pepper. Sometimes they also mix *sonti* (dried ginger). They believe that addition of pepper or *sonti* in the medicine make the drug therapeutically effective.

Drugs and Treatment

In case of acute and refractory diseases like Devarasattu (paralysis), the Koyas take treatment at least for three months. In case of general diseases like *Jvaram*-(fever), dysentery for diaarrhea, cough and cold etc. the normal treatment is for five to seven days. In case of minor ailments, scorpion bites etc., the treatment is for one or two days. The following are the common drugs and remedies.

(a) Rasam (Juice)

It is the extract made by squeezing the freshly collected parts of the medicinal plat. It is believed that the extract is highly effective. For the preparation of juice the medicine-men cut the required plants into pieces and crush then in mortar. The juice extracted by squeezing the crushed bits is administered orally or applies externally. The adults are prescribed about two tea spoons full and for children they use less than a teaspoon. Juice is always freshly prepared and is used within two to three hours.

(b) Paste (Makku)

Generally the Koya herbalists collect the required fresh and dried plants or plant parts and grind them into paste on a stone, and this is administered both internally and externally. For the preparation of paste they clean the dried material and make into a paste and it is applied all east twice in a day, according to medicine-man when paste is preserved for 24 hours.

(d) Mandu-Goli (Pills)

These are in the form of small globules. Plant parts and other dried-medicinal plants are pounded and mixed into a paste. This paste is made in to rounded pills about the size of a pea. The pills are to be taken thrice in a day; generally these pills are dried and kept ready by the medicine man.

(f) Kalchina Aakulu (Heated Leaves)

Generally fresh leaves of the medicinal plant are exposed to a flame in an open hearth and they are applied externally

Table 10.3: Treatment of diseases.

Diseases	*Treatment*
Thummulu (Sneezing)	The Koyas put one or two drops of *rasam* (Juice) of the leaves of *Padisam* plant in to the nose of the person affected (like nasal drops). This acts as a decongestant.
Rompa/Padisam/Jalubu (Cold)	*Treatment–1.* A few drops of *rasam* of the leaves of *Padisam* plant by are dropped in to the nose of the person affected, during sun light. This leads to continues sneezing. This medicine has to be taken three to four times and this aggregates sneezing and cures the person from colds.
	Treatment–2. This medicine is a combination of *Nimma* Chettu leaves, *Cheppata Chettu* leaves, *Vavila* chettu leaves, and a little bit of turmeric powder boiled together in a vessel. Inhalation of this steam three to four times a day gives complete relief from colds.
Daqqu (Cough)	The *rasam* extracted from *Karakkaya* is given internally twice or thrice a day until the cough is controlled. The Koyas also chew the bark of Maddu Vedama Chettu or, eat the curry of *Vakadukayalu* for controlling continues cough.
Talanoppi/Edekku (Headache)	*Treatment–1.* To control *edekku* the Koyas take the *rasam* made from the upper layer (*pottu*) of *Ballakayalu*, mixing it with sugar and this is taken at least 2 to 3 times during the time of headache.
	Treatment–2. To control *edekku*, the Koyas collect *alii cheemalu* from the forest and crush them and tie them in a cloth. The patient is made to inhale the crushed ants (all *cheemalu*) tied in the cloth and this treatment gives relief from the headache.
	Treatment–3. Paste of *Sonth* (dried ginger) is applied to the temples (*Kanatalu*) The patient gets relief within 15 minutes.

Contd...

Table 10.3–*Contd...*

Diseases	*Treatment*
Varagoba (Parsapunoppi)	*Treatment–1. Varagoba* is a kind of headache, which usually starts around 07.00 am in the morning and continues up to noon.To cure this problem the medicine-man takes the affected person to an open place and makes him stand with his back towards the sun. The medicine-man looks at the shadow of the man.He examines the shadow of the head and identifies the kind of headache. The medicine-man, makes the patient stand in that posture and look the shadow of his head. Then he puts a few drops of the *rasam* of the leaves of Chikkudu Chattu in this and the patient gets cured
	Treatment–2. To treat varagoba the *rasam* of the *Vempella* leaves and (Ontulli) *ontullipaya* is also used. The herbalist puts one or two drops of the Juice in to the nose of the affected person.

Table 10.4: Natural diseases are based on the heat concept.

Diseases	*Treatment*
Jvaram (Fever)	To control fever the pills made out of the paste of the *Nelavemu* leaves and Nela-tadi tubers are given for two or three days or until the fever is controlled. During the period of treatment the patient will also be given gruel (java) made of *Jowar.*
Banka Viraochanalu (Motions with Mucous)	*Treatment–1.* The bark of *morre* tree and gum of *Tumali* plant are chewed together for three to four days to cure motions with mucous.
	Treatment–2. The *rasam* made out of the bark of *jinne* plant mixed with a little bit of sugar is given for two to three days. For quick relief they also add some *Kaskas* while preparing the juice
Rakta-virochanalu (Motions with Blood)	To cure blood motions the juice made out of the bark of *Danimma* plant mixed with a little bit sugar is given for two to three days. For quick relief they also add some *Kaskas* and a little bit of dung of the cow while preparing the Juice.
Kalla Kalaka	To cure this eyes are wiped with drops of water falling from the leaves of *Aamudamu* plant for four or five days. Then the medicine-man makes an extract from *vempela* and *Balla Kayala* plant and pours two to three drops of this *rasam* in the eyes of affected person for three or four days or until it is completely cured.
Chali-Jvaram (Fever with Shivering)	In most cases this could be malaria. The *rasam* made out of *Nallamiriyalu* (black pepper) and leaves of the *Usiri* plant and *Kantula* plant is frequently given until the fever is controlled. A few Koya people are also taking modern medicines as the government health workers are providing the chloroquine tablets regularly to their houses.
Kancluka	This is attributed to the consumption of sweet toddy. The Koya herbalist makes the affected person vomit the sweet toddy by smelling *Velaga* plant leaves.
Dokkaubbu (Gas-trouble)	To control this gas-trouble, Koyas eat small quantities of *chiguru/iguru* (tender foliage) of the *rela* plant. At times the Koyas also eat food mixed with the paste of *sela* plant leaves to have a free motion ultimately, to control the gas-problem.

Contd...

Table 10.4–*Contd...*

Diseases	*Treatment*
Dokka- Vodchu/Neella virchanalu (Loose motions)	*Treatment–1.* The koyas take a little bit of the *rasam* made out of *Chinna* Pale plant's bark for one or two days until the motions are controlled. *Treatment -2.* To control loose motions the *rasam* made out of the barks of *Vepa, Mamidi, Chinta* plants is administered. The dose is one or two glasses daily.
Vantulu (Vomiting)	The paste made out of *Nela usiri* plant leaves and the of *kantula* plant in the medicine to control the vomiting. The Koyas also drink the *rasam* made out of *Danimma* fruit, or smell the juice of *velaga* plant leaves.
Chekkurupu/Chinna pundu (Sores)	To cure small sores the powder of *jowar* is mixed in water and applied to the sores. Sores are among the children of the Koyas of Bodugudem.
Sega-Kurupu/Gaddalu (Boils)	To cure boils the Koyas use the paste made out of the Mandara leaves. It is applied for four to five days.
Vatham-noppulu	To cure these pains curry made from the leaves of *Bachali, Mulaga*, and *Gongura* plants are eaten.
Keella-noppulu/Joint-pains (Rheumatism)	The Koyas preserve the dried meat of *Eluggoddu* and *Kondamuchu.* This meat is eaten as medicine to cure joint-pains and other bodily pains.
Ollu-noppulu (Body-Pains)	To cure these pains the Koyas use the *rasam* made out of *Karpuram Aavu* (cow) *gorojnam*, bark of *chinna virugudu* plant, and leaves of *Nelausirichethu* for two or three days.
Nadum-noppul (Backache)	The Koyas apply the *rasam* made out of *Nela-Tangedi* and *Nela- Usiri* plants, externally on the pain affected part of the body for two or three days to cure *Nadum-noppulu.*
Telakatlu/kallutiruguta/ murchalu/Reppa Kottatam (Epilepsy)	To cure this problem the Koyas put a garland made out of *Tella-Jivedu* tubers around the neck of the patient. Before tying to the patient it will be ritualized by putting *sambrani-dupam* in the name of the Goddess, Mutyalamma, the *kula devata* of the Koyas, in front of *gamamm* which is the place of worshiping the gods and goddesses.
Muttachula Noppulu (Pregnancy Related Pains)	To control this pain, the koyas uses Tamalapakulu and Gedevenna by grinding them as paste. This paste will be given to the pregnant woman when the labour pains start. This is given for a few days.To treat *muttachula-noppulu* the Koyas also give a mixture made out of Somichekka and *Sandi-Kayala-Kuppi* to the pregnant women for a few days.

Table 10.5: Diseases attributed to supernatural causes.

Diseases	*Treatment*
Ammatalli (Chickenpox)	To treat Ammatalli the Koyas spread small branches of *Neem* in the bed of the affected person. He will be given only non-oily foods. *Neem* leaves are also used in the special curry cooked for the patient. The patient will be given ritual bath by the *Pujari*, with boiled water. While boiling the water they also add *Neem* leaves and a little bit of turmeric powder. Later the *pujari* will reveal the name of the animal to be sacrificed to satisfy the Goddess. Mutyalamma as the cause of this disease is attributed to her.

Contd...

Table 10.5–*Contd...*

Diseases	*Treatment*
Korintadaqqu (Whooping Cough)	To prevent this cough the Koyas consult the *vejjollu* and get *Antram* (Amulet) which is tied around the neck of the patient. Besides *Antram* the Koyas also consume the *rasam* made out of black pepper and bark of Maddivedama plant.
Mondi Kurupulu/Narikuru-pulu (Dreadful Sorse)	The Koyas use the paste made out of the bark of Guppidi plant and leaves of *Uttareni* plant to cure Mondi *Kurupulu.* It is observed that most of these cases are not curable completely and at times they may even lead to deaths also. Therefore the cause for these *Kurupulu* is attributed as a punishment given by the *Konadevarlu.*
Sompidi-Endurogam (Emaciation)	To cure this, Koyas make a garland of tiny sticks of the *Somi* plant after sprinkling some turmeric powder on it. They also tie a leather garland around the neck of the person affected with *Sompidi.* This leather garland is being tied with some living creature like *Boddinka* (cockroach). It is hung around the neck of the patient, until the cockroach dies. It is believed that the person will grow normally after the death of the cockroach.
Devarasattu (Paralysis)	*Treatment–1.* To cure this disease the Koyas use rounded pills made out of tubers of the *Urimutyalu, Giritonda* and *Rimpa* plants and roots of *Vedama* plant, *Pedda* pale plants and the flowers of *Vami* plant. The patient has to consume these pills for three months. During the period of treatment the patient has to avoid consumption of oily-foods and rice, the patient will be given *gataka duda- java* (a kind of gruel) along with some buttermilk. *Treatment–2.* To treat this disease the Koyas also use the pills made out of the bark of *Pedelli* plant by mixing with a little bit of *karpuram*, and *Gavabestam.* These pills are to be taken daily until the disease is cured. Smoking should be avoided by the patient during the period of treatment. If any reaction arises due to the use of pills, the patient will be the Potlakayatokku to chew, and this neutralizes the side effects.

Table 10.6: Skin disesaes.

Diseases	*Treatment*
Dula/Durada (Itching)	A paste is made out of the leaves of *Vepa* (axadirachutur Indicia) and turmeric powder is applied externally on the effected part of the skin at least for one week.
Gajjipokkulu (Scabies)	The powder of *Ganuga* seeds mixed with coconut oil is applied externally on the affected part of the skin at-least for one week.
Tamara (Ringworm)	The *rasam* made out of the leaves of the Tamara plant mixed with lemon juice is applied externally on the affected part of the skin at-least for one week or longer till the skin disease is cured.
Soben (White Patches)	The Koyas use the ash made out of the *Potutadi chettu kayalu* mixed with coconut oil, externally on the affected part of the body, until the skin patches completely disappear.
Chaldi (Wounds)	To cure this disease *rasam* made out of the leaves of *Maredu* plant is applied externally on the affected part of the skin for two to three days

Table 10.7: Diseases attributed to weakness and lack of food.

Diseases	Treatment
Patchakamerlu (Jaundice)	*Treatment–1.*The Koyas eat food with milk, mixed with the paste made of *Vellulli* (garlic) and the tubers of *Jitti* plant at least for five to six days to cure *patchakamerlu.* *Treatment– 2.* Pills made of the leaves of *Golubu* mixed with a little bit of turmeric powder are given for at least a week to cure jaundice.
Tellakamerlu (Jaundice)	Garland made out of yellow, black and red threads is tied around the neck of the patient. This is kept for a week.
Yerrakamerlu (jaundice)	Pills made out of the tubers of the *Thadi-chettu* and the bark of *Chinnapale* plant are given for at least 5 to 6 days to cure Yerrakamerlu. There are no food restrictions during the period of treatment for this disease
Ubbukamerlu (Jaundice)	To treat *Ubbukamerlu* the affected person eats some *Chedukakara-kayalu* for four to five days. In addition the paste made out of the barks of the Ganuga tree is applied externally on the patient's body for at least five to six days.

Table 10.8: Minor ailments.

Diseases	Treatment
Kattigatlu (Cuts)	To cure cuts the bark of the *Gummadi* plant is used as a band-aid. The Koyas also use Thadi-poota to cure the cuts in general.
Debbalu (Wounds)	To cure wounds a paste made out of the bark of *Maddi* plant or leaves of the *Mulaga* plant is applied externally on the affected part of the body by tying with a cloth as a band aid.
Kalinagayalu (Burns)	To cure the wounds caused by burns, the Koyas use the leaves of *Amudam,* and *Tutaku* coated with *Amudamunune* (castor oil).
Pundlu (Sores)	To cure sores a paste made out of the *Guntakalakarses* leaves and urine of man is made. This paste is applied to the sores and a cloth is tied over it.
Emukaviru-gudu (Bone fracture)	A paste is made out of the leaves of *Bodde Kura.* It is applied on the fractured part of the bone externally. The fractured bone is kept in position by tying bamboo staves. The bamboo staves are removed every nine days, the paste is applied and bamboo staves are tied again. This process is repeated till the fracture heals.

Table 10.9: Bites of animals and other insects.

Diseases	Treatment
Pitchikukka Katu (Mad Dog Bites)	*Treatment–1.* The disease caused by the bite of a mad dog is called by the Koyas as *Verri-jabbu.* This is actually rabbis. To cure this disease the Koyas use a mixture made out of the fried heads of *Bangaru purugulu* (a kind of insect), turmeric powder and Jiggery This mixture has to be consumed for one week by the person bitten by the dog. Salt in the food should be avoided.

Contd...

Table 10.9–*Contd...*

Diseases	*Treatment*
	Treatment –2. The other treatment is pouring two to three drops of the *rasam* made out of *Korinda* tubers and garlic daily in the nose of the affected person at least for three to four days. *Treatment–3. rasam* is prepared by crushing *Neelatentam* tubers and tobacco leaves. Four to five drops of this is poured in the nose of the person bitten by the mad dog. *Treatment–4.* The Koyas also use a *rasam* made out of *Chedu-Kakara-Koyalu* mixed with a little bit turmeric powder. This is applied externally on the bitten spot and a cloth band (*Kattu*) is tied on the bitten spot. This *rasam* is also taken internally by the person bitten the by dog daily in the early hours of morning. During the period of treatment the patient should avoid eating of non-vegetarian foods, especially pork.
Pamukatu (Snake bite)	The Koyas use *Pamurallu* (snakestones) to neutralize the poison due to snakebites. Two types of snakestones are available to the Koyas: (1) *Pedda Pamurai* which is taken from the head of *Pinjaiu* and (2) *Chinna Pamurai* which is taken from Nagupamu (cobra) head. The Koya medicine-man keeps *pamurallu* on the snake bitten spot for a few minutes. After absorbing the poison, these snakestones will fall down from the snake bitten spot. Meanwhile the medicine-man makes a *rasam* out of the tubers of *Tellsvishal* and *Nallavishal* plants and roots of *Putta veduru* plant puts two to three drops of this juice in the eyes of the snake bitten person. After some time the affected person will become normal. The Koyas stated that the person bitten by snakes cannot identify the taste of *neem* leaves.
Telukatu (Scorpion bite)	*Treatment–1.* The Koyas use the *rasam* made out of the *Chenchela* plant leaves to prevent the pains caused due to scorpion bite. The Koyas stated that, pouring of the juice of *Chenchela* plant leaves on a scorpion results in breaking of its stomach and it dies. *Treatment–2.* To neutralize the poison and pains due to scorpion bite the Koyas use the mango gum which they preserve in a bottle. This gum is applied on the bitten place of the body externally two to three times.
Bites of other insects	Koyas use the *rasam* made out of the leaves of *Tagiredu* plant for two to three days to treat insects' bites. To neutralize the poison due to bites of insects like *Kandireega*, ant, and *centipedi* the juice of *Nallathumma* plant leaves mixed with turmeric powder is applied externally on the place of bite.

Table 10.10: Treatment of other health problems.

Diseases	*Treatment*
Constipation	The Koyas eat the paste made out of the *Rela* plant along with food to have free motions.
To have vomiting	The Koyas drink water mixed with soap or *Konkiri/Malamu* to have vomiting.

Contd...

Table 10.10–*Contd...*

Diseases	*Treatment*
To increase sexual desires	To improve sexual desire, the Koya men most frequently take the rasam made out of boiled leaves of *Ratna-pursha* plant by mixing with honey. Generally the Koyas eat the chiguru or *iguru* of the *Mulaga* plant with hot milk to have strength in sex. Most of the Koya men also eat the tubers of the *Magasirigadda* plant or *Elugoddinaramu* to increase their sexual power.
Garbasravam (Abortions)	The Koyas use a paste made out of the leaves of Amrutapalam and *Elagadam* plants by mixing with *Aavugoiejanam* (cow's fat) to cause abortion.
Tellabatta(White discharge)	The Koyas give a mixture of black-pepper, leaves of *amrutapalan* plant *ganneru-poota* and *meda-naram* of any animal along with food to the women to cure white-discharge.
Rommendu-jabbu (Milkless-ness in lacting mother)	In certain cases the Koya women may not have sufficient milk to feed her new born baby. To improve her breast milk, the women will be served a curry of *chedu-chepalu* and some *Boppai* fruits for several days. In addition the women will also be served a powder of dried earth-warms, garlic, and chili, in her daily food.
Disti (Evil Eye)	To avoid *Disti* the Koyas put *Dupam* before themselves and their cattle in general. The Koyas burn the leaves of *Durada* plant *llakommu* a piece of leather of *elugoddu* and *sambrani* together in a container as "Dupam".
Vendu (Heat in the Body)	To control this the Koyas boil *Tippateegapottu* and the leaves of *Neelavemu* in water. The person is given a bath with this water. The Koyas also consume toddy most frequently to control heat in the body.
Gali/Dooli (Evil Spirits)	To avoid *Gali/Dooli* or evil spirits the Koyas keep *Dupam* under the Tunika plant for their wellbeing and welfare of their cattle. *Gali/Dooli ate* attributed to supernatural spirits of the people who died in an accident or an unnatural death. It is believed that these spirits are roaming in the winds of hot -midday in the form of *Gali- Sudulu.* These may enter into the people's body and harm their health, if the people roam during hot-mid day.
Pedakadupula noppi (Appendicities)	The Koyas identify the pains of appendicitis. Now the Koya herbalists are sending the patients to the nearby PHCs or other government hospitals available at *Edugurallapalli*, or Bhadrachalam town, as they have failed to treat this problem with their traditional medicine.
Chevipotu (Pains in the Ear)	*Treatment -1.* To cure this pain the Koyas use the juice made out of the "*letathadichettumatta*" by pouring two to three drops in the pain affected ear for one or two days. *Treatment-2.* To cure this pain the Koyas also use the Juice made out of the leaves of "*Cheduanpakaya*" plant by pouring 2 to 3 drops in the pain affected ear for one or two days. *Treatment-3.* To cure this pain the Koyas use lemon juice mixed with Godavari Kachippa by pouring 2 to 3 filtered drops in the pain affected ear for one or days.
Mutrajabbu (Diabetes)	To cure these pills made out of a paste of the leaves of *Krishnatulasi* and *Podapatrapu* plants are used. These pills will be taken by the patient daily, until the disease is controlled.

Contd...

Table 10.10–*Contd...*

Diseases	*Treatment*
To Entice Women	The Koya men use the pills made out of a paste by mixing the roots of the Nallausiri and Tellausiri a little bit of Kunkuma and turmeric powder, urine of a male person and coconut-oil To entice the women and to have sex the men mix three to four of these pills in the food of the woman without her notice or knowledge with whom they like to have sex. This medicine is popularly known as Lanjamandu.
Neutralization of poison	The Koyas use the juice made out of *Uttareni* leaves to neutralize the poison due to snake or scorpion bite.
Enda debba(Sun stroke)	To treat this problem the Koyas take rest from physical work and consume sweet toddy often until they get strength and relief.

The Koya treatment thus entirely depends on the etiology rather than symptoms of disease. Through the use of medicines the Koyas have developed the positive aspect of treatment and the ways of strengthening the sick persons in increasing their power of resistance.

Conclusion

Ethno-medicinal practices among the Koyas have been evolved from centuries rooted into their socio-cultural realm and other value orientations. Many curatives are effectively useful in treating several ailments and other diseases. Even today, they have been using their own traditional medical practicing apart from other modern medical systems. Due to modernization, these health seeking cultures have been vanishing in the traditional societies, which have been developed from centuries. At this juncture, there is a great need to preserve and promote these valuable traditional knowledge systems, which provide treatment at very low cost.

References

Babu M. Ram, 2000. *'Ethno-Medicinal Practices of the Koyas of Bhadrachalam: A Case Study of Boddugudam Village, Khammam District, Andhra Pradesh'* (AnUnpublished M. Phil Dissertation submitted to the University of Hyderabad).

Haimendorf, C. Von Fuerer, 1945. *'The Reddis of the Bison Hills'*.

Rao, Mohana K., 1981. *'The Koyas of Andhra Pradesh'*. Hyderabad: TCR&TI.

Schltes, R.E., 1962. 'The Role of Ethno-Botanical in the Search of New Medical Plants'. *Lloyodia* 25: 257–366

Environment, Biodiversity and Traditional System, 2014 *Pages* 107-113
ISBN 978-81-7622-307-2
Editors: D.R. Khanna, G.S. Solanki and S.K. Pathak
Published by: BIOTECH BOOKS, NEW DELHI

Chapter 11

Biophysical Determinants Influencing the Adoption of Homegarden-Based Agroforestry in Western Himalaya

Kamal Kishor Sood

Division of Agroforestry,
Sher-e-kashmir University of Agricultural Sciences and Technology,
Main Campus-Chatha, Jammu, J&K (India)

ABSTRACT

Biophysical determinants play an important role in determining the type of vegetation of particular place. However, the role of these determinants on adoption of homegarden based agroforestry has remained ignored till date. Thus the role of these factors in adoption of homegarden based tree cultivation was studied in Western Himalayan province of Himachal Pradesh. Binary logistic regression was used to identify key factors influencing home-garden based agro forestry. Agrocliamtic zone, soil depth, soil texture, proportion of farm fenced, proportion of land adjacent to house and distance of house from nearest nursery, each individually, affected tree cultivation in homegardens. Based on logistic regression, out of seven variables, six appeared in the model. This implies that these six variables (agroclimatic zone, soil depth, soil texture, traditional fencing, distance of the housed from the road and nursery respectively) were key variables influencing likelihood of adoption of home-garden based agroforestry systems. The choice of tree-crop combinations which minimises the competition in high hill temperate dry zone would encourage the adoption of homegarden agroforestry systems. The study implies that the traditional fencing (bush+shrubs) and new nurseries needs to be developed to encourage tree cultivation in the form of homegarden agroforestry.

Introduction

There are many agroforestry practices across the world (Sood, 2003). Most are site specific and vary from one place to other even within the same country or region (Nair 1993). Homegarden is one of the agroforestry systems and is a very old tradition the world over (Soemarwoto, 1987). Trees in homegardens play an important role in livelihood security of the rural people. Trees provide timber and non-timber products to the people and also provide non-tangible benefits. For example, in Java (Indonesia), 60 per cent of all food comes from home-gardens (Hoskins, 1990). The homegardens of Kerala (India) (Chundamannil *et al.*, 1993; Nair and Krishnankutty, 1985; Soemarwoto, 1987), Bangladesh (Leuschner and Khaleque, 1989), Java (Christanty and Iskandar, 1985; Soemarwoto, 1987) and Kandyan gardens of Sri Lanka (Jacob and Alles, 1989; Soemarwoto, 1987) are well documented and play important role in the livelihoods of the people and assists in reducing the pressure on existing forests for tree products. The existence of homegarden based agroforestry systems has also been documented in Himalayan region (Mughal *et al.*, 2000; Nautiyal *et al.*, 1998; Singh *et al.*, 1998). These homegardens are adaptable to the biophysical and socio-economic conditions in their respective locations of their existence. However the empirical work on role of biophysical in adoption of homegardens is almost lacking. Therefore, current empirical work on role of biophysical determinants on adoption of homegarden based agroforestry systems was taken in Mandi district of Himachal Pradesh located in Western Himalaya.

Materials and Methods

The study was carried out in Mandi district of Himachal Pradesh where the agroforestry systems are existent. Data were collected using a prestructured schedule and personal interviews with the head of the household. Multistage sampling was used to select households. Out five forest divisions, two forest divisions, Joginder Nagar and Suket, were selected to obtain realistic representation of households from different agroclimatic zones. A list of villages into hill and foothill villages was prepared which were further subcategorised into programme (a village with participation Joint Forest Management) and non-programme. Two villages from each subcategory for hill and foothill villages were chosen in each selected forest division using simple random sampling with replacement method. In this way, there were eight sample villages in each of selected forest divisions and 16 villages in total. A list of households in each of the selected villages was prepared. A list of large (> 1ha) and small (<1ha) farmers from each selected village was prepared. Amongst large and small farmers 1/3rd of households were chosen for each category using simple random sampling. A total sample size of 401 households was selected out of which 52 which did adopt any agroforestry system and were dropped from the sample and remaining were divided into two categories: homegarden adopters and non-adopters. A chi-square test of association was used to test the association between adoption of home-garden based agroforestry and each independent variable at 0.01 level of significance. To identify key factors influencing agroforestry binary logistic regression analysis was used. The model specification for binary logistic regression is as:

Logit is defined as natural log of odds and the model was specified as:

$$Logit = Ln (P/1-P) = \beta_0 + \beta_1X_1 + \beta_2X_2 + \beta_3X_3 + -- + \beta_kX-1$$

where,

P is the probability of homegarden adoption

1-P is the probability of non-adoption

β_0 is the intercept term

β_1, β_2, β_3, and β_k are the coefficients associated with independent variables

$X_{1,}$ $X_{2,}$ X_3 and X_k are the predictors in the equation

Table 11.1 shows the coding scheme for the independent variables used in regression analysis.

The model was validated using model Chi-square and Hosmer-Lameshow goodness-of-fit and cases correctly classified. Standardised regression coefficients depicted the importance of the predictors in the model. Table 11.1 furnishes the variables used in the model, their abbreviations and statistics.

Table 11.1: Variables in model (Biophysical model)

Sl.No.	Variable Name	Abbreviation Variable Coding Scheme	Units	Mean	Standard Deviation
1.	Agroclimatic (Altitudinal) Zone	AZONE–[Mid-hill subhumid (601-1800m) = 1, High-hill temperate wet (1801-2200m), High-hill temperate dry (≥2201m)=3]	Scale	1.44	0.69
2.	Soil depth	SDEPTH (Shallow -<30 cm) =1, Deep-31-45 cm), Very deep–≥46 cm=3)	Scale	2.35	0.83
3.	Soil texture	STEXT (Clayey loam=1, sandy loam=2, gravelly=3)	Scale	1.76	0.77
4.	Distance of village from nearest road	DNRD	km	1.20	2.10
5.	Distance of villa nursery road	DNUR	km	2.84	2.47
6.	Proportion of farm fenced (Traditional fencing composed of live/dead bushes and shrubs)	PFF (Not fenced =0, Partial fencing =1, Fully fenced = 2)	Scale	1.02	0.79
7.	Proportion farm adjacent to household	PFADJ	Per cent	39.86	32.79

Results

The study revealed that agroclimatic zone, soil texture, soil depth, proportion of the farm fenced, proportion of the farm adjacent to the household and distance of the house from nearest nursery, each individually, influenced significantly the homegarden-based agroforestry. The distance of the house did not influence the

Table 11.2: Association between choice of homegarden-based agroforestry and biophysical factors.

Count Per cent of Row Per cent of Col	Agroclimatic Zone			Soil Texture			Soil Depth			Proportion of Farm Fenced		
	Mid-hill Sub-humid (601-1800m)	High-hill Temperate Wet (1801-2200m)	High-hill Temperate Dry (≥2201m)	Clayey Loam	Sandy Loam	Gravelly	Shallow (<30cm)	Deep (30-49 cm)	Very Deep (>49 cm)	Not Fenced	Small Portion	Major Portion
Non-adopters	68	24	36	39	67	51	57	38	62	70	58	29
	43.3	12.5	22.9	24.8	42.7	32.5	36.3	24.2	39.5	44.6	36.9	18.5
	29.3	31.2	90	25.3	54	71.8	71.3	57.6	30.5	66.7	43.3	26.4
Adopters	164	53	4	115	57	20	23	28	141	35	76	81
	85.4	33.8	2.1	59.9	29.7	10.4	12	14.6	73.4	18.2	39.6	42.2
	70.7	68.8	10	74.7	46	28.2	28.8	42.4	69.5	33.3	56.7	73.6
	χ^2= 73.475, d.f.=, p<0.0001, CFLF=0			χ^2=48.829, d.f.=2, p<0.0001, CFLF=0			χ^2=43.638, d.f.=, p<0.0001, CFLF=0			χ^2=35.514, d.f.=2, p<0.0001, CFLF=0		

Count Per cent of Row Per cent of Col	Land Adjacent to House			Distance from Nearest Road (km)			Distance from Nearest Forest Nursery		
	0-25 per cent	26-50 per cent	51 per cent and above	< 0.5	0.6-2 km	2.1-8	<2.0 km	2.1-5km	5.1-8km
Non-adopters	103	31	23	71	50	36	6	114	37
	65.6	19.7	14.6	45.2	31.8	22.9	3.8	72.6	23.6
	57.9	34.1	28.8	40.6	45	57.1	20	50.2	40.2
Adopters	75	60	57	104	61	27	24	113	55
	39.1	31.3	29.7	54.2	31.8	14.1	12.5	58.9	28.6
	42.1	65.9	71.3	59.4	55	42.9	80	49.8	59.8
	χ^2=24.836 d.f.=2, p<0.0001, CFLF=0			χ^2=5.140, d.f.=2, p<0.077, CFLF=0			χ^2=10.926, d.f.=2, p<0.004, CFLF=0		

adoption of the homegarden agroforestry. The proportion of homegarden adopters decreased from 70.7 per cent in midhill subhumid climate to 68.8 per cent in highhill temperate wet climate which decreased to 10 per cent in the households located in high hill temperate dry climate.

The farms will clayey loam soil had the highest proportion (74.7 per cent) of households with homegardens than that of with sandy loam (46.0 per cent) and gravelly (28.2 per cent) soils respectively (Table 11.2). With increase in thė soil depth the proportion of households adopting homegarden agroforestry increased from 28.8 per cent in case of shallow soils to 28.8 per cent and 42.4 per cent respectively in the households with deep and very deep soils respectively (Table 11.2). Similarly the proportion of homegarden adopter increased with increase in the proportion of the farm fenced (traditional fencing composed of live/dead shrubs/bushes) and proportion of the land adjacent to the household (Table 11.2). The percentage of the adopters of homegardens decreased with increase in the distance of the house from the nearest nursery, however, there no significant effect of distance of the house from the nearest road on homegarden agroforestry.

The binary logistic regression model is given in Table 11.3. Out of the seven variables tested in the model, six appeared in the model. Each variable in the model contributed significantly in predicting the adoption of homegarden (Table 11.3).

Table 11.3: Variables and their significance.

Step	*Variable In:*	*Coefficients (β)*	*EXP (β)*	*Improvement in-2LL (Chi-square)*	*D.F.*	*Significance of Change (p)*	*Standardised Coefficients*
1	AZONE	–2.627	0.072	77.938	1	<0.0001	–1.982
2	PFF	0.738	2.092	15.652	1	<0.0001	0.594
3	DNRD	0.207	1.230	8.133	1	<0.004	0.527
4	STEXT	–1.148	0.317	4.531	1	<0.003	–0.930
5	SDEPTH	–1.343	0.261	12.893	1	<0.0001	–1.173
6	DNUR	10.97	1.217	10.209	1	<0.0001	0.480
	Constant	7.700					

The model was significant in predicting the adoption of the homegarden agroforestry (Model Chi-square = 129.357, p<0.0001-Table 11.4).

Table 11.4: Model summary.

Statistic	*Value*	*D.F.*	*p*
Initial –2LL	402.362	–	–
Model –2LL	273.005	–	–
Model Chi-square	129.357	6	<0.0001
Hosmer and Lameshow Chi-square	13.428	8	<0.098
Nagelkerke R^2	0.414	–	–
N	349	–	–

The model provided a good fit as Hosmer and Lameshow Chi-square was non significant (Table 11.4). Based on standardized coefficients, agroclimatic zone, soil depth, soil texture, proportion of the farm fenced, distance of the housed from the road and nursery were respectively the key variables influencing the adoption homegarden agroforestry practices. The model predicted 78.2 per cent of the cases correctly (Table 11.5).

Table 11.5: Classification accuracy of the model.

Sl.No.	*Models*	*Type Household*	*Number of Observations*	*Predicted Membership*		*Overall Prediction*
				Adopter	*Non-adopter*	
1.	Biophysical	Adopter	192	160 (83.3%)	32 (16.7%)	78.2
		Non-adopter	157	44 (%)	113 (72.0 %)	

Discussion and Conclusions

At higher altitude light is a limiting factor for simultaneous cultivation of trees and crops in from of homegardens which discourages their adoption. With increase in soil depth, the competition between trees and crops and water in homegardens reduce which encouraged the adoption of homegarden based agroforestry in deeper soils. Farmers with finer textured soils would have more likelihood of adoption of homegardens as these soils are comparatively fertile and have more water holding capacity which reduces the completion for water and nutrients between crops and trees in such soils. Traditional fencing provides protection to trees as well crops from grazing and other inferences resulting in increased adoption of homegardens. Availability of the nursery nearer to the houses could ensure the supply of tree seedlings for planting in homegardens which would encourage homegarden adoption.

The study implied that strategies needs to be developed tree-crop combinations which minimizes the competition for light at higher altitude (high hill temperate dry zone) to encourage the adoption of homegarden based agroforestry. Traditional fencing could also encourage the homegarden adoption. Efforts to increase soil fertility would also increase the likelihood of choice in favor of homegarden based tree cultivation.

References

Christanty, L. and Iskandar, J., 1985. Development of decision making and management in traditional agroforestry: Examples in West Java. In: *Community Forestry: Socio-economic Aspects*, (Eds.) Y.S. Rao, N.T. Vergara and G.W. Lovelace. Regional Office for Asia and Pacific (RAPA), FAO, Bangkok, pp., 199–214.

Chundamannil, M., Krishnankutty, C.N. and Rajan, A.R., 1993. Socio-economic study of farm forestry: A survey of four villages in Kerala, *KFRI Research Report No. 90*, Kerala Forest Research Institute, India. pp. 66.

Hoskins, M., 1990. The contribution of forestry to food security. *Unsylva*, 41(1/160. (http:/www.fao.org/docrep/t7750e/t7750e02.htm)

Jacob, V.J. and Alles, W.S., 1989. The Kandyab gardens of Sri Lanka. In: *Agroforestry Systems in the Tropics*, (Eds.) P.K.R. Nair. Kluwer Academic Publishers in Co-operation with ICRAF, Dordrecht, The Netherlands. pp. 181–195.

Leuschner, W.A. and Khaleque, K., 1989. Homestead agroforestry in Bangladesh. In: *Agroforestry Systems in the Tropics*, (Eds.) P.K.R. Nair. Kluwer Academic Publishers, Dordrecht, The Netherlands. pp. 196–209.

Mughal, A.H., Ara, T. and Bhattacharya, P., 2000. Socio-economic aspects of agroforestry in rural Srinagar of Kashmir valley. *Indian For.*, 126: 234–240.

Nair, C.T.S. and Krishnankutty, C.N., 1985. Socio-economic factors influencing farm forestry: A case study of tree cropping in the homesteads in Kerala, India. In: *Community Forestry: Socio-economic Aspects*, (Eds.) Y.S. Rao, N.T. Vergara and G.W. Lovelace. Regional Office for Asia and Pacific (RAPA), FAO, Bangkok. pp. 115–130.

Nair, P.K.R., 1993. *An Introduction to Agroforestry*, Kluwer Academic Publishers in cooperation with ICRAF. Kluwer Academic Publishers, Dordrecht, The Netherlands. pp. 499.

Nautiyal, S., Maikhuri, R.K., Semwal, R.L., Rao, K.S. and Saxena, K.G., 1998. Agroforestry systems in the rural landscape: A case study in Garhwal Himalaya, India. *Agroforestry Syst.*, 41: 151–165.

Soemarwoto, O., 1987. Homegardens: A traditional agroforestry system with a promising future. In: *Agroforestry a Decade of Development*, (Eds.) H.A. Steppler and P.K.R. Nair. ICRAF, Nairobi, Kenya, Pages 157–170.

Singh, R.P., Rawat, D. and Jishtu, V., 1998. Agroforestry in cold desert areas of Himachal Pradesh, India. *Indian For.*, 124: 321–330.

Sood, K.K., 2003. Factors affecting Tree Growing in Traditional Agroforestry Systems in Western Himalaya, India. *Ph.D. Thesis* (unpublished), University of Aberdeen, Scotland.

Environment, Biodiversity and Traditional System, 2014 *Pages* 115–135
ISBN 978-81-7622-307-2
Editors: D.R. Khanna, G.S. Solanki and S.K. Pathak
Published by: BIOTECH BOOKS, NEW DELHI

Chapter 12

Tribal Culture and Sustainable Environment in the Era of Globalisation

Babu Lal Meena[1] *and Suwa Lal Jangu*[2]

[1]*M. Phil Student, Department of Geography,*
Delhi School of Economics, University of Delhi, Delhi
[2]*Assistant Professor, Department of Political Science,*
Mizoram University, Mizoram

ABSTRACT

It has now become imperative to the whole world to attain the sustainability of the environment not only through sustainable resource management but also by preserving tribal culture and identity. In addition, in attaining such a big goal we have to consider the importance of tribal culture and traditions. Hence, by protecting their natural habitat we can make the environment and the resources sustainable and can make a better earth with full of sustainable resources for coming generations. The present paper aims to explain the importance of tribal culture and identity in the sustainable resource management. The paper makes use of various research studies on the importance of tribal culture in sustainability of natural resources. The study is also based on the perception of the people on natural resource management and impact of globalization on environment. The paper discusses tribal ecological setting and other cultural experiences which have developed a style of life, a type of man-nature-spirit and give them unique identity which is the result of their long process of adaptation. Further, it explains the tribal subsistence economy which is based on the entire habitat including land, forest and water system. The paper demonstrates that the knowledge of tribal people regarding the preservation of forests, seeds, water and biodiversity. The knowledge about naturopathy and meteorology of tribal in their daily life has also been discussed in the paper.

Keywords: *Tribal, Globalization, Cultural life, Sustainable environment, Sustainable resources, Indigenous knowledge, Shifting cultivation.*

Introduction

In very certain terms, the present human world is more closely integrated and becoming a global village. The process towards a uniform socio-cultural and economic order is simply termed as globalization. Globalization and free-market are the basic principles dominant in this era. The issues of environment and sustainable resource management have also become important in the pace of globalization as the fast depletion of biodiversity and erosion have become common today. The issues of environment and sustainable natural resource management are very much linked with the original ethnic groups of the world. The indigenous people of the earth have been practicing the long cultural traditions with sustainable economic pursuits, social control mechanisms, techniques of production while ensuring their resources for several generations as their livelihood. They do this by using their resources judiciously and ensuring their renewability. Both the culture and nature seem to have played selective and reciprocal roles in the life patterns of tribal people. They have developed a sustainable model in their total life perspective of the environment. These they have learnt after a long struggle with environment. Sustainability of environment is an end result of their hard and enduring labour.

But in the pace of globalization and the present economic order, we have disturbed the tribal culture and identity by removing their natural forests and converting it into economic plantation. The tribal people are hardly in a position to take advantage of new opportunities in the current economic patterns as they lack required skills, unaware and untouched with modern science, technology and education. We have almost lost a huge cultural diversity in the form of erosion of tribal languages as one language is being disappeared every year. By clearing their forest, we have depleted huge biodiversity as well. Man is a phenomenon in which all civilization, all races, all customs pour out their streams as tributaries to the main stream. They can not afford to ignore, not only the loud proclamations but also the unheard whispers of truth from the great civilizations in advances races or even the tribal world, irrespective of their geographical locations that have something unique to contribute to the human spirit.

Tribal and indigenous peoples in India are facing enormous and complex challenges in today's globalised, modernised and rapidly altering world. They are socially, economically, and politically vulnerable and they are frequently exploited as the world's most isolated and marginalised groups. Government preventing them from exercising their civic rights and duties and keeping them from engaging in the much heralded civil society discourse of sustainable development. Moreover, tribal and indigenous groups are rapidly losing their cultural heritage, which includes customs, values, and indigenous knowledge, much of which contributed to sound, sustainable practices of biodiversity conservation in the past. Tribal and indigenous children every where are growing up in a world of diminishing cultural rootedness and a continuing lack of guaranteed national and human rights in a global environment of eroding sustainable development practices.

The work in this paper is based, in part, on the concept that the revival of cultural identity, including indigenous knowledge, is one of the determinants for a truly sustainable development. The paper suggests a new approach 'the affirmation of cultures and biodiversity conservation (ACBC)', as biodiversity and culture are the key elements of the survival of indigenous peoples in their inhibit region. The failures of government policy have led to their insecurity in land tenure, and industrial replacement projects in their areas result in a loss of their traditional self sufficiency and rising levels of debt, social stress in highland communities, and rising levels of migration to urban areas. The conservation policies encroach on community land by creating nature reserves restricting the traditional systems of swidden agriculture which used to contribute to their self-reliance and the cultural way of biodiversity conservation.

Communities following government sponsored development programmes have been the recipients of healthcare centres on the one hand, while their rights to land and any means of making a living are removed on the other. The net results of the insecurity and uncertainty of highland life are felt hardest by the women and young children in highland communities. Migration towards lowland urban centres appears to be increasing. The reasons underlying this migration are complex, although clearly one of the most important is the almost complete lack of a secure alternative that would allow people to continue living in the highlands.

Throughout the history of mainland North-East India there have been numerous indigenous groups residing in the mountainous areas. These ethnic highlanders lead lives very different from lowland dwellers. Through their adaptation to the environment and cultural exchange, indigenous knowledge has been accumulated through a long line of generations. The transmission of traditional culture and indigenous knowledge has continued within the family and villages of each ethnic group. It has been noticed that the traditional learning style has always revolved around various skills of indigenous livelihood as well as a certain set of social values based on the traditional belief system.

Tribals: Identity, Rights and Development

Adivasis are the most marginalized sections of Indian Society. This marginalization is social, economic and political. The process of commercial exploitation of natural resources and industrialization that began under colonial colonial rule and continued after independence threatened and gradually deprived the adivasis of their traditional means of livelihood. The situation has been further aggravated by economic reforms and liberalization initiated in the 1990s. The living conditions of the socially marginalised and economically deprived adivasis are further under threat from the new challenge of climate change. In India, nearly 70 million adivasis are directly dependent on climate sensitive secotrs (agricultures, forest and fisheries) and natural resources such as water, biodiversity, mangroves coastal zones and grasslands) for their sustenance and livelihoods.

The irony is that the adivasis, who are the least responsible for the cause of global warming due to their environment friendly lifestyles, are now the most vulnerable to the treat and impact of climate change process. The current model of development has seen a systematic and large scale violation of the adivasis' right to live with dignity and equity, their right of land, forest, water and other means of livelihood resources, their right to freedom and expression, etc in the name of national interest or public purpose. Common property resources (CP) are the most substantive part of their style of life but these are not always statutorily recognized by the state as belonging to adivasis. Tree plantations, promoted as one of the main solutions to climate change, are themselves resulting in further deforestation process where forests are being substituted by monoculture tree plantations.

Indigenous peoples are those who: (a) identify themselves as indigenous groups, (b) established their cultures and social institutions prior to European colonialism, and (c) continue to maintain those traditional ways of life to this day. They are "distinct from the dominant society that surrounds them, including practices such as communal ownership of land and resources, and a spiritual attachment to their territory". For indigenous peoples, all of these cleavages apply (language, religion, race, and nationality), the differences reinforce one another, and in combination, they more thoroughly separate indigenous peoples from others. Indigenous peoples have distinct pre-colonial cultures, spiritual ties to their lands, plus their own languages, religions, and racial characteristics, and all of these factors apply at the same time. Most indigenous tribes also suffer from more widespread and consistent inequalities in society and in the economy than do most minorities.

The International Labor Organization (ILO) has defined indigenous peoples in ILO Convention No. 169, which points to the "social, cultural and economic conditions [that] distinguish them from other sections of the national community" (ILO 1989). The social and cultural discrimination suffered by indigenous peoples exists in addition to the linguistic, racial, or religious cleavages that define ethnic minorities. The rights that are most important to indigenous peoples are those protections that help them to secure their unique cultures, resources, and habitats. The 2007 UN Declaration establishes numerous indigenous rights within its 46 articles. Some most important articles are following:

1. Article 7 of the DRIP proclaims a right to freedom from genocide and prohibits forced removal of indigenous children.
2. Article 12 identifies an indigenous right to develop and participate in traditional cultures.
3. Article 14 recognizes the right to transmit traditional indigenous cultures to future generations.
4. Article 5 defends the rights of all indigenous peoples to participate in both their traditional cultures and the surrounding dominant (nontraditional) culture.
5. Article 13 provides Cultural autonomy for indigenous peoples extends to the areas of language,
6. Article 14 regarding to education in their mother tongue,
7. Article 8 gives safeguard for freedom from forced cultural assimilation,
8. Article 12 protects religious freedom of tribals.
9. Many articles are regarding to land rights and rights to ownership of natural resources are spread throughout numerous sections of the UN DRIP.

10. Article 10 prohibits forced relocations of indigenous groups or relocation without informed consent.
11. Article 23 recognizes an indigenous right to development.
12. Articles 25 and 26 seek to protect the "distinctive spiritual" relationships between indigenous peoples and their territories, as well as their right to own the lands and resources where they have traditionally resided.
13. Article 28 calls for restitution or fair compensation in cases where indigenous peoples have had their land or resources "confiscated, taken, occupied, used or damaged without their free, prior and informed consent."
14. Article 3 of the DRIP proclaims an indigenous right to self-determination.
15. Article 19 elaborates political autonomy for indigenous peoples which require their input into any "legislative or administrative measure that may affect them."

However, Article 3's assertion of self-determination for indigenous peoples is limited by Article 46: Nothing in this Declaration may be interpreted as... authorizing or encouraging any action which would dismember or impair, totally or in part, the territorial integrity or political unity of sovereign or independent States.

The one-time 'lord of the forest' was reduced to the status of a subject and placed under the tutelage of the forest department. Tribals became a wage slave at the disposal of the forest department and forest contractor. Tribal India is an internal colony exploited by the rest of the country. The resource development decisions made by distant bureaucrats in the state capitals, in Delhi and in the boardrooms of the multi-national corporations in the West mean that even the renewable (forest) resources of tribal India are 'mined' to provide investment capital for the further development of the non-tribal plains and the developed world. The tribals receive few benefits from these activities. Given the environmentally dangerous rate at which some states are deforesting and the rapid depletion of many mineral resources in the south-eastern resources 'triangle'. It seems likely that the resources of the tribal areas will be largely exhausted without bringing any benefits to their original 'owners'.

Tribal Culture and Social Life

Tribal people are simply as 'primitive' societies. The economic organization of tribal communities varies but there are no great differences in wealth, and no exploitation of the many by a few. However, we as no-tribal peoples have tried to make the tribal feel ashamed of their way of life. It is painful to find them swallowed up and assimilated by a narrow, rather repellent and over-zealous form of Indian nationalism, unable to fulfill Nehru's ideal of 'unity in diversity'. Most of them are agriculturists; some are settled while others are nomadic, but hunting, fishing and gathering forest produce have remained an important part of their life. There are no capitalists or landlords. Although the chiefs have great power, most important decisions are taken collectively by the Village Council he presides over. Primitive communism prevails among many of them; the crops are still stored in communal granaries. In the context of pan-Indian culture and civilization tribals' culture is controlled and determined by their surrounding ecology and environment, isolated from the great traditions of Indian civilization in terms of objective reality and subjective awareness, based on value system of equality. The tribal people is called as Bhumiputras because they are ancestors of this land and their; they nurtured by this land and they nourished to nature.

The tribes are not permanent crystalline structures belonging to any particular stage of historical or social development. In our ancient scriptures they are known as Dasas, Dasyus, Nishad, and Kiratas. In our social system they lie on the opposite pole of the philosophical Varnashram dharm. They had lived in the ancient times in geographical isolation, away from the urban centres or the mainstream of the dominated social life, suffering from economic backwardness and social justice, maintaining a fair degree of cultural distinctiveness.

We get in their society a few distinguishable characteristic features like:

1. Homogeneity in culture
2. Less stratification
3. Segementary and kin-based social system
4. Less economic specialization with distinctive ecological isolation with backward food production and technology
5. Ethnocentrism (group sentiment) or communal solidarity through a participation in rituals and festival and sharing common beliefs and aesthetic of non-puritanical level.

Once you are accepted in the village, all doors are open to you. You acquire the status of a brother or an uncle and even on later visits, you will always be welcome. Tribal life is collective and not individualistic; it is a shared community life. There is little competitiveness based on acquisitive greed. The affection they have for each other is readily extended to a friendly stranger. Nothing is expected of you. They are touched if you buy them a pig and a few pots of their local brew to feast on. You are sure to be rewarded with the music of drums and flutes and singing and dancing till the small hours of the morning. Perhaps one of the reasons for the spontaneous joy of life in the tribal as compared to the rather morose Indian villagers is that they never have to feel smaller than others in their own communities; they do not judge people in terms of 'success' or caste. They have no sense of sin about sex, and little or no shame about their bodies in contrast to the Indian villagers' inhibitions. They preserved the physical expression of joy in their lively dances.

The dances always have the accompaniment of songs, and the music of flutes, drums and primitive string instruments which differ in design, but almost always have a sun-dried gourd-shell as a base. They make good music echoing the sound of nature. Tribal songs too have the melody of the wind and the rain, the rippling brooks, the cries of wild animals and of the vast silence. They sing about their joys and sorrows and even of rebellions- always with a slow, gentle and sensual rhythm and often with melancholy humour. But they do not work themselves up to the wild frenzy of African music or of western 'pop' music.

Wild animals also have great spiritual and symbolic value for the indigenous peoples. When the meat from a hunt is shared with friends and family, bonds of kinship and other social ties are reinforced. Communal hunting is also the vehicle through which tribal elders teach younger generations the values of courage, patience, and teamwork (Watt-Cloutier 2004). Without the group hunts, widespread social problems such as substance abuse and clinical depression escalate rapidly.

Tribal children learn dance as soon as they learn to work. The boys and girls grow up together free from any sense of guilt about sex. Sexual offences are rare in these tribal people or societies, although bawdy humour abounds in their conversations, songs and pantomime. Sex was never considered sinful by the ancient Hindus; on the contrary, it was treated as sacred and sculpted on temple walls as can be seen in the great temples of Konarak and Khajuraho for example. But successive invasions over the centuries influenced the Hindu mind, destroying its liberalism and preserving only it's puritanical, ascetic aspects, the superstitions and the fears. Rites for propitiating the gods and rules governing the conduct of men became the prime concern of popular Hinduism. The tribal, expect for a few converted to Christianity by European Missionaries, have remained largely unaffected by these encounters. They have their own gods, their rituals and their taboos.

Primitive societies look after both the young and the old. Even the very old are lovingly taken care of and are never left to themselves separated from their families, to will away in old people's home. They become honoured headmen, or members of the village council, and impart to the young their wisdom and experience which are no less valuable than the physical work they can no longer do. As grandfathers and grandmothers living in the same house, they also look after the children and teach them skills they will need when they grow up, while their parents are away working in the fields and forests.

Very often, the tribal chiefs will not allow their young to go to a forbidden mountain or to wash in prohibited waters. Nor will he do any of these things himself fear of landslides or of lurking crocodiles might have been the original reason for making these laws in some for off time. But these laws are easy to observe and do not lead to confusion and despair. Tribal tried to preserve their own culture and entity by retreating and taking shelter in some inhospitable regions of our country. Thus, they could maintain a cultural pattern of their own, and this can be easily distinguished from other social groups. In this context of pan-Indian culture and civilization their culture is controlled and determined by their surrounding ecology and environment. Village festivals are held as their response to the seasonal change and agricultural operations. Sacred specialists are the village priest, assistant priest, the sacrifice, and the medicine man.

All our tribal are either highlanders or forest-dwellers, and a few like Bhils live in the deserts, as do the sorry remnants of the 'Indian' and the aborigines in their 'reservations' in America, Australia and New Zealand. In India, however, the tribal inhabitants were not exterminated by ruthless genocide; instead they were allowed to languish away in poverty and isolation. Paradoxically, however, the tribal in remotest areas have survived the best, retaining a simple and indigenous way of life.

Tribal Culture and Sustainable Environment

Tribal people feel very comfortable to live in the jungle. The tribal occupations depend totally on environment settings. Job opportunity depends on the availability of local resources. In fact, cultural of tribal peoples are highly influenced by the opportunities and limitations offered by the physical environment of the country. So one can say that the local environment, that is, forest environment plays vital role in governing their economic life which again suggest their population number, size of hamlet, etc. due to their intimate exposure with the immediate forest environment, they have a very good knowledge of the qualities and varieties of all forest products and also know which of them are edible or which not. They know how to cope up with hot summer and severe winter. The economic activities are still closely in woven with forest and the economic utility of forest in their life can be understood in the following way: forest serves as a source of shelter, food and drink, game, and pasture-it also serves them in supplying indigenous medicine, and raw materials for domestic items and hut-making.

Primitive religion from the inception of the human civilisation the man continues to interact, adjust and adapt to his surrounding environment, both physical and social, to meet up the needs for his survival and livelihood. Man's interaction with environment is taking place through regulatory responses, *i.e.*, behavioural, social and cultural. These can be explained by three ideas *viz.* environmental determinism-that is determining effect of nature upon human society and culture; human adaptation to nature and nature as limiting factor to human possibilities. In other words, environment acts as determining factor in shaping culture depending upon nature and way of exploitation of the resources. Take the case of the Gond kings in the Gondwana area that comprises parts of Orissa, Chattisgarh, Uttar Pradesh and Madhya Pradesh. They were active promoters of surface water harvesting bodies. They gave rewards to villages with the best ponds, as the kings' prosperity was dependent on agriculture. Water added to their GDP. But, today, it is energy that adds to GDP, not water or even agriculture. Hence, everything is done at the expense of farm land and rivers. Unless water and food are accounted for in GDP, double digit figures would still mean deprivation.

In Boas' scheme, culture takes precedence over environmental settings. Cultures as the basis of human behavior have become the adaptive responses to the ecological setting. Cultures are human adaptation in most of its areas, latent as well as manifest. Material culture and religious rituals are the direct consequences of the interaction in the eco-system. A population is a specific eco-system adjusts to environmental conditions in specific past conditions. Agriculture, tribalculture, herdculture, globalculture, etc are examples of manmade environmental culture.

Man, through a part of nature, differ other components in the aspect that he has been culture which is a set of behaviours arising out of inter-action and inter-relation with the living and non-living components and is transmitted to the members from one generation to the next. This culture consisting of traits, complexes, institutions are various depending on the nature and types of the inter-actions with various parts of nature. The tribes, whatever, their homely synonyms, viz 'Simple', pre literate or

'Non-Literate society' or 'primitive society' are the autochthones or the early settlers in India. They are considered to be aboriginals or autochthones or Adivasi in common parlance.

Due to scarcity of food tribal people move from one place to another in small bands. Since they live in forest environment, their whole belief system is centered on the local ecosystem. To adjust with their natural environment, automatically their life has adopted various measures which have penetrated into their culture. Habitation in an uncertain environment creates a feeling of helplessness in their mind. To overcome such a feeling they have taken support of the supreme authority in his various forms, very much adjusting with their natural environment. The deities live in jungle, in hills, rivers, streams, in wind and even in some trees. For example, Sun is believed as the supreme authority, moon is his wife and the stars are their children. Sun gives light and warmth to the earth which ultimately created life on the earth. So he is regarded as supreme authority.

There is a close relationship between cultural and biological diversity all over the world, the two being interdependent on each other, co-exists together. There is a delicate interrelationship between the stability of the forest ecosystem the inherent biodiversity and the culture of the indigenous people. Disruption of forest ecosystem and biodiversity, results in degeneration of the indigenous culture.

Rituals and Religious

The tribal religion in India is described as 'Animism', 'Spiritism' and 'Bongaism'. It has several parts such as totemism, ritualism, taboo, magic, witchcraft and so on. The homogeneous sacred ecology. Idolatry is absent and anthropomorphism is ill developed. Animisam, naturalism, various beliefs and practices are part of the cultural life of the tribals. In tribals system plant, animals, hills, trees, rivers, lakes, ponds, stones, mountain tops, etc., are considered sacred. Nature worship is a form of belief and all nature's creations have to been protected. In fact, this sort of dependence on nature has greatly helped in the preservation and protection of many natural items of ecosystem in our country.

According to Ghurye, the tribal religion is a backward form of Hindu religion. Elwin opines that the religion of the tribals in India has a close affinity to the Saivaite section of Hinduism and points out the distinction between tribal religion and Hinduism, which is meaningless. Majumdar opined that the tribal religion represent today "marginal religion" a no-man's land between magic and religion, between pseudo-science and sciences. Most adivasis have a spiritual connect with the land they belong to, and look upon the forest as their mother- the source of life and livelihood. Each tribe has extensive religious and social norms that prohibit the exploitation of natural resources and the idea of separation and compensation are repugnant to their existence.

The tribal people's interaction with natural objects such as the sun, the moon, stars, trees, hills, birds, river, etc., became part of their great cultural heritage. The natural environment has an overall impact on the belief system of tribal people in a particular niche. Titiev (1955) has shown the interrelationship of the universally accepted patterns of culture through the model of bio-cultural triangle where the environmental relationship of man forms the base whereas the other two arms of the triangle comprise man-man and man-supernatural (man-social) interaction. Titiev opines that the three sides of the bio-cultural triangle are always integrated but the emphasis is not equal in all the societies (Dasgupta 1994). For example, the performance of the rituals and ceremonies depend on the nature. On every step of their agriculture operation, they perform various rituals. The offerings are mainly the newly harvested or produced crop or fruit. The indigenous systems of medicine to treat various ailments among Indian tribes depend upon their forest environment. Nature, especially trees, animals, birds, etc., has become part of tribal glorious mythology. Their dependence on the nature was total in ancient times when they were supposed to be "lord of forest".

Tribal Knowledge and Sustainable Environment

Living close to the nature, tribals are familiar with the natural resources. By empirical reasoning and trial and error, they have screened and developed a highly complex and very specific knowledge of local resources and its sustainable utilization. The traditional knowledge encompasses cultural products and practices along with folklore. Traditional knowledge is not just about medicines and methods of healing, although this is the area most vulnerable to experience and experiments. Due to traditional knowledge, tribals are poor but their quality of life is good. In a few words, traditional knowledge is as 'old is gold' and old wisdom'.

Indigenous and ethnic people learn to live in the most hostile environment from the pole to the equator, from the desert to the rainforest, and they have evolved a host of sophisticated strategies for doing so. The world's remaining areas of high biodiversity are often found on indigenous community's land and in their water bodies. These indigenous people have accumulated enormous knowledge about the sustainable use of plant species available to them in their native lands. The knowledge of ethnic people about the food, medicinal and other cultural value of plants has infused in them a sense of ethical responsibility to use these plant resources more sustainability and judiciously, conserve them in nature and pass on their traditional knowledge and wisdom with regard to their use to the posterity.

Tribal people have a vast knowledge of the medicinal plants available in the vicinity. They are very much reluctant to go to the hospitals and health centres and instead they depend on their local forest resources. They generally do not destroy those plants also. The people who live in any climatic conditions have an intimate knowledge of available food resources throughout the year. It is, therefore, clearly understood how forest-dwelling tribes are closely dependent on their forest resources for their survival. From such natural resources they not only draw the required energy for survival but also get sufficient protein, fat and minerals which are essential for existence.

From the forest, tribals derive their basic needs of survival for themselves. In turn they protect the forest with their folk knowledge and also enrich its fertility through their various cultural performances, beliefs and practices. The forest not only give livelihood to the tribals but also provide them social bondage and a culture reminiscent of their ancestors. Rajiv Sinha (1991), in his article "Ecosystem Preservation through Faith and Tradition in India", observes that ribal women think of forest as their *maika* or parental home and the trees are held sacred; he writes that although the tribal people cut some trees in the forest for economic and other purposes, side-by-side they take the vow of planting new trees in the vicinity of their habitat to keep up the ecological balance.

It has been observed that tribal people of Baster often collect fuelwood or timber for their houses but they never cut the whole tree. They allow it to grow and germinate in the next couple of years. Tribals are cognized their environment along with their ritual cycle. In fact, the environment knowledge can only be acquired due to intimate association with immediate environment which is otherwise known as cognizance of environment. The knowledge of the indigenous people must also be covered under 'Intellectual Properties Right' (IPR). Governments should recognize that indigenous lands need to be protected from environmentally unsound activities.

Environment and Biodiversity

Since diversity in nature is closely linked to diversity in culture, the loss of biodiversity also exhibits direct consequences on the fabric of societies, especially in countries where large proportions of the society life and survival depends on biodiversity. The link between climate change and biodiversity has long been established. Although throughout Earth's history the climate has always changes with ecosystems and species coming and going, rapid climate change affects ecosystems and species ability to adapt and so biodiversity loss increases. The basic unit of ecology rests on ecosystem which is again related with animals and plants of its physical environment. The religious pantheon plays a great role in the tribal life for preserving their geographical environment.

In fact, in their culture, plants, animals, trees, rivers, stones and hills are considered sacred and these are protected. The faith of these tribes has greatly helped in the preservation and the protection of many natural objects in the ecosystem. For instance, as per their customary law it is tabooed to kill or hunt or eat the totemic object by which fauna and flora are preserved in the forest. No one can easily understand about the rich natural resources and bio-diversities of our country in which Indian tribal people are closely interlaced in every sphere of their livelihood. Therefore, it is our duty not only to conserve the great cultural heritage of Indian tribes but also to conserve our rich biodiversities for our own sake.

The tribal belt is more often the center of high biological diversity and origin of our crop species. Some of the crop plants have their wild ancestors growing in the forest even today and have been survival the competition. These are resistant from pests, diseases and adverse weather conditions. There is a widespread belief that 'green medicines are healthier than synthetic products'. The indigenous people of North East India use varieties of plants and animals in medicines, dyes, flowers, essential oils, preservatives etc., which could form the basis of health care in this region.

The indigenous community of North East India practice crafts ranging from the making simple bow and arrows to the melting and casting of metals. Their scientific improvement will also boot the tribal economy sustainably. The tribal communities of North East India have a very rich knowledge about edible plants. The world remaining areas of high bio-diversity are often found on indigenous people's lands and in their water, and political conflicts have arisen over conservationists efforts to remove indigenous people from their lands. This has led conservationists to ask the questions: is bio-diversity conserved by indigenous people?

There are five different kind of evidence that indigenous peoples conserve bio-diversity:

1. The association of indigenous people's homelands with areas of high bio-diversity.
2. The existence of indigenous bio-diversity conservative institutions and techniques.
3. Indigenous peoples statements about their own actions and concerns.
4. Evidence of indigenous communities attempts to adapt new conservation practices in response to stresses, and
5. Comparison with activities by non-indigenous peoples in the geographical area.

All these evidence found in North East India. In Meghalaya the Khasis maintain sacred grove in many places in the Khasi Hills. These sacred groves are natural treasure house of plants and animals preserved through ages and are excellent manifestations of traditional wisdom. Tribals still love to live in their original, natural abode and practice their traditional way of living, rituals, customs, dance, music, and traditional belief system of health cure within their self-sustained forest ecosystem. In fact, tribals are an integral part of the forest ecosystem and like the other components of terrestrial ecosystem they have been playing a significant role in the operation of forest ecosystem and in the maintenance of the ecological balance in the area.

Since diversity in nature is closely linked to diversity in culture, the loss of biodiversity also exhibits direct consequences on the fabric of societies, especially in countries where large proportions of the society life and survival depends on biodiversity. The link between climate change and biodiversity has long been established. Although throughout Earth's history the climate has always changes with ecosystems and species coming and going, rapid climate change affects ecosystems and species ability to adapt and so biodiversity loss increases. It is known that loss of biodiversity impairs functioning of ecosystems. The biodiversity is an important attribute for the ecosystems, helping the soil maintain its multiple functions. Richness of perennial plants is more strongly tied to critical ecosystem services than other factors, such as annual rainfall or microbes in the soil.

Economic Life of Tribals

Like the material culture a close relation between the terrain and economy of the people can be observed. Their original means of agriculture was shifting cultivation.

Each village, usually inhabited by related families, owns the land surrounding it, where its ancestors had originally cleared the unclaimed Jungle. Thus in the hill areas, the land becomes the property of the clan inhabiting the village rather than that of the particular household as immovable property is not immediately divided even after the death of common ancestor. The economy of the tribal people of India can be understood in terms of ecology and environment and nature of their articulation with the greater economic system of our country.

1. The food-gathering economy
2. Pastoral or cattle rearing economy
3. Agriculture is one of the main economies of the ribal population.
4. A small section of tribal population practiced rudimentary forest-based industries
5. Rest of the tribal communities shifted from their traditional ways of life to the aoccuption of daily labourers, working in plantations and construction of roads/railways tracks, or as unskilled industrial labour.

Vidyarthi (1981) has classified the Indian tribal population into seven cultural strata basing on their ecosystem, traditional economy, belief in supernatural power and recent impact. These are as follows: (a) Forest hunting, (b) Primitive hill cultivation, (c) Plain agricultural type, (d) Simple artisan group, (e) Pastoral and cattle herders, (f) Industrial, and (g) Urban workers type.

The fate of economy of the forest-dwelling tribes is in many ways interlaced with the forest. Forest-based tribal development should be encourages. The development of these tribals should be thought in the perspective of local niche which means exploitation of local natural resources by which one can economically be benefited and at the same time balance of ecosystem may be maintained. The manner in which the indigenous people manipulate biodiversity for ecosystem functional integrity and through that for their own social function is interesting. The invaluable role of the indigenous people in the conservation of the plant genetic resources is now beginning to be recognized. However, economic benefits from the utilization of those genetic materials seldom accrue to them. Those who have conserved biodiversity tend to remain poor, while those who have converted such genetic into commercial products through biological technology are rich.

Generally people have the aptitude for learning the use of one or two herbs for self-therapy for cure of petty common ailments like bleeding from accidental cut or brushing, cough and cold, abnormal bowel movement, acidity, stomach trouble etc. Multifold utility of herbal plants is known to most Indian tribal and rural societies. The indigenous utility of herbs as medicines goes well with an understanding of the folk concept of illness in India. At the folk and tribal levels indicates that all of curative treatment were rendered by folk-medicine men who virtually exclusive dominated the field of medicine till before the modern mode of practices as professionals.

The native medical practitioners may well be put under two distinct categories: (1) The specialist practitioner who is adept in diagnosing illness, injury or disease and who renders treatment by examining the patient in a rational manner prescribing various folk herbal medicines, and (2) The practitioner who resorts to magico-religious methods in diagnosing the physical malady as being evil designs of spirits, demons or deities. The first one is a secular or profane minded practitioner who attempts rationally his treatment by administering one or the other type of herbal medicine following physical diagnosis of the malady. The folk-practitioner of the second category might rightly be called a witch-doctor commanding a congregation of spiritual beings, both benevolent and malevolent, with various whims and caprices. Since time immemorial both the rational minded practitioner and the witch-doctor have had intensive as well as extensive effective role as folk medicine men holding sway among the suffering tribal. The tribal are obsessed with the uncanny unearthly activities of the spirits, ghosts and deities. Diseases like diarrhea, malaria, goiter, gonorrhea, leprosy, head-aches, pneumonia, cholera, small-pox, lunacy indigestion or even the common cold and fever were all thought to be caused by one or the other watching spirit or deity.

Geographical Settings

In different geographical settings of tribal areas, different tribal people live and gather varied food materials from local natural resources for their survival. They have ultimately a symbiotic relation with their forest environment. The forest is also benefited from them through their various activities. Their cattle graze in the forest; cowdung becomes manure which helps to increase the fertility of soil and ultimately helps in forest growth. They burn the dried up leaves and bushes in the jungle. The ash becomes fertiliser and helps in forest growth similarly; there are several other examples where symbiotic relationship is well marked. It is also evident that close relation between geographical environment and food habits of tribal people exists. In fact, human needs are influenced by the environment and other ecological factors prevailing in the particular niche. This is reflected in their dietary habits.

Indigenous people are in addition also those who irrespective of their legal status live more in conformity with their social, economic and cultural institutions than with institution of the nation to which they belong. The adivasis consciousness and the articulation of indigenous people status is not so much about whether they are the original inhabitants of India as about the fact that they have no power whatsoever over anything (land, forest, water, resources) that lies in the territory they inhabit. This is despite being the original inhabitants of India in relation to others. The consciousness and the articulation are basically an expression of the yearning to have or to establish a special relation with the territory in which they live.

As appearance from the above description of the tribal territories, except on the North-East territories, where the so-called Aryans in ancient time had not penetrated, the tribes which are now euphemistically called adivasis (the original inhabitants) have been driven into inhospitable hills and forests and even deserts, which conquerors of different origins settled in the rich and fertile plains and river valleys. The vast jungle-clad of South-east Asia which spans the borders of India, China, Myanmar, Laos, Thailand and Vietnam is inhabited by people who belong to autonomous, state-resistant and subsistence- oriented societies of the hills. The colonial powers carved up these mountain regions into the nation states, each with its notions of territorial integrity but it was a division that wasn't accepted; they remained "ethnic minorities" within the state into which they were absorbed. This could be seen by their distinctive traits – segmentation, mobility, egalitarian social structures, flexible cultivation – which were simply a response to their new environment and "a strategic adaptation to avoid incorporation in state structures".

Tribal Environment and Development

All this does not, however, mean that if one wonders off to a tribal village in India one would find a Shangri-la. The tribals are certainly not free from human vices. They live in misery and poverty. One notices the marks of this poverty first- the cesspools of fifth, diseases, malnutrition, ill-nourished children and the adults who are shriveled and old at forty. Lack of sanitation and proper medical care are lead to epidemics. But belief in witchcraft practice to brand a sick child with a hot iron to drive away the evil spirit supposedly possesses him. Although most of them are peaceful, there are some who take pride in their machismo and murderous inter-tribal blood-feuds. They too have drunkards and social misfits who are often victims of the autocratic domination of their tribal chiefs against whom there can be no appeal.

Unfortunately, the tribals are ruthlessly exploited by everyone from the outside world who comes into contact with them- timber contractors, merchants and licensed distillers of the liquor they drink. Some petty official even make a regular habit of collecting chicken eggs, rice and forest produce from them without making any payment. The tribals are too poor to offer bribes and to avoid harassment; most of them run away from outsiders. This fear has often unfortunately consequences. They are unreasonable afraid of visiting government health department workers and vaccinators.

The tribals are bound by crushing burdens of debt to the money-lenders whose visits are dreaded he may take away a whole year's crop in a lean year. The fascinating part of it is that they take it all in their stride and remain redolent and kind. They never brood on their misfortune and are usually cheerful. The tribals have fought against economic exploitation again and again. Many tribal rebellions, fought with bows and arrows against guns, have been ruthlessly subdued and remain unsung sagas. The tales of their suffering and bravery remain buried in official files.

These people do not need industrial goods and moral education, but they should be provided with hospitals, medical care and a basic education aimed at removing harmful superstitions and creating an awareness of their rights. We have a lot to learn from them about living together in peace and joy. We need not divest them in their dignity and simplicity in order to grant them a meager share of our technological knowledge; nor need we make them give away their freedom for our dismal wage-slavery. The depletion of earth's forests, forest-dwellers and wildlife has been far greater in this century than during the millennia before the industrial age. It is only in the last decade that civilized man has begun to be seriously concerned about the ecological catastrophe caused by industrial progress. The tribal forest-dwellers are its immediate and most conspicuous victims. They are sustained by and are the natural custodians of the forests; they never harm nature nor do they needlessly slaughter its creatures. In the more technologically advanced world of tomorrow there may be no primitive men left to preserve the down of mankind.

The great problem is how they can pass from their present life of primitive simplicity into the modern world, not only without being destroyed in the process but also so that they may make their unique artistic and human contribution to the modern world. A free India has no need to be ashamed of her primitive populations rather she should treasure and use it, accepting with gratitude the priceless gifts of strength, honesty and beauty that it can bring. Dr Verrier Elwin had always felt about the inevitability of the tribals being totally assimilated in to our prevailing hybrid and very confused culture, or worse, being wiped out by the relentless march of technology and modern commerce.

Modern Education and Market Culture and Tribal Life

However, the impact of the formal school system on traditional indigenous lifestyle has disrupted indigenous and local communities. Through the school system the people of the younger generation no longer learn the knowledge, skill, and attitude of their indigenous society. In fact, indigenous school students are taught to look down upon their own traditional culture. As a consequence, the social relationship based on kinship commonly practiced among various indigenous communities has been broken. Natural resources are degraded through new agricultural practices and technology. Indigenous and local communities are encountering more serious social problems.

Towards the latter half of the twentieth century, it was realised that the formal school system had marginalised rather than modernized indigenous people. An indigenous movement for cultural revitalisation marked the beginning of the search by indigenous people and their advocates for alternative forms of education. Multi-ethnic education, multicultural education, and alternative education all implied respect for indigenous knowledge and culture. Indigenous education is regarded here, first, as alternative education emphasising indigenous learning styles. Thus, learning by doing in a lively environment is valued above classroom confinement.

Community-based education is preferred, for it allows more participation and local indigenous administration and control. Second, indigenous education focuses more on the local relation between land and people through indigenous knowledge. Third, it is also interested in interpersonal relations mediated by language, festivities and recreation. Last, indigenous education is primarily a rights-based approach where indigenous identity is both a human right and a social obligation. Indigenous education does not oppose new knowledge; it looks instead to integrate indigenous and new knowledge to bring about a better understanding of peaceful coexistence among diverse cultural identities.

Running Gamut of Politics, Government and State in Tribal Life

India's forest cover decreased by 367 square kilometers between 2007 and 2009, and it was primarily tribal and hilly regions that were to blame. The tribal and hilly regions are the last vestiges of India's forests. According to experts, the tribal movements are strengthening their base in tribal areas by extending open support to various social agitations of the tribal – such as those against the liquor menace and the land mafia. The reason why there's possibly so little afforestation carried out, as certain not-so-bashful state governments admit, is this: there's no more land left. Most of it has been sold or promised in the thousands of Memorandums of Understanding (MoUs) signed between government and private entities each year- to set up mines, industries, thermal power plants, Special Economic Zones that skip through the loops, ports and big dams, all valiantly opposed by indigenous communities who fear the destruction of their environment. None of these projects find any mention in the story as even possible causes of deforestation.

No space, similarly, has been given to the lapses in the land acquisition and environment clearance processes, which have been sped up exponentially since the liberalisation of the Indian economy in the early 902s. While it can't get ant clearer that the reporter has written this from the comfort of her desk, what makes it worse is that no attempt has been made to understand the traditional practices and knowledge systems of India's forest-dwelling communities. PESA along with Traditional Forest Dwellers Act of 2006 offer a huge opportunity to turn the tables in the scheduled areas where self-governance and community control over resources are the key empowering the marginalised forest dwell-ing communities who may be silent witnesses to the violence due to helplessness.

Most of the north-eastern Indian states (areas) where shifting cultivation, a practice where plots of fertile land are cultivated and then abandoned, are commonly practiced. The communities clear additional land as they move from one area to the next. As for podu or shifting cultivation, it is a traditional system of farming well-rooted in the principles of crop rotation; while one plot is cultivated over seasons, another is allowed to regenerate. The sheer bio-diversity and yield of indigenous crops grown in this sensitive system of farming is spectacular and compared to plantation and extensive cultivation practised in Punjab and Haryana, agriculture in tribal areas thrives alongside the forests that nourish it.

What a benevolent government that dubs wetland ecosystems as waste lands, while it acts as a front agency for mining companies with bauxite on their minds. In India's case (which has cornered up to 33 per cent of total Clean Development Mechanism projects worldwide) when native forests give way to mines and dams, compensatory afforestation is done with exotic cash-crops such as eucalyptus, which can either be sold as carbon credits and/or at an extremely high profit that is seldom shared with the forest communities that do the hard work. Businesspersons are as crime perpetrators against tribals and perpetrators as the saviours of India's forests: it is not indigenous communities that are razing India's canopy to the ground. It is because the Indian government refuses to recognise and continues to violate customary rights and every progressive law in the book to favour of an economic policy that puts profit over people and their environment.

A powerful rights legislation, the Forest Rights Act 2006 has been converted into a benefit scheme which is selectively and strategically used to distribute *pattas*, a connotation which best suit the dominant political interests of the State as distributor and allocator of resources. It is seen that implementing agencies of the government have reduced the law to merely an exercise of recording of rights in a mechanical and bureaucratic way and distribution of *pattas* only on limited rights on land.

The State forest departments/governments have abdicated their role of due diligence, mandatory under the Forest Conservation Act, possibly under political or other pressure and are routinely found approving even obviously damaging projects. Also that there is a lack of follow-up or a proper mechanism leading to violations pertaining to the case with forest, wildlife, water, catchment area treatment and soil conservation issues as well as with those pertaining to the Forest Rights Act and other related legislation. Since implementation of the FRA, such cases of forceful plantation in lands claimed under FRA have been reported from many states. Recently the Minister, Tribal Affairs in his public statements has admitted that Community rights are illegally being awarded to joint forest management committees, which are created by the forest department. These relocations are taking place despite the fact that the Forest Rights Act and the Wildlife Protection (Amendment) Act require that the process of recognition of rights has to be complete before initiating any relocation process.

Conclusion

The tribals have a 'culture' worth preserving. They are not literate but they have a rich oral culture- their poetry and their songs, myths, legends and dances. They have a storehouse of traditional skills and excellent handicrafts. As a result of the cultural invasions taking place now, the tribals feel that they should think and behave like their equally exploited but conforming neighbours. The trials are considered junglist, wild creatures, to be broken in, tamed and dressed up to look like everyone else. Tribal clothing is often considered indecent for these occasions (Republic Day-folk dance staged annually on Republic Day in New Delhi) and the organizers dare not present the participants in their authentic costumes. It is essential that tribals be understood and respected to further the steps taken to mitigate the negative effect of climate change. The concerns and interest of adivasis must be a part of the discourse on sustainable environment and in the discussions at governmental level. Industry is to blame for the increasing greenhouse gases, and it is the job of the government to restrict its emission.

Indigenous people are now under assault everywhere in world under the impact of modernization and they are fast declining. The tribals of India are also dwindling under the impact of industrial development and modernization. Large tribal population have been uprooted and displaced as 'ecological refugees' in Central India due to the construction of ambitious projects, sanctuaries, dams and so on. The traditional knowledge and resource management practices of the indigenous people should be applied in modern development strategies. The knowledge of the indigenous people must also be covered under 'Intellectual Properties Right' (IPR). Governments should recognize that indigenous lands need to be protected from environmentally unsound activities.

There is a strong need to recognize their traditional knowledge, innovations and practices on agriculture, agro-forestry, coastal and river management, use of medicinal plants, water management and harvesting, and disaster management, among others. Their adaptation processes contribute to mitigation and their mitigation measures are integral to how they continue to adapt to the ecosystems they inhibit. It is known that adivasis lifestyle is environment friendly and many of their practices and beliefs have protected and sustained the environment for centuries.

End Notes

1. The Declaration on the Rights of Indigenous Peoples was approved by the UN General Assembly on 13 September 2007. One hundred forty-three UN Members voted in favor and four states voted against the DRIP. The four "no" votes came from Australia, Canada, New Zealand, and the USA.
2. Year 1993 was also declared by the UN to be the International Year for the World's Indigenous People, and the years 1994–2004 were set as the First International Decade of the World's Indigenous People. We are now in the late years of the Second International Decade of the World's Indigenous People (2005–2015). The UN declares such years and decades for targeted groups (indigenous peoples, women, children, etc.) in order to bring more global attention to their plights.
3. Bolivia enacted the Law of Mother Earth in 2011. One of the clauses in Bolivian law accepts the definition of Mother Earth as the way it has been defined by the country's indigenous people. In 2008, Ecuador had passed the Right of Nature Law. UN Declaration on the Rights of Indigenous People (UNDRIP); Universal Declaration of the Rights of Mother Earth.
4. If we cut the entire forest down, where will we live?'- Muria adivasi, Warangal, Andhra Pradesh.
5. Forest Rights Act 2006 is not implemented in the Scheduled areas of North-East Indian states like Mizoram, Meghalaya, Nagaland, Assam and Arunachal. The Panchayat (Extension to the Scheduled Areas) Act-1996 under the 6th Schedule is still not implemented. The non-implementation of the Panchayats (Extension to the Sched-uled Areas) Act, 1996 (PESA) in the states with sched-uled areas (tribal dominated and forest areas) in Central India like Dandakaranya
6. Third World colonialism has replaced European colonialism as the principal global force that tries to subjugate indigenous peoples ... In-vasion and occupation of indigenous nations once done by foreign white expansionist are now done by foreign brown expansionist power ... Most Fourth World indigenous nations have maintained the quality of lands, waters and resources while Third World States have not. Systems that do work are being destroyed to prolong systems that don't work. Over one-half of the world's conflicts are being fought over Fourth World geography, not East-West politics, or North-South economics. The development and modernisation of Third World States is heavily dependent upon the

invasion and annexation of Fourth World nations (Nietschmann 1999: 1-2).

7. In the late 1950s, Nirad C. Chaudhari in his the Continent of Circe saw the disappearance of tribals as an inevitable outcome of industrialization.
8. Political scientist and social anthropologist James C Scott has provided a tentative answer in his recent book, The Art of Not Being Governed: An Anarchist history of Upland South East Asia. He says the vast jungle-clad massif like that of South-east Asia which spans the borders of India, China, Myanmar, Laos, Thailand and Vietnam is inhabited by people who belong to autonomous, state-resistant and subsistence- oriented societies of the hills.
9. Chronicler of India's tribals, Dr Verrier Elwin says in one of his *Twaru* (Tribal Welfare and Research Unit) newsletters: "The tribesmen cannot be isolated any longer; they cannot be left alone."

References

Dash, Tusha, 2012. Forest Rights Act, Right Act is not working right. *India Together*, 13 Feb.

Debnath, Debashis, 2005. *Ecology and Rituals in Tribal Areas*, p. 18–25.

Down to Earth, Science and Environment Fortnightly, February 1–15, 2012, p. 14.

Frontline, April 20, 2012, p. 32.

Glancey, Jonathan, 2011. Nagaland's tribal complexities, Bookmark, Nagaland: A Journey to India's Forgotten Frontier, *Business Standard*, 17 September.

Janah, Sunil, 2003. *The Tribal of India*. Oxford University Press, p. 1–21.

Jones, Steve, 1987. Tribal underdevelopment in India. *Development and Change ISAGE*, London and Beverly Hills, 9: 41–70.

Kafila Blog: Run from big media/February 17, 2012. What the Wall Street Journal Can't See in India's Forests, Adivasis, Forest Rights Act, India deforestation, land grab, *Wall Street Journal*, Shivam Vij/Aruna Chandrasekhar.

Mahanta, K.C., 2008. *North-East India: The Horizon of Anthropology*. Kalpaz Publication, Delhi, p. 248–71.

Menon, Sreelatha, 2011. Worshiping Varuna. *Business Standard*, 27 March.

Meyer, William H., 2012. Indigenous Rights, Global Governance, and State Sovereignty. *Human Rights Review*, DOI 10.1007/s12142–012–0225–3, Business Media. B. V. Media June 2012.

Pattnaik, Sandeep Kumar, 2012. Adivasis and climate change. *National Centre for Advocacy Studies (NCAS)*, Pune.

Roy, Arundhati, 2012. Capitalism: A ghost story. *Outlook*: A Weekly Magazine, 26 March.

Saenmi, Sakda and Tillmann, Timmi, 2006. ACBC: Affirmation of cultures and biodiversity conservation, *International Social Sciences Journal (ISSJ)*, 187, UNESCO, Blackwell Publishing Ltd., 9600, Garsington Road, Oxford University, UK.

Sarkar, Amit and Dasgupta, Samira, 2000. *Ethno-Ecology of Indian Tribes*. Rawat Publication, Jaipur, p. 47–102 and 124.

Sarthak, Sengupta, 2002. *Tribal Studies in North-East India*, Mittal Publication, pp. 109–115, 229–233.

Scheduled Areas Need a Fresh Legal Perspective by Sanjay Upadhyay, *Economic and Political Weekly*, October 9, 2010, Vol XLV No 41.

Sinha, Rajiv K. and Sinha, Shweta, 2005. *Ethnobiology*. Surabhi Publications, Jaipur, p. 9–20.

Xaxa, Virginius, 1999. Tribes as indigenous people of India. *Economic and Political Weekly*, December 18.

Environment, Biodiversity and Traditional System, 2014 *Pages* 137–145
ISBN 978-81-7622-307-2
Editors: D.R. Khanna, G.S. Solanki and S.K. Pathak
Published by: BIOTECH BOOKS, NEW DELHI

Chapter 13

Conservation of Bidiversity for Sustainable Livelihoods of Rural People of India

*D.K. Belasre**

Advisor,
ENVIS, Disaster Management Institute, Bhopal

Biodiversity as Life-Supporting System

The values of biological diversity and their importance for development indicate why biological diversity conservation differs from traditional nature conservation. Biological diversity conservation entails a shift from a reactive posture–protecting nature from the impacts of development–to a proactive effort seeking to meet peoples' needs from biological resources while ensuring the long-term ecological sustainability of biotic wealth. It involves not only the protection of wild species and their habitats, but also safeguarding the genetic diversity of cultivated and domesticated species and their wild relatives. It maintains the life-support system provided by nature in all its variety, and the living resources essential for ecologically sustainable development.

Ethnic Social System in India

The ethnic social systems of India consist of **4635** identifiable communities which are diverse in biological traits, dress, language, forms of worship, occupation, food habits and kinship patterns. They speak 325 languages and use 25 scripts derived from various linguistic families such as Indo-Aryan, Tibeto-Burman, Dravidian,, Austro-Asiatic, Andamanese, Semitic, Indo-Iranian, Sino-Tibetan and Indo-European. There are thousands of dialects. 65 per cent of communities are bilingual. The tribals

* *Corresponding Author:* E-mail: d.belsare@yahoo.com

are tri-lingual. 85 per cent of Indian communities are rooted in local ecosystems on local natural resources. They derive their livelihood, occupation, dress, songs and settlement patterns from these resources.

Their culture and beliefs are diversified that flourished in order to protect biodiversity of nature. These concepts are sacred trees, sacred groves, sacred gardens etc., followed through generations. The faith and beliefs of rural people played significant role in biodiversity conservation for centuries using indigenous knowledge systems. There are

4635 identifiable communities which are diverse in biological traits, dress, language, forms of worship, occupation, food habits and kinship patterns. They speak 325 languages and use 25 scripts derived from various linguistic families such as Indo-Aryan, Tibeto-Burman, Dravidian,, Austro-Asiatic, Andamanese, Semitic, Indo-Iranian, Sino-Tibetan and Indo-European. There are thousands of dialects. 65 per cent of communities are bilingual. The tribals are tri-lingual. 85 per cent of Indian communities are rooted in local ecosystems on local natural resources. They derive their livelihood, occupation, dress, songs and settlement patterns from these resources. They worship animals, plants, forests, rivers, mountains and lakes in the form of incarnation of the Gods' creation. Nearly 50,000 sacred groves are conserved *in situ* and are worshiped by communities and provide a safe refuge to several endangered and threatened species of plants and animals.

Ethnonyms of Indian Social System

An analysis of the ethnonyms of 1018 communities suggests a close relationship of the nomenclature of a community with its occupation based on locally available resources, place of origin of village or territory and deity of religious association. 55 per cent of these communities derive their names from the traditional occupation. Peasants derive their identity from land, *i.e.* Bhuiyar, Bhuinhar, Bhumikara and Bhumji (Bhu means land). These names still prevail among tribals of Madhya Pradesh. Other names are Angaria (iron smelter), Alvan (Salt maker), Angarok (Body guard), Atishbaz (Explosive manufacturers), Badhik (Slaughterer), Bafand (Cotton dresser) Bahurupi (Mimic), Baiyakar (Musical instrument), Bakarwal (Goat herder), Bansphor (Bamboo basket maker), Buna (Weaver), Bajandar (Musician), Bandukkar (Gun maker)Chitrakar (Scroll pinter), Chunari (Line maker), Churihar (Bangle maker), Lohar (Blacksmith, Malakar (Garland maker), Sapera (Snake charmer) etc.

Environmental ethics of Bishnoi community of Rajasthan forbid felling of trees like *Prosopis cineraria* and tolerates damage of crops due to wildlife.

Festivals and Biodiversity

Various festivals are organized by agrarian and ethnic communities to celebrate biodiversity of animals, plants and crops in their environment. The Lord Ganesha is worshiped by offering 101 different kinds of leaves in Patra Puja. In Southern India a particular vegetable or leaf is a must for many festivals. In Telengana during 'Endlagatte Punnam' festival biodiversity of agriculture is celebrated. 'Banashankari Jatra' is organized for three days in Badami town and temple of Banashankari (Goddess of forest), during which hundreds of varieties of leaves and flowers are used for decoration of town and temple. In Jaganath Puri during each day of the year an offering of freshly harvested rice is made in order to awaken one to a different understanding of agriculture. Thus vast cultural beliefs and biodiversity of India are based and nurtured on one another.

Use of Biodiversity of India for Sustainable Growth and Reduction in Rural Poverty

Biodiversity and Livelihoods

A livelihood comprises the capabilities, assets and activities required as means of living. It comes from a variety of sources and activities, variable over time and comprises of several different activities for each given household–even for each working member, and even within a year. It is sustainable if it can cope with, and recover from, stress and shocks, maintain or enhance its capabilities and assets. This also provides net benefits to other livelihoods locally and more widely both now and in the future without undermining the natural resource base. It implies systems of how people make a living and whether their livelihoods are secure or vulnerable over time. Its security means "secure ownership of access to resources and income generating activities including reserves and assets to offset risk, shocks and meet contingencies. Thus it is a livelihood security, rather than just food security, because achievement of food security is just one of the objectives of livelihood security (Maxwell and Smith1995).

Aquaculture and Fisheries: The freshwater fish diversity can serve as a pathora of livelihood for 20 million fishermen communities of India. The aquaculture production still comes mainly from extensive and traditional freshwater farming systems, but due to climate change the water bodies are suffering from reduced water area and thus capture fisheries are depleting. Organized farming of other species on a commercial scale has yet to be developed and this appears to be a significant constraint to diversification of aquaculture in this sub-region (Ganapathy, 1996). The prospetcts for marine culture are not promising. Brakishwater fish ponds created health problems due to environmental pollution.

Lac cultivation on Palas trees can serve as income generation for the poor. Jharkhand has potential to feed 3000 tribal families. Each tree gives annual income of 3000 rupees from stick lac.

The *sericulture* is a source of income generation in rural India. There are five varieties of silk worms reared in this country. *Bombyx mori*, the silkworm, feeds on the leaves of *Morus*, to produce the best quality of fibre. *Antherea assama* produces the famous muga silk (confined only to the Brahmaputra valley of India in the world). Tasar silk is a product of *Antherea mylitta*, which feeds on *Terminalia tomentosa* grown in the thick jungles of Bihar, Madhya Pradesh and Orissa. *Antherea royeli* and *Antherea perniyi* has been recently introduced to produce the oak tasar silk, *Phylosamia ricini*, the eri silkworm, which feeds on *Ricinus communis*, is raised in Assam and Orissa.

There are five varieties of bees, *Apis laborosia* (tarai region), *Apis cerana indica* all pains and hill varieties) *Apis dorsata* (Rock-bee), *Apis florea* (little bee) and *Tragona* sp (stingless bee). The honey collectors sale about 1800 tonnes of honey/per ann from forests of India and derive their livelihood. Bee-keeping in India has not taken up at large scale, but looking to the potential of forests to support millions of bee colonies, it would serve as important source of livelihood.

Wild plants used as food in the form of tuber, leaves, and seeds. The leaves of *Eremures himalaicus, Sedum sp, Origanum vulgare, Ardisia sp, Meliosma pinnata, Urtica hyperborea* are used as food in Ladakh region. In coastal region *Sesuvium portulacastrum* is eaten as spinach. In Madhya Pradesh, Andhra Pradesh and Santhal Parganas of West Bengal, lichen species like *Heterodermia tremulans, Everniastrumcirrhatum, Ramalina sp, Usnea longissima, Rocelia montagnei* are used as spices and flavouring agents. They are sold in local market to earn money also

In trans-Himalyan region the traditional rearing of livestock is an integral part of the overall agricultural system and is a major source of livelihood for most residents. Cow, sheep, goat, pig and yalk are the most important animals. The livestock provides wool, meat, milk, hide, organic manure, draught power, and local transportation. Farmers cross yak with cow for agro-domestic use. The original yak is segregated in the fifth generation.

Medicinal plants collected from forests are processed and sold in the market. In Madhya Pradesh the forest department has created a center called "Sanjivini" to purchase these raw medicinal products on cooperative basis. Women rural folk earn cash income directly without any middle agency.

Faunal diversity of camel, goat, sheep, buffalo, cow and chicken species are the direct source of livelihood for the rural poor, *e.g.* Karaknath variety of chicken found in Jhabua district of M.P. can fetch double the price as compared to poultry breed product.

In order to achieve goals, we require more advanced technology, trained and experienced manpower, and much-needed capital expenditure. We have to develop organized marketing system so that the poor is not exploited by middle men and traders.

The globalization, which is referred to greater freedom and higher internal and external integration of commodity and capital market, has direct impact on sustainability of the livelihood of the poor people through its impact on production and employment conditions. The livelihoods refer to the multiple ways in which people meet their 'material' requirements of food, shelter, and clothing through natural resources. Strategy for management of biodiversity for sustainable livelihoods of the rural poor is to understand the range of community traditions, knowledge, practices and livelihoods related to biodiversity, strengthening extension services to the poor for their livelihoods, and formation of village assembly as functional unit for taking critical decisions regarding use and conservation of biodiversity through proper management.

India has rich floral diversity which is used for primary health, food security and insecticides by local communities for their livelihood. Some of the medicinal plants used for health are *Myristica malabarica, Garcinia indica, Utleria salicifolia, Vateria indica* (Western Ghat region), *Pterocarpus santalinus, Decalepis hamiltonii, Terminalia pallida, Shorea tumbuggaia* (Deccan Penninsula), *Holarrhena pubescens, Mallotus philippensis, Pluchea lanceolata, Peganum harmala* (Gangetic Plain), *Aqularia malaccensis, Smilax glabra, Ambroma augusta, Hydnocarpus kurzii* (North-East region), *Ephedra gerardiana, Hippophae rhamnoides, Arnebia euchroma* (Trans-Himalayan region), *Aconitum heterophyllum, Ferula jaeschkeana, Saussurea costus, Nardostachys grandiflora, Taxus wallichiana, Rhododendron anthopogan, Panax pseudoginseng* (Himalayan region), *Convolvulus microphyllum, C. pluricaulis, Cressa cretica (Desert), Commiphora wightii, Caesalpinia bonduc, Balanites aegyptiaca, Tribulus rajasthanensis* (Semi-Arid region), *Rhizophora mucronata, Acanthus ilicifolius, Avicennia marina, Sonneratia caeseolaris* (Coastal regions) and *Calophyllum inophyllum, Adenanthera pavonica, Barringtonia asiatica, Aisandra butyracea* (Island region).

Singh and Arora (1978) reported wide variety of tubers, edible grasses, flowers, fruits and seeds which are used as food. The leaves of *Eremures himalaicus, Sedum sp, Origanum vulgare, Ardisia sp, Meliosma pinnata, Urtica hyperborea* are used as food in Ladakh region. In coastal region *Sesuvium portulacastrum* is eaten as spinach. In Madhya Pradesh, Andhra Pradesh and Santhal Parganas of West Bengal, *lichen species like Heterodermia tremulans, Everniastrum cirrhatum, Ramalina sp, Usnea longissima, Rocelia montagnei* are used as spices and flavouring agents.

Plant species used as food during femine are: flowers of *Utrica diocia, U. parviflora*, rhizomes of *Diascorea* sp, fruits of *Aisandra butyracea, Calamus erectus, Elaeagnus conferta, Heracleum lanatum, Podopphyllum hexandrum, Machilus edulis, Melia dubia, Morus australis, Terminalia chebusa, Zanthrozylum acanthopodium* and shoots of *Rheum nobile, Diplazium esculantum, Polygonum molle, Ficus virens*. In desert, the femine food species used are: *Prosopis cineraria, Cenchrus biflorus, Calligonum polygonoides, Citrullus colocynthis, C. lanatus, Dactyloctenium aegyptium, Acacia jacquemontii, A. leucophloea, Ziziphus nummularia, Z. mauritiana,, Indigofera cordifolia, Tamarindus indica,, Capparis deciduas, Salvadora oleoides, Achyranthus aspera, Eleusine coracana, Sesamura indicum* and *Cyperus rotundus*. Seeds of *Indigofera glandulosa, I. linifolia, I. cordifolia, Echnochloa* sp, *Panicum* sp, *Eleusine* sp and grains of *Bombusa sp, Vigna copensis*, and *Phasiolus sublobatus* are also used during famine.

The insecticidal property of plants was recognized by the farmers and they use it to control insect pests like aphids using *Ipomea sp, Calatropis gigantean* and *Clerodendrum multiflorum*, blister beatle pest on buck-wheat by *Fagonia indica schweinfurthii, Leucas indica* for storage of chilli, *Euphorbia sp* for pigeon pea, *Eleusine coracana* for termites, *Bidens biternata* for insect pest on paddy, *Azadirachta indica* for storage of millets, oil cake of *Madhuca indica* and crushed fruits of *Catunaregam spinosa* for storage of grains etc (Dastur 1964).

The freshwater fish diversity can serve as a pathora of livelihood and biodiversity of conservation values (Belsare, 2010). They serve as source of protein and food for increasing human population in developing countries (Belsare 1984). The extensive farming of fish is more sustainable for livelihood of the rural poor (Belsare 1986, 2005), because it serves as security of food, health and income generation. When natural disasters like flood and earthquake occur, our first response is obviously to address the immediate needs of the survivors in terms of employment or earned income and food. In the longer term, however, the loss and recovery of livelihoods require in depth analysis by a wide range of expertise, but driven by the real needs of the local people. The utilization and proper management of freshwater fish diversity resource serve as good source for livelihood of the people.

Indian water bodies are much altered by anthropogenic influence on them, such as clustering of human population around them, intensive culture of a few species like major carps (rohu, catla and mrigal) and exotic carps (Grasscarp, Common carp, Silver carp, Bighead etc), eutrophication, soil erosion on the adjoining land where natural vegetation is destroyed or altered and pollution created drastic ecological imbalance, especially, in aquatic food web. The fisheries departments and corporations claim that the freshwater fish production has gone up (based on major carps and common carp and in few cases of catfish), the actual fish landings are depleted. The indigenous fish species are replaced by a few species and bottom fish fauna is drastically altered. Exactly similar situation occurred in African natural and man-made lakes (Worthington and Lowe-Connell 1995). Pandey (2003) pointed out that the North Bihar has high potential of fish production that may feed the poor, but his presumption lacks the proper analysis of ecological changes in water bodies due to human influence. The poors of North Bihar still depend on *Amphipnus cuchia* (Cuchia eel) and catfishes for their food and earnings by transporting human population across the Ganges as well as trapa farming. *In situ* conservation of freshwater fish diversity and livelihoods of the people in India and elsewhere requires in depth analysis of aquatic ecology, if we want to avoid this disaster and think of the future generation's welfare.

22 million fisher-folk of India depend on fish resource for their livelihood. The strategy of micro-credit, gender empowerment, chains of communication, needs-versus-rights ('Bio-rights') based approach and cooperation with the national government is the present need. This approach is now applied to the Tsunami victims in coastal areas where fast recovery for livelihood of the poor is the priority. The hill stream fish fauna of the eastern Himalayan region has diverse species that are used by the local people for food and sports. Exploration of fish diversity in hill streams of the peninsular region of India is not taken seriously and awaits discovery of new species. The recent report of ZSI (Jabalpur Station) has brought out list of freshwater fishes of Madhya Pradesh, but could not identify new species. There are several villages on the bank of Narmada, whose inhabitants are mainly fisherfolks. Our major task is now to survey their livelihood depending on local fish catch. The Central Institute of Fisheries Education's (CIFE) Center at Powarkheda is giving training to them for aquaculture of major carps and catfish and educating them regarding latest technologies for more income generation from culture fisheries.

Modern biotechnology has increased the value of biodiversity both within crop species and among their wild relatives, because it has greatly enlarged the pool of potential sources of useful genes for crop breeding. Genes can now be moved from completely unrelated species into crop plant varieties (GM species).

Many of the natural resources found on Indian lands have become more valuable in the context of the modern global economy. Several factors have spurred renewed interest in natural resources in India. These factors are the mobility of capital, ecological limits to growth in developed countries, lenient environmental restrictions in underdeveloped nations, lower transportation costs and advances in biotechnology, cheap third world labour, and national privatization policies. Limits to logging in developed countries have led timber trans-nationals overseas. Increased demand and higher prices for minerals have generated the reopening of mines and the proliferation of small-scale mining operations. Rivers are coveted for their hydroelectric potential, and bio-prospecting has put a price tag on biodiversity. The lands unsuitable for productive activities are now resources of the future!

Biodiversity also provides inputs for agriculture without which production, either would not occur or would be greatly decreased. These include soil fertility, pollination, pest control, and water for irrigation and livestock consumption. Pest control depends heavily on the mosaic of natural and semi-natural ecosystems that border agricultural fields and serve as a reservoir of biological control agents. These inputs from biodiversity depend largely on the services provided by nearby natural ecosystems. All have traditionally been provided at no cost by the natural environment. Biodiversity actually boosts ecosystem productivity where each species, no matter how small, all have an important role to play and that it is this combination that enables the ecosystem to possess the ability to prevent and recover from a variety of disasters. This is obviously useful for mankind as a larger number of species of plants means more variety of crops and a larger number of species of animals ensure that the ecosystem is naturally sustained. In economic terms, natural ecosystems may be considered capital assets–a form of "natural capital." Like other forms of capital, if properly nurtured, ecosystems provide a flow of valuable goods and services over time. Compared to record-keeping of human, financial, and physical capital, little account has been taken of the natural capital stocks (ecosystems and their biodiversity) that supply ecosystem services. If ecosystems deteriorate to an unsustainable level, then the problems resulting can be very expensive economically.

A Business Model of Small Holder Aquaculture to Reduce Poverty

A viable business needs to be at the heart of the small holder aquaculture enterprise. These models for improvement should be considered as part of a commercial enterprise with business arrangements that deliver value to farmers as owners. The business model should consider opportunities that create more value for smallholder's right along the value chain, not only in production, but also in a financially sustainable manner.

Investments in farming services and farmer groups and organizations may take some time to realize returns and require different approaches than investments in farm operations. Different types of investment are appropriate at different points along any "improvement pathway". Introduction of the right skill sets at the right time into the pathway is key. Similarly, building market connections often requires a sufficiently high quality product to sell in sufficiently large quantities. Successful investments in improving productivity and organizational assistance in M.P. for example preceded investments in improving market connections. Where opportunities for improved productivity and farm efficiencies may be limited, yet the product is available in sufficient quantities, investments in improved market connections might be more appropriate for generating value to smallholders.

Connecting farmers horizontally and vertically requires intermediary organizations to facilitate capacity to provide professional assistance for enterprises, and for investment in capacity development of both small holders and their organizations. Competent intermediary organizations can also be channels for effective delivery of finance and other services to smallholder aquaculture farmers, but sound business models are needed. Investment in the business of farmers' organisations has received minimal attention to date.

Business strategies need to differentiate different types of farm and different products in different places, and the role of public and private actors in the process should, therefore, be facilitative rather than prescriptive. Opportunities for "catalytic" investments in smallholder aquaculture should be sought out to address gaps or inefficiencies in value chains and should be careful to assess whether the elements needed to correct them, such as functioning farmer organizations, are in place. Grants and subsidies alone can never be the basis of business planning and should only be used to temporarily address gaps on the way to a self-sustaining business.

Investments should be considered within the context of an economic vision or investment goal, using business principles as a foundation. Duration and exit strategy should be considered where external assistance is required. The goal should be to leave behind a functioning enterprise operating as a business. Investment in developing the capacity of smallholders and local people to run small business enterprises is necessary to sustain the business as well as to contribute to the long-term improvement of health and employment opportunities in communities.

Partnerships are important in increasing the likelihood of success with smallholder aquaculture enterprises. Public agencies, private investors and smallholders and their organizations bring different skills and resources and these complementary partnerships can reduce risk and increase the overall chances of success. Selection of partners with suitable levels of motivation, financial capacity and appropriate complementary skills is required. Investments in production assets, working capital and infrastructure would be most suitable for private and public banks, but many banks are wary of smallholders because of difficulties in dealing with large number of customers and concerns over risk. Bank concerns can be mitigated by productivity and organizational improvements such as partnering with farmer groups or intermediary institutions, and supporting technical organizations which reduce risk and transaction costs. Once engaged, banks may also bring further skills to enterprise development, such as business models and financial management. Investments in market access and value chain cooperation are also important for creating sustainable opportunities for further value creation. Improving market access and marketing may require bank financing–such as for large investments in improving product quality through better post-harvest facilities–but there are also opportunities for building partnerships with large domestic and international customers. Such stakeholders can be important partners in connecting smallholder aquaculture enterprises to markets and making skills available to support improvements and developing long term contracts that meet the needs of both parties. Such coalitions of partners that together provide the right combination of investments and skills are rare to date, but provide real opportunities for delivering social and environmental benefits.

Suggestions for Success

1. Creating partnerships for investment in small holder aquaculture and ensuring such coalitions have access to the resources and skills they need to cover all categories of investment and to manage risk more effectively.
2. Business models which create sustainable intermediary and farmer organizations also need particular attention.
3. Investment in intermediary organizations and farmer organizations and cooperatives that can deliver professional services to smallholders and enable strength, security and bargaining power through cooperation.
4. Wider networking for knowledge sharing of experiences in smallholder aquaculture investments, both among smallholders and investors. Networking arrangements can produce greater economies of scale and the sharing of experiences across wider networks can be used to advocate for greater awareness of investment and change in smallholder aquaculture.
5. Improvements in rural internet access and mobile communications platforms are opening opportunities for innovations that could connect large numbers of smallholder aquaculture enterprises at scale with the services needed to facilitate improvements and business sustainability.
6. Private corporate social responsibility investments could be used to bring more smallholders into supply chains, and build and test smallholder business models. The capital investment in a small feed mill by a smallholder aquaculture group and finance from member savings and a government grant for equipment would help to reduce feed costs for cooperative members, leading to improved profitability of fish farming.
7. There are number of investment funds – from government and philanthropic sources – being considered for smallholder aquaculture and these developments are welcome. Their impact can be increased through partnerships and packaging of the investments in ways that cover the spectrum of needs across the "improvement pathway". In future, measures which bring about environmental improvements for smallholder aquaculture may gain access to emerging climate change finance, and such opportunities should be explored for smallholder aquaculture.
8. Further investigations should be conducted at State level to enable more focused approaches to small holder aquaculture business investments.

Environment, Biodiversity and Traditional System, 2014 *Pages* 147–157
ISBN 978-81-7622-307-2
Editors: D.R. Khanna, G.S. Solanki and S.K. Pathak
Published by: BIOTECH BOOKS, NEW DELHI

Chapter 14

Indigenous Knowledge on the Utilization of Plant Resources among Bawm Tribal Community of South Mizoram (North East India)

R. Lalfakzuala[1]* *and F. Lalnunmawia*[2]

[1] *Department of Botany,* [2]*Department of Forestry*
Mizoram University, Aizawl – 796 009 (India)

ABSTRACT

The present study revealed the indigenous knowledge relating to the cultivation and use of cultivated and wild plants among the Bawm people in Mizoram (North East India). This paper is based on the spot investigation into the use of forest resources for the livelihood as well as ethno-healing practices among the Bawm tribes in Mizoram. Some of the plants used for medicine, wild edible plant species and wild edible fruits traditionally used by them are listed with how they are being utilized. It has been observed that most of the medicinal plants are used for the treatment of minor diseases such as diarrhoea, stomach pain and cold etc. The Bawm tribe utilizes 21 plants species for herbal medicine, 22 plants species are taken for vegetable, 30 wild plants species are consumed as fruits, 19 plants species are used for timber, 5 plants species are used for wrapping material, 5 plant species are used for thatching and 8 plants species are used for pig fodder. The people in the study area have their own system of preservation and utilization of their valuable bio-resources. However, it was observed that the Bawm tribes do not have any proper document on the use of herbal

* *Corresponding Author:* E-mail: fmawia@rediffmail.com

medicine and the name of plants used by them in the form of scientific name. Documentation of indigenous knowledge system and local knowledge and practices are important to biodiversity management and conservation. Greater understanding of the utilization of appropriate indigenous knowledge would help in the sustainable utilization of bio-resources.

Keywords: ***Indigenous knowledge, Bioresources, Traditional use, Documentation, Conservation.***

Introduction

Indigenous knowledge is being increasingly utilized to identify useful species and methods for preservation, processing and application of those species. Research and development costs can be significantly reduced using traditional knowledge that has been extracted or mined from an IKS (Posey, 1997). An IKS may be commonly held within a community or indigenous society or it may be known only to specialists, tribal elders, lineage or gender groups (Warren *et al.*, 1995).

Despite increasing international recognition, indigenous peoples' roles in biodiversity conservation have been underestimated. However, as scientific documentation and understanding improve wild resources and areas are increasingly found to be the result of co-evolutionary relationships between indigenous peoples and nature (Posey, 1997).

Limited works have been done on the traditional and indigenous knowledge systems in Mizoram. Lalnundanga and Jha (2000) have stated that ethno-botanical work helps in the search of plants which are used by local people as folk medicine. Some of the most notable contributions, based upon actual ethno-botanical survey, are made by Lalramnghinglova (1996); Lalramnghinglova and Jha (1997). Lorrain had mentioned some traditional medicines used by the Lushai people in Mizoram. Vailinga (1991) has also documented 165 diseases and their ethno medicine while Rozika (2001) documented 204 plants along with their uses.

Study Area and Methodology

Mizoram is located at the Southern part of North-Eastern India surrounded by the Burma in the East, Manipur and Cachar Dist. of Assam in the North, Tripura and Bangladesh in the West and again, Burma in the South. It lies between 22°19′ N and 24°19′N latitude and 92°16′E and 93°26′E longitude and the Tropic of cancer passes through the middle of the territory and has a total geographical area of 21,087 Sq.Km. They mostly distributed in the villages of the southern portion of Mizoram. The study area enjoys a sub-tropical climate with temperature ranging from 11°C and 26°C during winter and 19°C to 39°C during summer whereas it receives an annual rainfall of from 25.5 to 30 cm. A field survey was conducted in the year 2004 to identify the plant species traditionally used by the Bawm tribal people. Information was collected through interview with the local people and practitioners. Locally available literatures on the use of traditional herbs were also collected for proper documentation of the traditional knowledge of the people of the study area.

Results and Discussion

The Bawm tribal community has their origin from Chhinlung in Myanmar where a section of the Bawm people still lives. They first migrated to the western part of Burma and then around 1800 A.D. they came to Mizoram (India) crossing the Kolodyne river in Southern part of Mizoram and have occupied 16 villages in the southern part of Mizoram. They speak in Bawm dialect among themselves but do not have their own script. The Bawm people are non-vegetarians and rice is their main staple food. Land and forest are the major economic resources for the Bawm people. They generally practice jhum cultivation in the hilly region of South Mizoram.

It has been seen that most of the medicinal plants are used for the treatment of minor diseases such as diarrhea, stomach pain, cold etc. They have utilized 21 plants species for herbal medicine, 22 plants species as vegetable, 30 wild plants species are consumed as fruits, 19 plants species for timber, 5 plants species for wrapping material, 5 plant species are used for thatching and 8 plants species are used for fodder. Name of the plants used for medicine, wild edible plant species, wild edible fruits, timber, wrapping materials, thatching and fodder traditionally used by the Bawm people are listed in Tables 14.1–14.7.

There has been a relatively intact nature of Bawm communities with the natural environment. The people in the study area have their own system of preservation and utilization of their valuable bio-resources. However, it was observed that the Bawm tribes do not have any proper document on the use of herbal medicine and the name of plants used by them in the form of scientific name. However, they do not have any proper document on the traditional use of forest resources and name of plants in the form of scientific name.

Different tribes in Mizoram have their own indigenous utilization of herbal medicine to solve their health problem. Well documentation and scientific exploration of those medicinal plants is needed to find out their own active principle action. Some indigenous peoples have established databases, which they themselves control, ensuring their ability to control access and use of their knowledge and related information. Documentation of indigenous knowledge system and local knowledge and practices are important to biodiversity management and conservation. Greater understanding of the utilization of appropriate indigenous knowledge would help in the sustainable utilization of bio-resources. Indigenous knowledge may contribute to improved development strategies in several ways such as by helping identify cost-effective and sustainable mechanisms for poverty alleviation that are locally manageable and locally meaningful; by a better understanding of the complexities of sustainable development in its ecological and social diversity; and by helping to identify innovative pathways to sustainable human development that enhance local communities and their environments.

However, indigenous knowledge continues to be largely disregarded in many regions like Mizoram, it plays only a marginal role in biodiversity management and its contribution to society, in general, is neglected. Furthermore, indigenous knowledge is being lost under the impact of modernization and of ongoing globalization processes. Thus, there is a need to protect and further develop the

Table 14.1: Plant used for medicine.

Sl.No.	Scientific Name	Bawm Name	Mizo Name	Family	Habit	Status	Part Used	Mode of Utilization
1.	*Adhatoda vasica*	Kawldi	Kawldai	Acanthaceae	S	A	Lf	A leaf is boiled and the water is used for bathing against fever.
2.	*Ageratum conyzoides*	Dura	Vailenhlo	Asteraceae	H	VF	Lf	Leaves are crushed and applied on the head for the remedy of headache.
3.	*Azadirachta indica*	Neem	Neem	Meliaceae	T	A	Lf	Decoction of leaves is drunk against for dysentery.
4.	*Callicarpa arborea*	Naikiah	Hnahkiah	Verbenaceae	T	A	Br	Decoction of bark is drunk for the treatment of stomach pain.
5.	*Chromalaena dispermus*		Belhnam	Pholeng	Compositae	H	F	LfLeaves are crushed and applied for antiseptic.
6.	*Clerodendrum colebrookianum*	Phuihnam	Phuihnam	Verbenaceae	S	F	Lf	Decoction of leaves is taken for the treatment of urinary problem and high blood pressure.
7.	*Curcuma longa*	Aieng	Aieng	Zingiberaceae	H	Cu	Rh	Rhizome is crushed and the water taken for the treatment of cold.
8.	*Emblica officinalis*	Suanlu	Sunhlu	Euphorbiaceae	T	F	Br	Decoction of bark is taken for treatment of urinary tract problem and diabetes.
9.	*Eugenia jambolana*	Lenmui	Lenhmui	Myrtaceae	T	F	Br	Powdered of bark is taken for treatment of diarrhea.
10.	*Gingiber officinalis*	Sawhthing	Sawhthing	Gingiberaceae	A	Cu	Rh	Rhizome is crushed and taken along with sugar to remove worm.
11.	*Girardinia palmate*	Zawngkasai	Kangthai	Urticaceae	H	A	Lf	Leaves is grinded and paste on the head against headache.
12.	*Trevesia palmata*	Kawhtebel	Kawhtebel	Anacardiaceae	T	F	Br, Lf	Decoction of bark is taken against cholera and decoction of leaves is applied for treatment of rheumatism.

Contd...

Table 4.1–*Contd...*

Sl.No.	*Scientific Name*	*Bawm Name*	*Mizo Name*	*Family*	*Habit*	*Status*	*Part Used*	*Mode of Utilization*
13.	*Lagerstroemia speciosa*	Thado	Thlado/ Chawnpui	Lythraceae	T	A	Br	Decoction of bark is taken against for dysentery.
14.	*Mangifera indica*	Theihai	Theihai	Anacardiaceaea	T	F	Fr	Powdered of fruit skin is taken against for diarrhea.
15.	*Mikania macrantha*	Zunring	Japan-hlo	Asteraceae	C	A	Lf	Leaves is crushed and applied for antiseptic and clotting of blood.
16.	*Musa* spp.	Nahchang	Changel	Musaceae	T	A	St	Water from stem is applied for antiseptic and clotting of blood.
17.	*Nicotiana tabacum*	Vaihlo	Vaihlo	Solanaceae	A	Cu	Lf	Leaves is crushed and applied for antiseptic.
18.	*Psidium guajava*	Kezen	Kawlthei	Myrtaceae	T	A	Sh, Lf	Young shoot and leaves are taken for treatment of diarrhea.
19.	*Spondias pinata*	Tuaithlawh	Tawitaw	Anacardiaceae	T	F	Br	Decoction of bark is taken for treatment of diarrhea.
20.	*Tamarindus indica*	Thingtherela	Tengtere	Caesalpiniaceae	T	A	Br, Lf	Bark is boiled and water part is applied for treatment of skin diseases. Leaves are crushed and applied on head to cure fever, headache and dizzy.
21.	*Vitex peduncularis*	Thinguimilek	Thingkhuaihlu	Verbenaceae	T	R	Br	Decoction of bark is taken against for fever.

Plate 14.2: Wild edible plant species.

Sl.No.	*Scientific Name*	*Bawm Name*	*Mizo Name*	*Family*	*Habit*	*Status*	*Part Used*
1.	*Acacia* spp.	khangmawk	Khanghu	Mimosaceae	S	F	Young leaves
2.	*Adiantum caudatum*	Makawk	Chakawkria	Adiantaceae	H	F	Young leaves
3.	Amomum dealbatum	Airiatil	Aidu	Zingiberaceae	H	F	Tuber
4.	*Arisaema leschenaultia*	Tualrawng	Telhawng	Araceae	H	A	Rhizome
5.	*Bambusa tulda*	Rawsing	Rawthing	Poaceae	T	A	Young shoot
6.	*Bambusa spp*	Tuai	Mautuai		T	A	Young shoot
7.	*Cajanus cajan*	Kawkleng	Behliang	Leguminoseae	S	A	Pod
8.	*Caryota mitis*	Meibu	Meihle	Palmae	T	F	Pith of the stem
9.	*Caryota urens*	Thangtum	Tum	Palmae	T	F	Pith of the stem
10.	*Cephalostachyum capitatum*	Ranal	Rawnal	Graminae	T	A	Tender shoot
11.	*Clerodendrum colebrookianum*	Phuihnam	Phuihnam	Verbenaceae	S	F	Leaves
12.	Cocasia esculenta	Bachek	Dawl	Araceae	H	A	Young leaves
13.	Colocasia spp.	Baibing	Baibing	Araceae	H	A	Spadix
14.	Dysoxylum procerum	Thingthupui	Thingthupui	Meliaceae	T	F	Young shoot and leaves
15.	*Garcinia lancifolia*	Pel	Pelh	Clisiaceae	H	F	Young leaves
16.	*Musa glauca*	Saisuah	Saisu	Musaceae	T	A	Stem
17.	*Musa paradisiaca*	Balhla	Balhla	Musaceae	T	F	Young compressed inflores-cences
18.	*Parkia roxburghii*	Zawngehah	Zawngtah	Mimosae	T	F	Pods and seeds
19.	*Solanum* sp.	Vairemte	Tawkte	Solanaceae	S	A	Fruit
20.	*Spilanthes acemella*	Ankasa	Ankasa	Asteraceae	H	A	Leaves
21.	*Trevesia palmata*	Tlangkamkar	Kawhtebel	Anacardiaceae	T	F	Fruit
22.	*Zanthoxylum rhetsa*	Chingit	Chingit	Rutaceae	H	A	Young shoots and leaves

Plate 14.3: Wild edible fruits.

Sl.No.	Scientific Name	Bawm Name	Mizo Name	Family	Habit	Status
1.	*Amomum dealbatum*	Airiatil	Aidu	Zingiberaceae	H	F
2.	*Artocarpus chama*	Tatpawng	Tatkawng	Moraceae	T	F
3.	*Artocarpu. Heterophyllus*	Lamkhuang	Lam-khuang	Moraceae	T	F
4.	*Artocarpus lakoocha*	Theitat	Thei-tat	Moraceae	T	F
5.	*Baccaurea ramiflora*	Theipangkai	Pang-kai	Euphorbiaceae	T	F
6.	*Bruinsmia polysperma*	Theileitirpawng	Thei-pa-ling-kawh	Styraceae	T	F
7.	*Carallia branchiate*	Theiria	Theiria	Rhinophoraceae	T	F
8.	*Carcinia panniculata*	Theirawmfarah	Theivawmva	Gutiferaea	T	F
9.	*Dillenia pentagyna*	Aitlang	Kawrthindeng	Dilleniaceae	T	A
10.	*Elaegnus caudata*	Sarzuk	Sar-zuk	Elaegnaceae	S	F
11.	*Emblica officinalis*	Suanlu	Sun-hlu	Euphorbiaceae	T	F
12.	*Eugenia jambolana*	Lenmui	Lenhmui	Myrtaceae	T	A
13.	*Euphoria longan*	Theifeimung	Thei-fei-mung	Sapindaceae	T	F
14.	*Ficus prostrata*	Theitit	Thei-tit	Moraceae	T	F
15.	*Ficus semicordata*	Theipi	Theipui	Moraceae	T	F
16.	*Flacourtia jangomar*	Riamsang	Sakhithei	Binaceae	T	F
17.	*Garcinia cowa*	Chengkek	Cheng-kek	Guttiferae	T	F
18.	*Garcinia sopsopia*	Vawmva	Vawmva	Guttiferae	T	F
19.	*Garuga pinnata*	Burbu	Bungbu-tuairam	Burseraceae	T	F
20.	*Glochidion arborescens*	Tuaitit	Tuai-tit	Euphorbiaceae	T	F
21.	*Mangifera indica*	Theihai	Thei-hai	Anacardiaceae	T	F
22.	*Memecylon celastrinum*	Theiawrak	Theikawrak	Melastomaceae	T	F

Contd...

Table 4.3–*Contd...*

Sl.No.	Scientific Name	Bawm Name	Mizo Name	Family	Habit	Status
23.	*Musa paradisiaca*	Balla	Balhla	Musaceae	T	A
24.	*Protium seratum*	Theidil	Bil	Burseraceae	T	F
25.	*Psidium guajava*	Kezem	Kawlthei	Myrtaceae	T	A
26.	*Rubus acuminatus*	Theite	Theihmu	Rosaceae	T	F
27.	*Spondias mangifera*	Tuaitaw	Tawi-taw	Anacardiaceae	T	A
28.	*Tamarindus indica*	Thingterela	Tengtere	Leguminoseae	T	A
29.	*Willughbeia edulis*	Theibak	Vuakdup	Apocynaceae	T	F
30.	*Ziziphus mauritiana*	Marei	Borai	Rhamnaceae	T	F

Table 14.4: Plant used for timber.

Sl.No.	Scientific Name	Bawm Name	Mizo Name	Family	Habit	Status
1.	*Albizia procera*	Vairu	Kangtek	Mimosaceae	T	F
2.	*Albizia stipulata*	Vangsik	Vang	Mimosaceae	T	F
3.	*Anogeissus acuminata*	Zairum	Zairum	Combretaceae	T	F
4.	*Artocarpus chama*	Tatpawng	Tatkawng	Moraceae	T	F
5.	*Bombax insigne*	Tumpang	Pang	Bombacaceae	T	F
6.	*Castanopsis tribuloids*	Siathing	Thingsia	Fagaceae	T	F
7.	*Dipterocarpus indicus*	Vailawng	Lawngthing	Dipterocarpaceae	T	F
8.	*Duabanga suneratioides*	Parzuang	Zuang	Sonaratiaceae	T	F
9.	*Gmelina arborea*	Thanvuang	Thlanvawng	Verbenaceae	T	F
10.	*Laegerstroemia speciosa*	Do	Thlado	Lythraceae	T	F
11.	*Mahonia nepalensis*	Thingaieng	Pualeng	Beberidaceae	T	F
12.	*Mesua ferrea*	Hersa	Herhse	Guttifereae		F
13.	*Michelia champaca*	Ngiau	Ngiau	Magnoliaceae	T	R
14.	*Sapium baccatum*	Thingbawkpi	Thingvawkpui	Euphorbiaceae	T	F
15.	*Schima wallichii*	Khiang	Khiang	Theaceae	T	F
16.	*Tectonia grandis*	Teak	Tlawr	Verbenaceae	T	F
17.	*Terminalia chebula*	Thingkhuailul	Reraw	Verbenaceae	T	F
18.	*Terminalia myriocarpa*	Char	Char	Combretaceae	T	F
19.	*Vitex penducularis*	Thingnimilit	Thingkkhawilu	Verbenaceae	T	F

Table 14.5: Plant used for wrapping materials.

Sl.No.	Scientific Name	Bawm Name	Mizo Name	Family	Habit	Status
1.	*Amomum dealbatum*	Airiatil	Aidu	Zingiberaceae	H	F
2.	*Callicarpa arborea*	Naikiah	Hnahkiah	Verbenaceae	T	F
3.	*Musa paradisiaca*	Balhla	Balhla	Musaceae	T	A
4.	*Musa* spp.	Nahchang	Changel	Musaceae	T	A
5.	*Phrynium capitatum*	Nahthial	Hnahthial	Scitamineae	H	A

Table 14.6: Plant used for thatching.

Sl.No.	Scientific Name	Bawm Name	Mizo Name	Family	Habit	Status
1.	*Bambusa* spp.	Mau	Mau	Poaceae	T	A
3.	*Calamus guruba*		Thilthek	Palmae	S	F
4.	*Imperata cylindrical*		Di	Graminae	H	A
5.	*Licuala peltata*		Laisua	Palmae	S	F

Table 14.7: Plant used as fodder.

Sl.No.	Scientific Name	Bawm Name	Mizo Name	Family	Habit	Status
1.	*Bambusa spp*	Tuai	Mautuai		T	A
2.	*Colocasia esculenta*	Bachek	Dawl	Araceae	H	A
3.	*Ipomea batatas*	Bara	Kawlbahra	Convolvulaceae	H	A
4.	*Manihota esculenta*	Pangbal	Pangbal	Euphorbiaceae	H	A
5.	*Mikania macrantha*	Zunring	Japan-hlo	Asteraceae	C	A
6.	*Musa glauca*	Saisuah	Saisu	Musaceae	T	A
7.	*Musa* spp.	Nahchang	Changel	Musaceae	T	A
8.	*Spilanthes acemella*		Ankasa	Asteraceae	H	A

H: Herb; T: Tree; S: Shrub; C: climber; VF: Very frequent; A: Abundant; F: Frequent; Cu: Cultivated; R: Rare; Lf: Leaf; Br: Bark; Bb: Bulb; Wp: Whole plant;

knowledge generated and perpetuated by local communities through awareness-raising, training programmes, international property rights arrangements, and validation procedures.

Acknowledgement

We are thankful to the farmers of Chawngte Village in Saiha District of Mizoram for providing valuable ethnobotanical information and sharing their traditional knowledge on the utilization of plants. Thanks are also due to the North Eastern Biodiversity Research Cell (NEBRC), NEHU, Shillong for financial support to R.Lalfakzuala for conducting the field study.

References

Lalnundanga and Jha L.K., 2000. Medicinal plants of tropical semi evergreen and montane sub-tropical Pine Forest of Mizoram. Paper presented in *National Seminar on Ethnobotany in North Eastern India: Past, Present and Future*, 12th April, Aizawl.

Lalramnghinglova. J.H., 1996. Ethnobotany of Mizoram: A preliminary survey. *J Econ. Tax. Bot.* Addl. Series, 10: 439.

Lalramnghinglova, J.H. and Jha L.K., 1997. Ethnomedicine from Mizoram–Northeast India. *Ethnobotany*, 9(142): 105–111.

Lorrain J.H., 1940. *Dictionary of the Lushai language*. The Asiatic Society, Calcutta. pp. 1–356.

Posey, Darrell D., 1997. Wider use and applications of indigenous knowledge, innovations and practices: Information systems and ethical concerns. In: *Biodiversity Information: Needs and Options*, (Eds.) D.L. Hawksworth, P.M. Kirk and S. Dextre Clarke. CAB International, pp. 69–92.

Rozika, R., 2000. *Ramhmul Damdawite (Medicinal Plants)*, Revised. Published by the Medicinal Plant Board of Mizoram, pp. 1–457.

Vailinga, Rev. M., 1991. *Mizo Pipute Sulhnu leh Pipute Damdawi*(Local Language). Christian Book Store. Chanmari, Lunglei. pp. 1–27.

Warren, D.M., Slikkerveer, L.J. and Brokensha, D. (eds.), 1995. *The Cultural Dimension of Development: Indigenous Knowledge Systems*. Intermediate Technology Publications, London.

Environment, Biodiversity and Traditional System, 2014 *Pages* 159–174
ISBN 978-81-7622-307-2
Editors: D.R. Khanna, G.S. Solanki and S.K. Pathak
Published by: BIOTECH BOOKS, NEW DELHI

Chapter 15

Pandit Loishang and the Folk Medicine among the Meiteis of Manipur

Ch. Sarajubala Devi

Assistant Professor,
NERIE – NCERT, Shillong
E-mail: sarajubala@yahoo.com

ABSTRACT

Like any other elements of folklore, folk medicine plays a major role in every traditional community. This body of knowledge among the Meiteis of Manipur had been developed and institutionalized in Pandit Loishansg[1], which was patronized by the king. The establishment of such institution was to response to the problem of health, disease and sickness of the community. The institution of Pandit Loishang consisted of magico-religious physicians and other functionaries known as Maibas[2] and Maibis who are well versed in diagnosis and treatment of different diseases by using a wide range of herbal medicines,

1 Pundit Loishang is the socio-religious institution of the Meiteis which could give guidance not only to the masses but also to the king in every sphere in every moment of sad and sorrow. Therefore, the king as well as all the Meiteis abided by the advice of the members of this institute. At present it has three departments namely—*Maiba Loishang* (Council of traditional Meitei Literaties and priests), *Maibi Loishang* (Council of traditional Priestesses) and *Pena Khongba Loisang* (Council of traditional musicians).

2 Sircar made a distinction between two categories of maibas, whom I prefer to address as 'traditional knowledge experts' of the Meitei society—the first category functions as 'traditional priests'; the second '... variety of specialist...are not traditional priests, but act as Shamans and curers'. For details see, Manjusri Chaki Sircar, *Feminism in a Traditional Society*, (New Delhi: Shakti Books), 1984, p. 119.

stones, animal substances, *etc*. Apart from this, in many cases, the process is also ritualized through some magico-religious process like chanting mantra and incantation. Hence folk medicine is an important organ of collective folk mind, belief and superstition of the Meiteis who inhabit the valley of Manipur. Though the importance of this age old practice of folk medicine has been lessened among the educated youths because of awareness and preference of modern science and rationality, common people still have the faith in this system as the real needs. Apart from the institutionalization of folk medicinal practice the development of other cult as a new healing centre in Manipur known as 'Ibudhou-Phaoba' (the upcoming/ rise of the ancestor god) and heavy rushing crowd which one sees around this new place is a testimony to the fact that their relevance for healing the human psyche has not declined in the mist of modern medicine. In many rural areas they still find a place as it is only available means in the absence of hospital. With this background the present paper is an attempt to understand the knowledge of the Meitei community in terms of folk medicine and their utility at past and present. And such study of the Meitei folk medicine will certainly highlight some important aspects of the Meiteis in terms of their wider knowledge about health and environment.

Keywords: ***Pandit-Loisang, Maiba-Maibi, Umanglai-Haraoba, Folk medicine, Ecology, Environment, and Ritual.***

Introduction

Like any other society, good health and prosperity of the individual and community has been a major preoccupation of Meitei society. The celebration of a long festival of Forest deity commonly known as Umanglai-Haraoba[3] (pleasing of forest deity) by every village every year is a sure sign of their effort to protect their community from any pain and other suffering.[4] The study of this Umanglai-Haraoba reveals that Meitei community gives a lot of emphasis on their environment and ecology for bringing good health and prosperity of the community. The sacrificial rituals and other associated practices show a great reverence towards this deity.[5] The preservation of many plant and animals having medicinal values thought this practice is a clear understanding of the environment and ecology towards the development of

---

3 The festival of Umanglai-Haraoba is the longest festival of the Meiteis which normally performed for 10 days every year in all the villages. However, the Tangjing Lai-Haraoba at Moirang usually is performed for 30 days. The main theme of this festival is to protect the population, crops, and cattle from disease and any harm caused by the evil spirit. As the ritual is conducted for fertility and prosperity of the community, the ritual is very much related with their understanding of health and its relation with environment and forest.

4 Khuraijam Bijoykumar Singh, The Sacred Grove of Umanglai: A Traditional Way of Bio-diversity Conservation and Environment Management in Manipur, Seeds of Hope, Lokayan, 2004.

5 A study of four sacred groves in Manipur revealed therapeutic application of nearly 120 plants species representing 106 genera and 57 families out of which 42% of trees and 33% of herbs found in the groves has medicinal values which can be used for skill disorder, ulcer rheumatism, bronchitis, etc. For further detail see Ashalata Devi Khumbongmayum, M.K. Khan & R.S. Tripathy, "Ethnomedicinal plants in the sacred groves of Manipur, Indian, Journal of Traditional Knowledge, Vol. 4(1), January, 2005, pp. 21

community health. Ashalata Devi, M.K. Khan and R.S. Tripathy shows the significance of this sacred grove as the repository of herbal medicine of the community.[6]

Based on a shared understanding of human nature and the causes of suffering, Meitei society has developed its own healing institutions and practices. And the conceptual understanding of 'body', 'soul' and 'shadow' finds its way in the 'traditional medicine' and the ordering of social relationships among the Meiteis.[7] An observation of the relationship between these three component elements as expressed in the discursive uses as well as in many literatures, myths, rituals and the cosmological world of the Meiteis reveals the matrix that binds different levels of thought and practices of a particular health care system. How the body has been conceived by the people has a lot of bearing on how they respond when the body is in some 'critical' situation which is called 'sick' or 'illness.[8]

Methodology

Use of both secondary and primary sources are the basis of this paper. Secondary sources are mainly drawn from the archival material composed by the *Pandit Loisang* and other folk healers of Manipur. Since some of the practices are not yet documented but orally transmitted from the person to other disciples focus interviewed were also taken with few healers locally known as *Maiba* (male healers) and *Maibi* (female healers). In addition to these key informants, 10 elderly men and women from Keirak village, Kakching Municipality area and Wabagai villages of Thoubal District Manipur were also interviewed in order to capture the community knowledge about the folk medicinal practices.

Cultural Context of Health, Disease and Medicine in Meitei Society

Every culture, irrespective of its simplicity and complexity, has its own belief and practices regarding health and disease and the way of treatment, and it is also true in the case of Meitei community. Therefore, the phenomenon of disease and health are universal experience, which are as old as human community is. However, the concept of health, disease and treatment vary according to the culture of a particular

6 The concept of body, soul and shadow represent important component of life in Meitei society. The understanding the philosophy of these three elements and their interrelationship constitutes an important part of folk medicine in Meitei society. The healing practitioners not only to understand the condition of these three component through pulse reading, manifested symptoms and divinity but he also has to find out the suitable remedial measures by applying herbs and other complementary rituals.

7 In Meitei language, the term fundamentally used for physically sickness or illness is *hakchang naba, masha-ngamdaba, masha-nungaitaba,* etc. The variation in the terms though related to the statement of the condition of human body do not mean differently.

8 As informed by an Maiba, in regard to the rite and rituals of the Meiteis there are two types – one is the performance of periodic rites and rituals on regular basis to honour and respect their ancestor goods who are believed to be the guardian and protector of the land, community and cattle and another is the rites and ritual for immediate effect to appease the angry evil spirit which are not periodic in nature.

community. In one way it is a system of community's response to their environment as health is always influenced by one's environment at the same time cultural patterns are also always seen as a means of responding to its environment. For human being in a given set up, disease, accident and disaster not only threatens one's state of well being but it also affects the integrity of the community as a whole.

Fear of being dead is universal for everyone and therefore, the phenomenon for pain and death involves an emotional tension for every community. This emotional tension induces a permanent fear in the mind of each and every individual in the society. This particular fear of occurring death enables Meiteis to plan and act, to challenge and resist obstacles and calamities, and finally to understand the nature and causes of these phenomenon and the means to remedy it. The development of socio-religious institution of *Pandit-Loishang* in Manipur Valley is the response of Meitei society towards this problem. And in many communities the best means for dealing health and disease is to link the phenomenon with one's faith and belief. This process enables them to legitimize the entire health knowledge by the religious institution so that people can response to the problem collectively.

There are two main activities in regard to the relationship between health and environment in Meitei society *i.e.* – using environment for medicinal purpose (extracting medicine) and act of honoring and respecting the environment. In regard to the first activity Meitei community had a fair control and command over their environment in terms of utilizing many herbs and minerals for treatment of different diseases. And the second activity is their understanding and realization the interrelationship between health of the community and their immediate environment. The development of both 'periodic' and 'immediate'[9] rites and ritual to honour and respect the environment is their collective effort to maintaining a harmonious relationship with nature. Therefore, the understanding of Meitei culture, religious belief will help to understand the nature and characteristics of their total health care system. In Meitei society it is observed that religion not only serves the function related to one's faith and belief in god but it also helps to legitimize and regulate the knowledge and activities related to health.[10]

Metaphors of Health and Meitei Folk Medicinal Life

In general, Meitei divide illnesses into two categories. Firstly, those that trace their etiology to natural phenomena which includes the change of climate, weather, seasons, etc. And secondly, those attributed to sorcery (the malevolent actions of other people) and extra-human agency. Both the factors affect both physical and mental well-being of individual and society. In the first case, the seasonal change that produces extreme climatic variation and availability of new local fruits and

---

9 Meitei religion serves a remarkable function in terms of health by imposing hygienic purposes such as avoiding the visit to a house where someone has passed away, arrangement of the disposal of death, cleaning the house where death occurred, birth occurred, etc.

10 Personal interview with Kh. Yaiskul Singh, Maiba of keirak village.

excessive consumption of the same by the people affect the health.[11] Use of herbal folk medicine is the common practice in such cases. It hardly needs any kind of training, and remains more or less a family lore, men, women and children are fairly well acquainted with this kind of medicine. Self or home treatment is usually the first step in medical care, consisting primarily of herbs, barks of trees, flowers, roots, leaves, seeds etc. Many of these are commonly known to Meiteis. Some grow wild in the area, some are cultivated in home gardens and others are bought from the market. And for each and every disease there is a special herb to cure. So far, about 1200 plants (FRLHT Database) and 300 animals including birds, fishes, insects, etc. are reportedly used by Maibas-Maibis in making of traditional remedies used by them for treatment of patients.[12] For example, for cold, cough and flu *Nongmangkha* (*Phlogancanthus thyrsiflorus*) leave are eaten or decoction of it is taken orally. Likewise, for constipation rose leaves are taken for urinary tract infection juice of crushed *lam ikaithabi* (*Mimosa Pudica*) is used.

Regarding second cases, Meitei theory of sickness describes a different source of evil caused by invisible spirits that exist outside their social boundaries. These spirits inhabit in trees, rivers, lakes, mountains and deserted places around the habitation. It is alleged that spirits and ghosts cause various kinds of suffering and are agents of illness and fatality. Diseases like cholera, plague, and small-pox were, in the past, attributed to divine anger. Similarly, even leprosy was believed to be a curse. Moreover, diseases like possession, madness, hysteria and a number of other similar diseases were believed to be caused by evil spirits, evil eyes, witches and black magic. The treatment of these diseases in many cases requires prayers, offerings, rituals, incantations, mantras, holy words, sacrifices etc. to appease the angry gods, exorcize the evil spirit, or punish the evil doer or at times to transfer the disease to animals, trees or objects. Therefore, a pluralistic method of dealing the health problem *e.g.* both ritual and herbal medicine is the approach that Meitei adopted. Therefore, a traditional health specialist always looks at a sick individual to determine whether the illness can be ascribed to natural causes or sorcery.

The *Pandit Loishang*: The Institutionalization of Folk Medicine in Meitei Society

Since folk medicine is part of the entire socio-religious life of the people, there is the process of institutionalizing the knowledge related to health of the people and community in Meitei society. The development of medico-philosophical literatures like *Maibalon, Hidaklon, Shingligi- Maram, Laimuron, Taoroinai- Yangbi, Thepalon, Laihatlon* etc. by the *Pandit Loishang* enabled them to strengthen the knowledge and belief revolved around health and disease in Meitei society. In order to make full control and command over the common people in terms of various knowledge and

11 *Report on Local Health Traditions of Manipur*, Manipur State Biodiversity Board Government of Manipur Imphal, Manipur, 2011.

12 Unlike Maibi, he Maiba does not receive a calling, and chooses to become a healer. They are the learned persons.

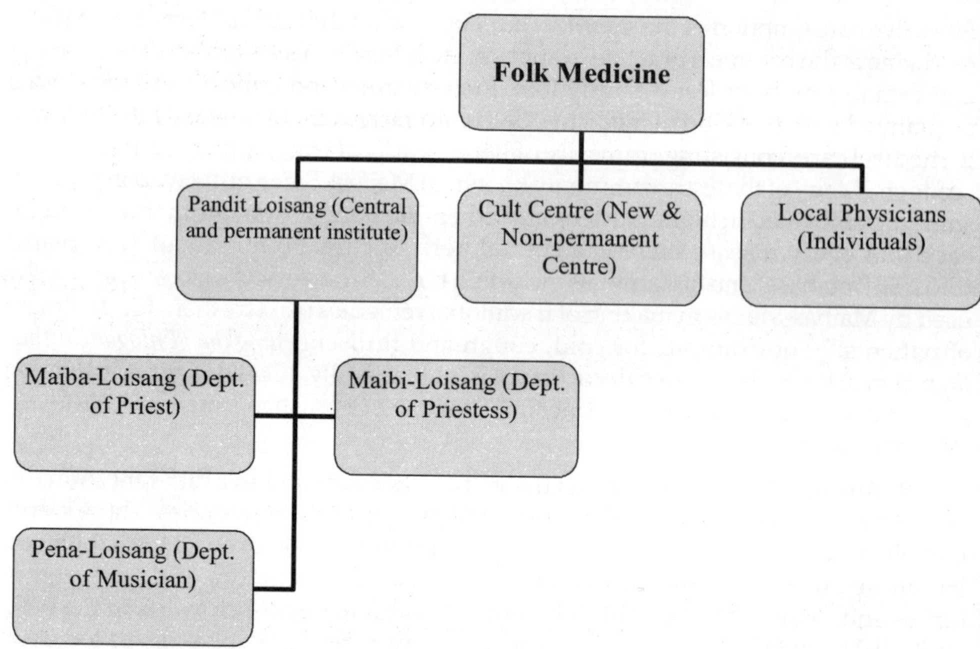

Figure 15.1: Structural organization of functionaries of Pandit Loisang.

practices related to health of the community, they also develop myth and belief according to which the knowledge and information written in the philosophical books of *Maibalon* etc. cannot be shared with common people. Only the *Maiba* and *Maibi* can use and apply the knowledge. This helps the emergence of professionalism in traditional folk medicine in the form of institutionalization of *Pandit Loishang* in Manipur. The institution is the highest institution which not only responsible for the entire socio-religious life of the people but it is also institution to protect people and animals from various diseases.

This institution is the highest institution which not only responsible for the entire socio-religious life of the people but it is also institution to protect people and animals from various diseases. The rituals of Kongba-Laothong Phatpa, *Saroi-khangba, Kwak-Tanba* performed by the *Maiba-Maibi* in the community are their responsibility to protect the land, people and various other domesticated animals from epidemic and other accident and other harms caused by evil spirits. Therefore, the establishment of such central institute enables to regulate and control the health in the community. One important point to be noted here is that without the support of the community such ritual cannot be performed. Every household has to contribute items for sacrifice in this ritual of *Saroi-Khangba*.

Types of Meitei Folk Medicinal Practitioners

Traditional healers do not perform the same functions, nor do they all fall into the same category. Each of them has their own field of expertise, with their own methods of diagnosis and their own particular medicine. They are as follows.

Maibis: The Diviners

Maibi or diviner, who is usually a female, is the most important authority in Meitei socio-religious life because of their important and prominent position in the Meitei religion. She is a person who can predict causes of different kind of diseases and epidemics as well as the welfare of the community. However, a *Maibi* may or may not have knowledge of medicinal herbs. Their specialization is divination where she operates within a traditional religious supernatural context and acts as a medium with the ancestral spirits. They concentrate on diagnosing the unexplainable diseases. They analyse the causes of specific events and interpret the messages of the ancestors or the spirits and god. Although the focus is on divination, they often also provide medication for the specific cases they have diagnosed. Training to become a *Maibi* is not a personal choice but is a calling bestowed by god or ancestors. After she is called she undergoes training under a senior *Maibi* during which she learns to use magico-religious practices and became a professional.

Maiba: The Traditional Doctor or Herbalist

A *Maiba*[13] who is male is a knowledgeable person in the philosophy of life and death. He particularly specializes in the use of herbal and other medicinal preparations for treating disease. He possesses an extensive knowledge of curative herbs, natural treatments and medicinal mixtures of animal origin. Their comprehensive curative expertise includes preventive and prophylactic treatments, rituals and symbolism as well as preparations for luck and fidelity. Some treat only one disease and become renowned experts on that disease. These include specialists in diseases of specific organs (for example heart, kidney or lung disease consultants).

Folk Medicinal Practitioners' Approach to Curing Diseases

The theory underlying traditional medicine in Meitei society is essentially similar with many tribal societies who viewed disease as the both natural and supernatural phenomenon governed by a different force of nature. And the traditional Meitei healing process follows different stages which can be classified as follows.

1. Identification of the cause or discovery of violation of established order through supernatural divination.
2. Removal of the hostile source by neutralization of the sorcerer or seeking the ancestors' forgiveness with sacrifices and rituals to appease their anger or by prescription of certain medication. In this regard, patients are treated as integral components of a family and of community at large rather than as isolated individuals. Family members are almost always involved and the healing process is facilitated by the broader involvement of members of the community.

---

13 A new Cancer therapeutic herb, locally known as 'zanlung damdei' (in Hmar tribal language) was discovered by an 82-year old villager named Chawlien Hmar at Saikot village in Churachandpur district. With the tale of Chawilien's miracle plant gaining significance every hour, the daily turnout at Chawilien Cancer Medicine and Research Agency now witnesses a crowd of not less than 10,000 every day. (source: Manipur daily, *The Sangai Express*, June 18, 2008).

Therefore, as mentioned above the health in Meitei society does not only to protect from disease, but to develop a capacity to do things successfully, therefore, ill-fated is even treated as bad health and always seeks the help from traditional healers commonly called as *Maiba* and *Maibi*. Both the *Maibas and Maibis* or Manipuri shamans have been part of Manipuri culture since pre-history. They have played a major role in Manipuri society as curer, priest, magician, and sorcerer, and basically one that can help to alleviate or eliminate both physical and psychological problems. Therefore, the *Maiba and Maibis* are the important figures linked with the mysterious world of spirits and mysticism. They are practitioners in the areas of alternative healing, sorcery, and traditional ceremonies and have played a major role in Manipuri society.

Therapeutic Process and Methods used by Meitei Folk Medicinal Practitioners

Traditional diagnosis is both an art and a method of seeking to discover the origins of the disease and determining what it is. The diagnostic process not only seeks answers to the question of how the disease originated (immediate causes), but who or what caused the disease (efficient cause), and why it has affected this particular person at this point in time (ultimate cause). The types of medications used by Meitei traditional healers can be classified into three categories:

Preventive and Prophylactic Medication

Most of the work of traditional healers concerns protecting patients from possible afflictions. This can be achieved in various ways, for example by performing ceremonial acts, using medicine against disequilibrium, or wearing amulets called Jantra, Kurao, etc. The traditional term for such kinds of affliction is called *'Minu-Changba'* (casting of evil eye), the patient of such cases have multiple symptoms like fever, incessant vomiting etc. another form of this kind is *'Hingchabi-Changba* (spirit possession), symptoms being abnormal pulse rate, fever, talking nonsense subconsciously etc. for treating such diseases the practitioner will find out the cause and appease the related spirit with appropriate rituals and sacrifices.

Following is case history of two patients treated by Kh. Yaiskul Singh and L. Maibiren Singh of Keirak village, Thoubal district (in july2011) respectively:

Case–1: Name of the patient: H. Usharani Devi, 16 years, female.

Name of the Disease

'Hingchabi Changba' (the entering of evil female spirit)

Symptoms

The patient suffers from fever, frequent nervous movements of the body, abnormal, talking nonsense, abnormal pulse rate,(pulse rate is feel on different fingers, sole, palm and back of the ear), constant desire to eat something, weeping, suffocation, etc.

Diagnosis

Taking eventual history of the person like where did s/he go, with whom s/he talked, what did s/he eat in front of whom and also analyse the words s/he speak in the subconscious state. Through this the *Maiba/Maibi* identifies the kind of spirit.

Treatment

1. Ritualistic observation of '*thou-touba/offering chaktha*'(respond to the spirit which is spirit specific). The *Maiba* by using the leaves of *Tairen* (*tona ciliata*) and *Tingthou* (a kind of grass) sprinkle some holy water to the patient's body.
2. Asking the patient to disclose the spirit's identity, which is a very tough job, many times the *Maiba* has to threaten the spirit to burn or thrash the body of the victim with *Tairen* (*tona ciliata*), while chanting the mantra.
3. Asking the patient about what s/he wants to eat or to dress when s/he is in a subconscious state ad offering the same to the spirit afflicting the individual.
4. The patient is allowed to pass across the family hearth which is considered to be the abode of the ancestors.
5. Sprinkling of the mixture of charcoal and turmeric in and around the homestead.
6. Sacrificing of *Ngakra* (a kind of local fish) along with the objects she tells in the subconscious state on a banana leaf at the gate of the house of the victim.
7. The patient was given an amulet containing some mantra to be worn in the left arm.

The patient whom he treated was recovered from all the complications within one day of the treatment.

When interviewed with the girl she mentioned the she was first taken to Kackching Rural Hospital, Kackching, the doctor gave few medicines which did not help her in any way. Her parents on the advice of the other villagers she was taken to Regional Institute of Medical Sciences, Imphal, for treatment. The medicine the doctor prescribed did not help her. In this condition, one of the villagers again suggested her parents to take her to the *Maiba* of the village and she did accordingly and surprisingly she was cured in one day.

Case–2: Name of the patient: U. Mangi Singh, 43 years, male.

Name of the Disease

' *Semjinba*' (sending of an evil spirit by someone)

Symptoms

The patient suffers from fever, vomiting, no appetite, constant complains of fear (that someone is following him all the time), suffocation, weeping, abnormal pulse rate,(pulse rate is feel on different fingers, and back of the ear), suffocation, etc.

Diagnosis

Taking eventual history of the person like where with whom he had a bad relation, did he eat something given by any person, if so what did he eat, and let him tell all his problems. After hearing from the patient the maiba chant some mantra in holy water and see the face of the person who is sending the evil spirit to the victim.

Treatment

1. Ritualistic observation of *'thou-touba/offering chaktha'* (respond to the spirit which is spirit specific). The *Maiba* by using the leaves of *Tairen* (*tona ciliata*) and *Tingthou* (a kind of grass) sprinkle some holy water to the patient's body.
2. The patient is allowed to pass across the family hearth which is considered to be the abode of the ancestors.
3. The *Maiba* made an anthropometric figure of the spirit by wax and gave him some miniature clothes to wear, three variety of deformed fruit, arum, ginger, some amount of puffed rice and offer it in the
4. Sprinkling of the mixture of charcoal and turmeric in and around the homestead.
5. Sacrificing of *Ngakra* (a kind of local fish) along with the objects she tells in the subconscious state on a banana leaf at the gate of the house of the victim.
6. The patient was given an amulet containing some mantra to be worn in the left arm.

The patient whom he treated was recovered from all the complications within one day of the treatment.

When interviewed with the man (the patient) he mentioned the he went to Kackching Rural Hospital, Kackching, the doctor there gave few medicines which

Figure 15.1: The diviner of newly emerging Cult Centre Called Panthou of Wangkhei in Imphal East surrounded by help seekers.

(*Courtesy*: Manipur daily, The Sangai Express)

did not have any impact on him. His friends suggested him to go to Imphal for treatment but since he did not have money he stopped the idea of going to Imphal, rather he decided to meet the local *Maiba* who ultimately helped him to get rid of the disease in a cost less than Rs.100/-. He is happy and has great respect for Oja Maibaren (the *Maiba*).

From the above few cases it is known that the efficacy of folkmedicinal practice is still a puzzle and is often doubted by health professionals, scientists and rationalists. At present though we do not have any conceptual and methodological tools to establish a causal connection between folk medicine of the Meiteis and their well-being, many people across the classes still fall into this practices as the last means. The present available evidence of heavily crowded in few local healing centers also

Figure 15.2: People, mostly from Imphal staying overnight waiting to collect the herbal plant called 'zanlung damdei' (in Hmar tribal language) which was discovered by an 82-year old villager in Manipur.

(*Courtesy*: Manipur daily, *The Sangai Express*)

shows a great belief of people in this practice.[14] From this it is also known that there is a renewed interest in religion and spirituality and their role in the psychotherapeutic process in Manipur. Traditional folk medicine therapies of the Meiteis have much to contribute to this end. The rich know-how and techniques employed in traditional therapies to bring about attitude change can provide new insights to evolve more relevant psychotherapies. Modern health practices in Manipur are very poor and frequently referred their patients of both minor and major problem to private hospitals.[15] Obviously, these services are perceived by the general population to be highly inadequate to manage the present health problems in Manipur. Mutual learning from these diverse perspectives can significantly enhance our ability to alleviate the suffering of the people. For this we need a new conceptual framework which can integrate the finer elements of this folk medicinal practice of the Meiteis. There are indeed many positive developments perspectives if we give more attention and support to this system as alternative practices.

Treatment for Ailments

These are prepared in different forms such as cold and hot infusions, decoctions, powders, and pastes of host of herbs and plants, and a variety of earthy ointments that comprise animal fat, clay and sometimes ashes. These formulations are made into different medicine mixtures. The recipes range from general (known to many people) to a secret and are part of the knowledge that the healer will pass onto his apprentice. There are specific herbs, plants and roots, some of which examples are given below.

Examples of Folk Remedy

1. Management of cold and cough with fever.

Sl.No.	*Name of the Herbs/ Shrubs*	*English Name*	*Part Used*	*Method of Preparation of Formulation*	*Mode of Adminis-tration*	*Avoidance*
1	Nongmangkha	Adhatoda	Leaves	Decoction	Internal	Milk and milk products
2	Shing	Ginger	Stem	Decoction	Internal	Milk and milk products
3	Uchithi	Black-pepper	Seed	Decoction	Internal	Milk and milk products
4	Laipung-khanga	Indian wormwood	Leaf	Decoction	Internal	Meat, milk, sour fruits

14 The frequent referring of patient of rural areas from Primary health centers and rural and district Hospitals to Government and private Hospitals at Imphal and latter on to outside Manipur is sure sign of poor response of modern medical facilities in Manipur. This situation sometimes leads people to choose alternative practices most probably the help of tradition folk medicinal practitioners. This is why some of the traditional healing centers are heavily rushed by patients.

15 Some *Maiba* informants said they regularly consult people seeking supernatural powers of invulnerability. Persons such as political candidates and those in the armed forces sought help to get them elected into positions of power or to make them invulnerable to physical dangers presented in their field of work or travel. He also further said that He said he has many clients from the armed forces asking for safe passage.

2. Management of measles.

Sl.No.	Name of the Animal	English Name	Part Used	Method of Preparation of Formulation	Mode of Adminis-tration	Avoidance
1.	Sa-om	Bear	Gall-bladder	Dried and crush into powder and mixed with molasses	Internal	Soar fruits and oil, and taking bath

3. Management of ring worm

Sl.No.	Name of the Plant	English Name	Part Used	Method of Preparation of Formulation	Mode of Adminis-tration	Avoidance
1.	Ashi-heibong	Hairy fig	Leave	Pluck the leaf and apply on the infected part	Rub the hairy leaf on the infected part 2/3 minutes for five days	Do not use soap on the infected part during the treatment

Sources: *Report on Local Health Traditions of Manipur*, Manipur State Biodiversity Board Government of Manipur Imphal, Manipur, 2011.

Appeasing the Ancestral God or to Destroying the Power in Others

Many Meiteis believe that if the ancestors withdraw their protection and gift of good fortune, the descendant is left vulnerable to all sorts of misfortunes and diseases. The wrath of the ancestors is usually evoked by discord in the home, the violation of customs and traditions or non-observance of certain taboos. They performed an important ritual usually to three important household deities namely, *Sanamahi, leimarel* and *Phunga-lairu* to reduce patients' anxiety and relieve the feelings of guilt. Completion of the rituals often has a calming effect on the patient. A large part of the Meitei traditional healer's practice is also devoted to *counseling* individuals. Treatment is comprehensive and has curative, protective and preventive elements. It can either be natural or ritual, or both, depending on the cause of the disease. It includes among others, ritual sacrifice to appease the ancestors, ritual and magical strengthening of people and possessions, steaming, purification. The rituals which are common for such kind of disturbances occurring in a family and individual are '*Apokpa Khurumba*' (prayer to the ancestor deity) '*Sagei Lai Khurumba* (prayer to the lineage deity), '*Lam lai Khurumba*' (prayer to the deity of the locality) in an elaborate way.

Importance of Folk Medicinal Practitioners

Traditional healers' availability and accessibility, their familiarity with culture bound syndromes and traditions, their relationships with patients and their families place them in a position to serve as an alternative to mainstream health providers. Visits to such healers are not only for health problems but also for other problems. The belief that illness arises from supernatural causes and indicates the displeasure of ancestral gods or evil spirits or is the effect of black magic always linked the disease and the problems to the gods and need to appease them reinforces the existence and retention of the system of *Maiba* and *Maibi*.

As per the information given by some *Maibas* there are five main areas in which people believe *Maiba* can help: in love matters; business matters; strange illnesses believed to be caused by sorcery; political relationship[16], social relationships such as disputes and problems with neighbours, friends or family; and desire for physical strength and invulnerability.

Illness is another reason with most informants saying that if they became ill they would first go to a doctor to be healed before consulting a *Maiba*. If the doctor's treatment proved unsuccessful or another illness emerged in place of the original illness, then the patient would seek the help of a *Maiba*. *Maiba* still have the power to combat sorcery and many informants testified to this. Some informants claimed that doctors sometimes admit they cannot help because they also believe sorcery to be the cause.

Conclusion

The Meitei traditional health care system is well defined and regulated by their culture and belief in nature. Despite the availability of modern health care, the older generation still has inclination towards herbal medicine and indigenous methods of treatment. They found this method of treatment quite convenient and affordable especially in rural area. And even in urban areas it is an alternative health practices for some special cases. Many people also classified the health problem as domain of doctor (doctor-gi-laina) and domain of folk medicinal practitioners (Maiba-gi-laina) by seeing the symptoms according to which they prefer either one.

Regarding the magico –religious therapeutic practices as well as the new emerging cult and heavy rush in some of such centers also clearly shows people's option for such megico-reigious practice as an alternative when the modern medicine fail to care. Therefore, the utility of such practices may be scientifically relooked from the psychological perspectives as another option for few specific cases. Moreover, the practice is not counterproductive to the modern health care and because of its accessibility and affordability the practice can still be retained.

The growing disinterest in the use of the folk medicinal plants and its significance among the younger generation of the Meiteis may lead to the disappearance of this practice and the host of knowledge incorporated in the environment and the belief system. The local knowledge on various medicinal plants, folk medicinal preparations are useful resources, which may be scientifically evaluated and disseminated for efficacious drug development and improved health status. The protection and

---

16 Some *Maiba* informants said they regularly consult people seeking supernatural powers of invulnerability. Persons such as political candidates and those in the armed forces sought help to get them elected into positions of power or to make them invulnerable to physical dangers presented in their field of work or travel. He also further said that He said he has many clients from the armed forces asking for safe passage.

conservation of bio-diversity are not only desirable but are also necessary for the development health. As the practice of folk medicine of the Meitei has a component of environmental and ecological concern, it could be encouraged as positive aspect ecological and biodiversity conservation.

In this juncture, consideration on traditional medical systems that survived in this area which has proved to have sustainability can be upgraded through education, licensing and incorporation in to state health planning as a complementary component. The state sponsored medical system does not look at 'indigenous medicine' as whole and fail to see the socio-cultural basis of its usage. However, this knowledge could be profitably used as an integrated Health care.

References

_______________ "The sacred Grove of Umanglai: A traditional way of biodiversity conservation and environment management in Manipur. *Seeds of Hope*, Lokayan, New Delhi, 2004.

Ashalata Devi Khumbongmayum, Khan, M.K. and Tripathy, R.S., 2005. Ethnomedicinal plants in the sacred groves of Manipur, India. *Journal of Traditional Knowledge*, 4(1).

Brara, N. Vijaylakshmi, 1998. *Politics, Society and Cosmology in India's North-East.* Oxford University Press, New Delhi.

Brown, R., 1975 (Reprint). *Statistical Account of the Native State of Manipur and Hill Territory Under Its Rule*, (Calcutta, 1873). K.M. Mittan Sanskaran Prakashak, Delhi.

Durkheim, E., 1976. *The Elementary Forms of the Religious Life*, George Allen and Unwin, London.

Govt. of Manipur, 2011. *Report on Local Health Traditions of Manipur*, Manipur State Biodiversity Board Government of Manipur Imphal, Manipur.

Hodson, T.C., 1986 (First Published 1908). *The Meitheis*. B.R. Publishing Corporation, Delhi.

Malinowski, B.K., 1978 (First published in 1925). *Magic, Science and Religion and Other Essay*, Souvenir Press, London.

McCulloch, W., 1980 *An Account of the Valley of Manipur*, (Calcutta 1859), Gain Publication, Delhi (Reprinted).

Nettleton, 1995. *The Sociology of Health and Illness*. Polity Press.

Parratt, Saroj Nalini, 1980. *The Religion of Manipur: Belief, Rituals and Historical Development*. Firma KLM Private Limited, Calcutta.

Shakespear, J., 1911. A note on the rain stopping in Manipur. *Folklore*, Vol. 22, London.

Singh, H.V., 2011. The Maiba and the Maibee of Manipur. *The Oriental Anthropology*, Vol, 11.1.

Singh, Kh. Bijoykumar, 2005. Sanamahi movement among the Meiteis of Manipur: A sociological study of a socio-religious movement. Unpublished *Ph.D. Thesis*, JNU.

Singh, L. Bhagyachandra, 1991. *A Critical Study of the Religious Philosophy of the Meiteis Before the Advent of Vaishnavism in Manipur*. The Paradise Publishing House, Imphal.

Singh, L. Romeo, 2011. Importance of plants in the medication of ailments: An ethno-medicinal perspective from Karang Island, Manipur. *Souvenir*, Y.K. College, Manipur.

Singh, L. Warjeet, 2011. Traditional medicinal plants of Manipur as anti-diabetics. *Journal of Medicinal Plants Research*, 5(5): 677–687.

Sircar, Manjusri Chaki, 1984. *Feminism in a Traditional Society*. Shakti Books, New Delhi.

The Sangai Express (Manipur English daily).

Turner, B.S., 1992. *Regulating Bodies: Essays in Medical Sociology*. Routledge, London.

WHO, 2001. *Legal status of Traditional Medicine and Complementary/Alternative Medicine– A world wide review.*

www. Cdc.gov / tb / publications / guidestoolkits / … / Burma / … / chapter 2.pdf.

www.ejud.uib.no / ejhdv 16–n1 / v 16 no1–page 19 PDF

Environment, Biodiversity and Traditional System, 2014 *Pages* 175–181
ISBN 978-81-7622-307-2
Editors: D.R. Khanna, G.S. Solanki and S.K. Pathak
Published by: BIOTECH BOOKS, NEW DELHI

Chapter 16

A Preliminary Study on the Culm Growth of *Melocanna baccifera* in Mizoram

Zonuntluangi Bawitlung and Ramachandra Laha*

Department of Botany, Mizoram University, Mizoram – 796 004

ABSTRACT

Mizoram is endowed with rich bamboo forests, which cover about 49.10 per cent of the total geographical area. Bamboo area of Mizoram is the highest in the country. Bamboo is used for a variety of purposes and has gained renewed importance in the present day context of shortage of wood due to its fast growing character. From the study area, five quadrats of *Melocanna baccifera* were selected and five newly culms from each quadrat were taken for growth measurements. The culms complete their growth within three to four months after the emergence of rhizomes from the ground but their diameter and height do not increase after the growth is over. Monsoon dependency of the plantation was prominent since the maximum recruitment of culms seems to be monsoon dependent rather than intermittent showers. With increase population pressure, natural stands of bamboo are being indiscriminately cut for fuel wood and furniture and for converting cultivable lands. As such bamboo are ideally suited for such efforts in view of availability and fast rate of growth and multifarious uses.

Keywords: Bamboo, Culms, Quadrat, Growth, Height, Plantation.

Introduction

In tropical countries like India, bamboo form the largest single renewable raw material resource for the pulp industries. India is rich in the distribution of bamboo,

* *Corresponding Author:* E-mail: zntboitlung@gmail.com

both by area and number of species, next only to China. About 10.03 million ha. comprising 12.8 per cent of total forest area in India is occupied by bamboos (Tewari, 1992). The North eastern region of the country accounts for 50 per cent of the bamboo production. Out of the 150 bamboo species in India, more than 58 species occur in the North-east India, of which 22 species have been recorded in Mizoram (Lalramnghinglova and Jha, 1997; Jha and Lalnunmawia, 2001). Bamboo area of Mizoram is the highest in the country. In Mizoram non-clump forming bamboo species (*Melocanna baccifera*) contributes almost 98 per cent of the total bamboo crop while a meager 2 per cent (*Dendrocalamus hamiltonii etc.*) is shared by clump forming species (Anon, 2002). Bamboos are woody monocots and represents one out of six or seven main natural groups of the grass family Poaceae. Culms of all the bamboo species complete their growth within 3-4 months after the emergence of rhizomes from the ground, but their diameter and height do not increase after the growth is over (Mc Clure, 1967; Banik, 1980; Ueda, 1981; Shanmughavel, 1995). In *Bambusa bambos*, culms may reach their maximum height in about 4-5 months analogous to *Gigantochloa levis* and *G.ligulata* (Othman, 1992).Under cultivation, the average daily height growth may reach approximately 30.0 cm at sixth year which may vary considerably between species depending on the soil conditions and mineral nutrition (Kondas, 1973). The maximum growth per day is found to be in the vigorous growing period. According to records, the growth per day was 91.3 cm by *Bambusa bambos* observed at Kew Garden in England (1855) and was 88.0 cm by *Phylostachys edulis* which Shibata (1898) observed at Koishikawa Botanical garden in Tokyo. Ueda (1981) recorded the growth of 119 cm and 121 cm per day in *Phylostachys edulis* and *P.reticulata*. New culms generally appear during the rainy season. Ueda (1960) suggested that 1-3 year old bamboo culms must always be left in reserve while 4 year old culms increase shoot production, reduce shoot mortality and deformed culms (Ordinario, 1978). Studies on the growth of *Phylostachys edulis* (Hsiung *et al.*, 1980) have revealed the sequence of internodal elongation and the formation of the new nodes.

With increased population pressure, natural stands of bamboo are being indiscriminately cut for fuel wood and furniture and for converting cultivable lands and extensive forest fires. Sustained availability can be ensured only by plantations. Some bamboo species are planted in villages by farmers only for own use. The uncontrolled exploitation of these resources, however, has resulted in reduced yields and deterioration in quality. Bamboo is used for a variety of purposes and has gained renewed importance in the present day context of shortage of wood due to its fast growing character. Growth and yield of bamboo depends upon source of planting materials, method of planting and spacing (Gupta and Sood, 1978). Organic matter is important for the growth of bamboo. It helps to improve the physical and chemical properties of the soil by stabilizing the moisture content and nutrient retention. These conditions promote the growth of rhizome and bamboo culms (Sun and Ye, 1991, Abd. Razak, 1992)

Study Site

The study site is confined to Khawzawl, Champhai District, which is situated between 23° 32′04.05" N latitude and 93° 10′59.97" E longitude with an elevation of

3962 ft. The area enjoys a sub-tropical humid climate with a temperature ranging between 11° C-32° C. The area has a loamy soil with a pH 4.4-5.6.

Materials and Methods

From the study area, five quadrats (10mx10m) of *M. baccifera* were selected from the fifth year plantation and identified with paint marking. New culms sprouting from the selected clumps were taken for growth measurements. In each quadrat, five newly sprouted culms were marked with paint totaling 25 observations. On alternate days, the height of each culm was measured and after it had reached 1.37m height, the diameter at breast height (DBH) was recorded. All observations were made for the basal area girth, DBH and growth in height. Every month, the number of newly sprouted culms and their survival were recorded.

Results

Table 16.1: Monthly recruitment of culms/quadrat.

Month	1		2		3		4		5	
	C	S	C	S	C	S	C	S	C	S
Jan	–		–		–		–		–	
Feb	–		–		–		–		–	
Mar	–		1		–		–		–	
Apr	–		–		–		–		1	
May	–		–		–		–		–	
Jun	5		4		2		4		4	
July	6		4		7		3		6	
Aug	8	26	7	21	6	22	7	19	8	22
Sept	5		5		3		4		5	
Oct	3		2		1		–		1	
Nov	1		–		2		1		2	
Dec	–		–		–		–		–	
Total	28		23		21		19		27	

*C: New culms; S: Culms survive.

Table 16.2: Growth measurement of culms of *Melocanna baccifera*.

Days	Melocanna baccifera		
	Height(m)	DBH (m)	BA (m)
1	0.09		
2	0.13		
4	0.19		
6	0.30		

Contd...

Table 16.2–*Contd...*

Days	*Melocanna baccifera*		
	Height(m)	*DBH (m)*	*BA (m)*
8	0.50		
10	0.71		
12	0.90		
14	1.12		
16	1.37	0.095	0.127
18	1.60		
20	1.85		
22	2.90		
24	2.33		
26	2.36		
28	2.39		
30	2.64		
32	2.88		
34	3.12		
36	3.34		
38	3.58		
40	3.84		
42	4.50		
44	4.26		
46	4.57	0.140	0.175
48	4.81		
50	5.40		
62	6.44		
72	7.52		
82	8.60		
92	9.63		
102	10.65		
122	12.86		
132	13.50		
134	13.50		
136	13.60		
138	13.80		
140	15.00	0.165	0.187
142	15.20		
144	15.50		
146	15.50		
148	15.50		
150	15.50		

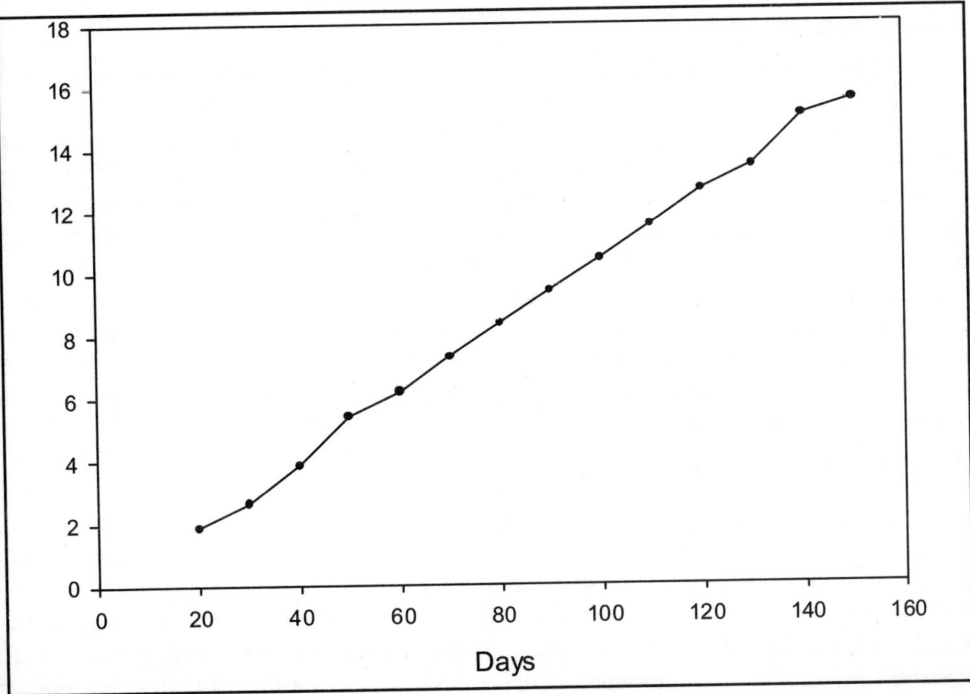

Figure 16.1: Culm growth pattern of *Melocanna baccifera* (Each point is the average of five quadrats).

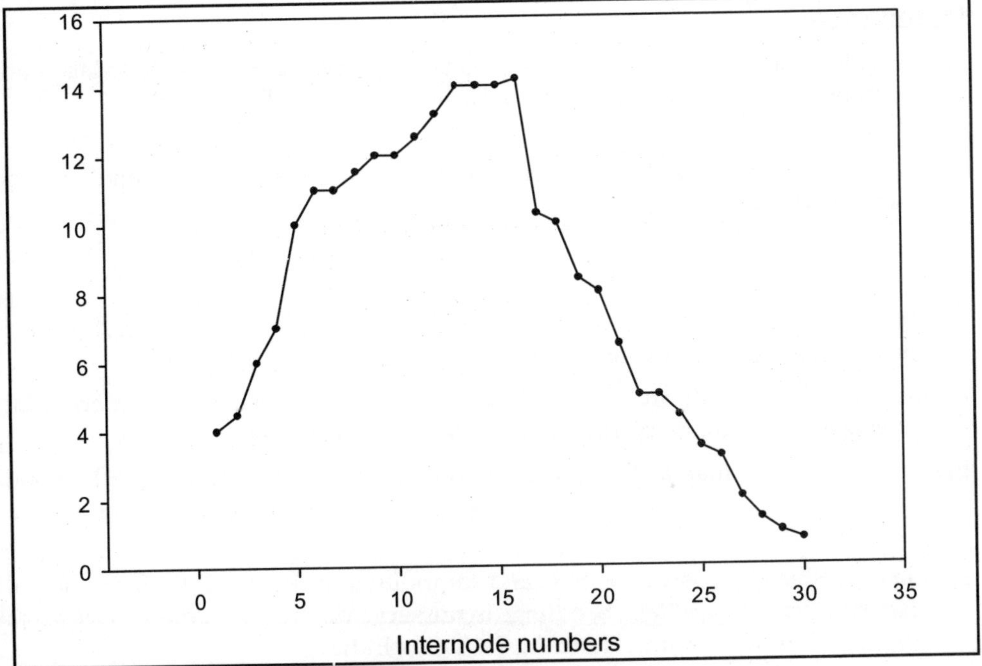

Figure 16.2: Variation in internode length within the culm.

Discussions

The appearance of the culm was very slender and the growth of the sprout was initially slow. The average increase in height of culm on alternate days was approximately 23cm. The culm is strictly erect with a pendulous tip. In the initial stage the culm appears as a cone wrapped up in a series of culm sheaths of decreasing sizes towards the tip of the cone. The height growth of the culm is caused by successive elongation of the internodes. The daily increase in length of the internodes during this period ranges from 1.5 cm – 7.5 cm. Although the culms grow to their full height in 3-4 months, they remain immature till about next 2-3 years. On maturity the culms gradually lose their culm sheaths. They do not bear any branches for a considerable height during growing stage. An average of 23 culms were produced on one quadrat in a year. Maximum recruitment of culms took place between June-September which was the monsoon season.

Conclusion

Bamboo is the most versatile gift of nature. It can be easily grown in forest and farms. Bamboo are as highly useful to pulp industries as they are to cottage industries, agricultural and housing needs, soil and water conservation, rural employment etc. Considering the economic importance and problem of continuity of bamboo supply there is a need to establish bamboo plantation both in the forest and on private land. Bamboo are ideally suited for such efforts in view of availability of provenances and species suited to diverse climatic and edaphic conditions, fast rate of growth, repeated quick harvests and multifarious uses both to the rural poor and the industries.

References

Abd. Razak, O. and Hashim, M.N., 1992. Vegetative propagation of selected Malaysian bamboos. Paper presented at *National Bamboo Seminar*. FRIM, Kepong, 2–4 Nov., 1992.

Anon., 2002. *The Bamboo Policy of Mizoram*. Department of Environment and Forests, Govt. of Mizoram.

Banik, R.L., 1988. Investigation on the Culm Production and clump expansion behavior of five bamboo species of Bangladesh. *Ind. For.*, 114(9): 576–583.

Gupta, B.N. and Sood, O.P., 1978. Storage of *Dendrocalamus strictus* Nees, Seed for maintenance of viability and vigour. *Indian Forester* 104: 688–596.

Hsiung, W.Y., Ding, Z. and Li, Y., 1980. Intercalary meristem and intermodal elongation of bamboo plants. *Scient. Sily. Sinic.*, 16: 81–89.

Jha, L.K. and Lalnunmawia, F., 2001. Study on growth and management of bamboo species of Mizoram, Annual Tech. Report, Ministry of Environment and Forest, Govt. of India. Pp1–25.

Kondas, S., Sree Rangaswamy, S.R. and Jambulingam, R., 1973. Performance of *Bambusa arundinacea* Retz. Seedlings in nursery. *Madras Agricultural Journal*, 60: (9/12): 1719–1726. Tamil Nadu Agric. Univ., India.

Lalramnghinglova, J.H. and Jha, L.K., 1997. *Natural Resources Management Vol. 1: Mizoram.* APH Publishing Corp., New Delhi, pp. 257–269.

McClure, F.A., 1966. *The Bamboos: A Fresh Perspective.* Hary. Univ.Press, Cambridge, Mass, 347 pp.

Ordinario, F.F., 1978. Sustained yield treatment of *Bambusa blumeana.* Schult. Progress Report. PCARRD Project 283. PCARR, Los Baflos, Laguna, Philippines.

Othman, R., 1992. Production and culm growth of two commercial bamboo species in Peninsular Malaysia, *BIC–India Bulletin,* 2(1): 5–7.

Shanmughavel, P., 1995. Studies on the growth of transplanted seedlings of *Bambusa bambos* at Chickally, Sathayamangalam Forest Division, Tamil Nadu, *Van Vigyan* 35(1): 14–16.

Tiwari, D.N., 1992. *A Monograph on Bamboo.* International Book Distributors, Dehradun, India.

Sun, T.R. and Ye, S.C., 1991. Relationship between soil, available nutrients and rhizome distribution on *Phyllostachys heterocloda.* In: *Selected Paper on Recent Bamboo Research in China,* (Eds.) A.N. Rao, X.P. Zhang and S.L Zhu. Bamboo Information Centre, China, pp. 116–120.

Ueda, K., 1960. Studies on the physiology of bamboo. *Bull. Kyoyo Univ. Forests,* 30: 167.

Ueda, K., 1981. Bamboo industry in Japan: Present and future. *17th IFURO World Congress,* 5: 244–256.

Environment, Biodiversity and Traditional System, 2014 *Pages* 183–198
ISBN 978-81-7622-307-2
Editors: D.R. Khanna, G.S. Solanki and S.K. Pathak
Published by: BIOTECH BOOKS, NEW DELHI

Chapter 17

Conservation of Wildlife of India with Special Reference to India's Wild Cats

D.K. Belsare

Advisor,
ENVIS, Disaster Management Institute, Bhopal
E-mail: d.belsare@yahoo.com

Introduction

India's wildlife is distributed in Palaearctic, Ethiopian and Oriental bio-geographical realms and constitutes 26 endemic species of mammals which are critically endangered as per IUCN list. Three species *i.e.* Indian Wild Ass, Leopard and Red Fox are threatened with extinction. Many of its wild cats, which balance India's ecosystems, are on the verge of extinction.

The first cats evolved in India some 70 million years ago and are changed little over the last 30 million years. This period gave them plenty of time to adapt to varied habitats.

They have big feet and a long tail for adaptation of arboreal life, partially webbed feet to make aquatic work more efficient and long legs for running down swift prey in open habitats etc. The wide range of skin pattern plays a special role in their survival. Dark fur and spots serve as camouflage in the forest. Stripes are good for hiding in the grass lands. Light fur blends with snowy background. More than half of the world's cat species live in India and many of them are found in tiger habitat.

Present report is focused on conservation of 14 species of wild cats of India.

Biology of Indian Cats

There are twenty species of cats, each of them is distinct by its characteristics. They are grouped under six genera; *Panthera, Felis, Neofelis, Pardofelis, Prionailurus and Acionyx*. The species are *Panthera tigris tigris* (Indian Tiger), *Panthera leo persica* (Asiatic Lion), *P. pardus* (Leopard), *P. uncial* (Snow Leopard), *Neofelis nebulosa* (Clauded Leopard), *Felis bengalensis* (Leopard Cat), *Prionailurus viverrina* (Fishing Cat), *F. silvestris* (Desert Cat), *Pardofelis mammorata* (Marbled Cat), *Prionailurus bengalensis* (Leopard Cat), *F. manul* (Pallas Cat), *Catopuma temminckii temminckii* (Golden Cat), *F. rubiginosa* (Rusty Cat), *F. caracal* (Caracal), *F. lynx* (Lynx). *Acinonyx jubatus* (Asiatic Cheetah) is now extinct from this continent. Of all the Indian cats, the Tiger received more attention because of being esteemed adversary of hunters, an important advertising symbol, a central theme in the mythology and folklore of the region and subject of one of the largest international conservation efforts ever undertaken. It is majestic, cunning, brave and big cat which inhabits almost all types of ecosystems of India. The Madhya Pradesh has the largest number of tiger population (300 tigers) which is 19 per cent of India's and 17 per cent of the world's tiger population. This State is called 'The Tiger State'.

1. The Asiatic Lion (*Panthera leo persica*) or Persian Lion

Asiatic Lion is a subspecies of the Lion which survives today only in the Gir Forest of Gujrath, India (Ravi and Johnsigh 1993). They have less swollen tympanic bullae, shorter postorbital constriction with divided infraorbital foramen. The colour ranges from reddish-brown to a highly mottled black to sandy cinnamon grey. In adult males, the maximum skull length is 330-340 mm, while that of females is 266-277 mm. They reach a weight of 160-190 kg in males and 110-120 kg in females. The scientific record for the longest male is of 292 cm, while the maximum height to the shoulders reported is of 107 cm. They are highly social animals living in units called prides. Their lion prides are smaller than those of African lions, with an average of only two females. The Asiatic males are less social and only associate with the pride when mating or on a large kill. The Gir Forest Nationnal Park of Gujrath has 411 lions in 1,412 km^2 sanctuary covered with scrub and open deciduous forest habitats (IUCN 2006).

2. Leopard (*Panthera pardus*)

The Leopard is the most perfect cat among the big cats because of its beautiful appearance and graceful movement. Its great adaptability has allowed it to occupy virtually all habitats of India ranging from the desert to the forest and from sea level to more than 3000 msl elevation of the country (Henschel *et al.*, 2008). Compared to other members of the Felidae family, the leopard has relatively short legs and a long body with a large skull. It is similar in appearance to the jagaur, but is smaller and more slightly built. Its fur is marked with similar to those of the jaguar, but the leopard's rosettes are smaller and more densely packed, and do not usually have central spots as the jaguar's do. Body is completely dark.

The species' success in the wild is in part due to its opportunistic hunting behaviour, its adaptability to habitats, its ability to run at speeds approaching 36 mph, its unequaled ability to climb trees even when carrying a heavy carcass, and its

Figure 17.1: Leopard (*Panthera pardus*).

notorious ability for stealth. The leopard consumes virtually any animal it can hunt down and catch.

Its body is comparatively long for a cat and its legs are short. Head and body length is between 125 and 165 cm (49 and 65 in) and the tail reaches 60 to 110 cm (24 to 43 in). Shoulder height is 45 to 80 cm (18 to 31 in). The muscles attached to the scapula are exceptionally strong, which enhances the leopard's ability to climb trees.

Leopards show a great diversity in size. Males are about 30 per cent larger than females, weighing 30 to 91 kg (66 to 200 lb) compared to 23 to 60 kg (51 to 130 lb) for females. Large males of up to 91 kg have been documented in Kruger National park in South Africa; however, males in the South Africa's coastal mountains average a much smaller 31 kg. This wide variation in size is thought to result from the quality and availability of prey found in each habitat.

They also show a great diversity in physical appearance, particularly because of the wide variations in colour coat and rosette patterns. The leopard's yellow coat tends to be more pale and cream colored in desert populations, more gray in colder climates, and of a darker golden hue in rainforest habitats. Overall, the fur under the belly tends to be lighter coloured and of a softer, downy type. Solid black spots in place of open rosettes are generally seen along the face.

3. Snow Leopard (*Panthera uncial*)

The Snow Leopard is confined to high elevations of the Himalaya, Palmir, Altai, Hindu kush mountains (Fox *et al.*, 1991). It purrs rather than roar. They show several adaptations for living in a cold mountainous environment. Their bodies are stocky, their fur is thick, and their ears are small and rounded, all of which help to minimize heat loss. Their feet are wide, which distributes their weight better for walking on snow, and they have fur on their undersides to increase their traction on steep and unstable surfaces, as well as to assist with minimizing heat loss. Snow leopards' tails are long and flexible, helping them to maintain their balance. The tails are also very thickly covered with fur which, apart from minimizing heat loss, They have a short muzzle and domed forehead, containing unusual large nasal cavities that help

Figure 17.2: Snow Leopard (*Panthera uncial*)

the animal breathe the thin cold air of their mountainous environment. In summer, they usually live above the tree line on mountainous meadows and in rocky regions at an altitude from 2,700 to 6,000 msl. In winter, it comes down into the forests to an altitude of around 1,200 to 2,000 msl They prefer broken terrain and can travel without difficulty in snow up to 85 centimetres deep, although they prefer to use existing trails made by other animals. It leads a largely solitary life, although mothers may rear cubs for extended periods of time in dens in the mountains. An individual snow leopard lives within a well defined home range but does not defend its territory aggressively, when encroached upon by other snow leopards. Home ranges vary greatly in size. In Nepal, where prey is abundant, a home range may be as small as 12 km^2 to 40 km^2 (Oli *et al.*, 1994) and up to five to ten animals are found here per 100 km^2, whereas in habitats with sparse prey, an area of 1,000 km^2 supports only five individuals (Mallon, 1984).

Like other cats, they use scent marks to indicate their territory and common travel routes. These are most commonly produced by scraping the ground with the hind feet before depositingurine or scat, but they also spray urine onto sheltered patches of rock. They are most active at dawn and dusk. They are known for being extremely secretive and well camouflaged.

They are opportunistic feeders, eating whatever meat they can find including carrion and domestic livestock (Oli *et al.*, 1994). They are capable of killing animals three times their size but will readily take much smaller prey such as hares and birds. Unusually among cats, snow leopards also eat a significant amount of vegetation, including grass and twigs. Their diet varies across its range and with the time of year, and is dependent on prey availability. In the Himalayas they prey mostly on Bharals(Himalayan blue sheep), but their main prey consists of Siberian Ibex andArgali Himalayan Tahir, deers, boars and various rodents and birds.

Snow leopards prefer to ambush prey from above, using broken terrain to conceal their approach., and can leap as far as 14 m. They actively pursue prey down steep mountainsides, using the momentum of their initial leap to chase animals for up to 300 m. They kill with a bite to the neck, and may drag the prey to a safe location before feeding. They consume all edible parts of the carcass, and can survive on a single bharal for up to two weeks before hunting again.

Snow leopards usually mate in late winter and have a gestation period of 90–100 days, so that the cubs are born between April and June. Oestrous typically lasts from five to eight days, and males tend not to seek out another partner after mating, probably because the short mating season does not allow sufficient time to reliably do so. Paired snow leopards mate in the usual felid posture, from twelve to thirty six times a day.

The mother gives birth in a rocky den lined with fur shed from her underside. Litter sizes vary from one to five cubs but two or three is more usual. The cubs are blind and helpless at birth, although already with a thick coat of fur, and they weigh from 320 to 567 g. The eyes open at around seven days, and the cubs are able to walk at five weeks and are fully weaned by ten weeks. The cubs leave the den at around two to four months of age, but remain with their mother until they become independent after 18-20 months. They disperse over considerable distance to seek out new hunting grounds. They become sexually mature at two or three years of age. Their normal life-span is of 15-18 years.

4. The Clouded Leopard (*Neofelis nebulosa*)

The Clauded Leopard is a rain forest cat and lives in eastern border of Nepal-India at 2,000 msl height (Choudhury, 1993). It is also called 'Mint Leopard', because of similarity in its coat with mint leaves. It has a tan or tawny coat, and is distinctively marked with large, irregularly-shaped, dark-edged ellipses which are said to be shaped like clouds (Choudhury 1996). It weighs between 15 and 23kg and has a shoulder height of 25 to 40 cm. It has the longest canines (5 cm), which are used for tearing primary prey like arboreal and terrestrial animals, particularly gibbons, macaques and civets supplemented by other small mammals, deers, birds, porcupines etc. It is an excellent climber because of its short, flexible legs, large paws and sharp claws. The tail is as long as its body and is used for balancing the body. It may hang upside-down and descent tree trunks with head-first. It is a solitary creature, but remains with females during courtship. It reaches sexual maturity at two years of age. Femalegives birth to a litter of 2 to 4 cubs after a gestation period of about 85 to 93 days Initially, the young are blind and helpless. Unlike adults, the kittens' spots are "solid"–completely dark rather than dark rings. The young can see after 10 days of birth, are active within 5 weeks, and probably become independent at about 10 months of age.Its average life-span is 11 years.

5. Fishing Cat (*Prionailurus viverrinus*)

The Fishing Cat ((*Prionailurus viverrinus*) is a medium sized grayish, civet like cat with low hind quarters and a long muzzle. It occupies Himalayas and Malabar Coast of India. It is always found in the vicinity of water body and hunts fish, crabs, water birds and small mammals. It fishes by deeping its slightly webbed paw into water and throwing fish on the bank of water body (Haque and Vijayan, 1993).Its fur has an olive-grey colour with dark spots arranged stripe-like running along the length of the body. The face has a distinctly flat-nosed appearance. The size varies between locations. The Indian specimens grow to 80 cm plus 30 cm tail, weigh up to 11.7 kg They are stocky of build with medium short legs, and a short muscular tail of one half to one third of the length of the rest of the animal. The Fishing Cat lives along rivers, streams andmangrove swamps. It also hunts other aquatic animals such as frogs orcrayfish, and terrestrial animals such as rodents and birds.The inter-digital webs on its paws help the cat gain better traction in muddy environments and water, like other mammals living in semi-aquatic environments. Gestation period is 65-70 days (Bhattacharya 1992). The young fishing cats become independent at the age of 10 months. Their life span is of 12 years,

The Fishing Cat is endangered due to its dependence on wetlands, which are being converted for agriculture, and also due to human over-exploitation of local fish stocks (Belsare, 1994).

6. Indian Desert Cat (*Felis silvestris ornata*)

The Asian Desert Cat lives further east, from Pakistan and north-west India in a band through central Asia as far as Mongolia. The main threats to the survival of this species are hybridization with domestic cats, disease transmission, and competition with feral domestic cats. Other significant threats are ongoing habitat loss, fragmentation and degradation in some areas (Nowell, 2008). It is like the European Cat and inhabits drier parts of Central India. It feeds on rodents, birds, beetles, grasshoppers, small mammals and snakes.

7. Marbled Cat (*Pardofelis mammorata*)

The marbled cat is primarily an animal of moist tropical forest, but it has been reported from mixed deciduous-evergreen forest areas, mountainous evergreen forest, secondary forest, clearings in northern India and Nepal (Suinsquit and Sunsquit 2002). It is about the size of a domestic cat. Its head and body length is 45–62 cm and it weighs 2–5 kg. Its tail is extremely long and bushy. The background color of its fur varies from dark gray-brown through yellowish gray to red-brown. The flanks and back are strikingly marked with large, irregular, dark-edged blotches. The legs and underparts are marked with black dots, and the tail is marked with black spots and rings. There are spots on the forehead and crown, which merge into narrow longitudinal stripes on the neck and irregular stripes on the back.

It has been found at 3000 msl height. Birds and/or rodents have been reported as likely forming a major part of its diet. Although previously the marbled cat was considered to be primarily nocturnal, but recent studies have shown that it is arboreal and terrestrial in habits (Grassman *et al.*, 2005).

Because of its dependence on forest habitat, the major threat to the marbled cat is habitat destruction caused by felling of trees and the traditional, shifting, "jhum" method of local cultivation. The marbled cat is also thought to be intolerant of human disturbance, abandoning a forest that is even moderately disturbed. Poaching for skins, bones and meat may also be a threat (Nowell and Jackson 1996).

8. The Pallas Cat (*Felis manul*)

The Pallas Cat is a small central Asian species having wide head with large forward facing eyes and low rounded ears. Its short stout legs support a chunky body which is covered with luxuriant fur in order to protect from cold. It inhabits Ladakh region and hunts pikas along stream beds. Its small round ears help to observe prey from behind rocks. It is about the size of a house cat at 60 cm long, not including its 25 cm tail, and an average weight of 3.6 kg. Its fur is ochre in colour with vertical bars, which are sometimes not visible due to the thick fur Its eyes have round pupils rather than the slit-like pupils of other cats; its legs are proportionately shorter than those of other cats; and its fur is long and thick. The combination of its stocky posture and dense fur makes it appear stout and plushy. Its coat changes with the seasons. The winter coat is greyer and less patterned than the summer coat. The ears are set low.

Pallas's Cat inhabits the Asian steppes up to heights of 4000 msl. They are thought to be crepescular hunters and feed on small rodents, pikas and birds. Like other species of exotic felines, Pallas's Cat has been hunted for its fur.

9. The Asian Golden Cat (*Catopuma temminckii temminckii*)

It is also called Temminck's Golden Cat and "Fire Cat". It is a medium-sized wildcat of SouthernAsia'. It inhabits the forests of southeast Asia ranging from tropical rainforests to temperate broadleaf and mixed forests up to an altitude of 2,517 msl (Bashir *et al.*, 2011). Little is known about this enigmatic felid in the wild, due to the lack of research programs relating to this species.It has a cat-like appearance, but is more similar to a small leopard in size and behaviour. Its height is 56 cm at the shoulder, and the length from nose to tail is 75 to 105 cm.The weight ranges from 9 to 16 kg which is about two or three times the size of domesticate Cat. The long and flexible tail is about 50 cm long. Its life span is seventeen years.

Its pelage colour is uniform, but highly polymorphic ranging from red to golden brown, dark brown to pale cinnamon, gray to black. Transitional forms among the different colorations marked with spots are also found. Its vocalizations include hissing, spitting, meowing, purring, growling, and gurgling. Other methods of communication observed include scent marking, urine spraying, raking trees and logs with claws, and rubbing of the head against various objects.

Female cats sexually mature between 18 and 24 months, while males mature at 24 months. Females come into estrus every 4 months, at which times they leave markings and seek contact with the male by adopting receptive postures. During intercourse the male will seize the skin of the neck of the female with its teeth. After a gestation period of 78 to 80 days, a litter of 1 to 3 kittens are born in a sheltered place.

The Asian Golden Cat is a territorial and solitary species. Previous observations suggested that it is primarily nocternal, but a field study on two radio-collared specimens revealed arrhythmic activity patterns dominated by crepiscular and diurnal activity peaks, with much less activity late at night. The male's territory is 47.7 km^2 whereas the female's territory is 32.6 km^2.They travel between 9 to55 m a day and are more active in July.

The Asian Golden Cat is threatened species due to destruction of its habitat and the decline of prey (Misra *et al.*, 2006). This felid is poached for its fur and bones; latter are used in Chinese medicine as a substitute for tiger bone. Since it attacks livestock it comes in conflict with farmers, who poison or shoot it on sight ranges from Tibet to India and is becoming rare over most of its range because of habitat destruction (Choudhury, 2007). It lives in forests and hunts small game birds.

10. The Rusty-spotted Cat (*Prionailurus rubiginosus*)

The rusty-spotted cat is one of the smallest cat species, with a head and body length of up to 48 cm and a weight of up to 1.6 kg. The rusty-spotted cat's coat is a short, soft fawn-grey with a rufous tinge, patterned with transverse lines of small rusty-brown spots which form solid stripes along the back of the head. The belly, chest, and throat are white, marked with large dark spots and bars. The tail, which averages about half of the head-body length, is faintly marked with dark rings.

The rusty-spotted cat is found in moist and dry deciduous forests, tropical thorn forest, scrub forest, grasslands, arid shrublands, rocky areas, and hill slopes of India (Dubey, 1999). It feeds on small birds, rodents, frogs, insects and, possibly, small lizards, as well as domestic fowl. In the wild, rusty-spotted cats are thought to be mainly, but not completely nocturnal and to spend the daytime resting in a hollow log or dense cover (Suinquist and Sunquist, 2002). One female rusty-spotted cat reached sexual maturity at 68 weeks of age. Gestation period is 66–70 days.

11. The Leopard Cat (*Prionailurus bengalensis*)

It is a miniature version of the Leopard and is found in western India. It is also called 'Money cat' because of its spots resemble the Chinese coins. It is one of the few cats that form permanent pair. Its prey species are large birds, hares and Mouse Deer. It inhabits tropical forest, scrubland, pine forest, second-growth woodland, semi-desert, and agricultural regions, especially near water sources; and may also be found at heights up to 3,000 m. Leopard Cat bears black markings that may be spotted, rosette, or even forming dotted streaks, depending on the subspecies. There are two dark stripes running from the eyes to the ears, and smaller white streaks running from the eyes to the nose. The backs of the ears are black with a central white spot.

It feeds on feed on variety of small prey, including mammals, lizards, amphibians, birds, and insects. In most parts of their range, small rodents such as rats and mice form the major part of their diet. It has no fixed breeding period. Gestation period is of 60–70 days, two to four kittens are born in a den. They usually pair for life and raise their kittens together for about 7 to 10 months (Sanquist and Sanquist 2002)

12. The Eurasian Lynx (*Lynx lynx*)

The Eurasian Lynx is of the size of the Clouded Leopard weighing upto 25 kg and has short tail and plain brown coat. Its long ear tufts are distinctive. The Eurasian lynx has a very broad distribution from western Europe through the boreal forests of Russia, and down into central Asia and the Tibetan plateau (Nowell and Jackson 1996, Sunquist and Sunquist 2002). Populations in the southeast of its range (Europe and southwest Asia) are generally small and widely separated, whereas the bulk of its historic range from Scandinavia through Russia and Central Asia is largely intact. preys on hares, small deers, large birds sheep and goats. Home range size varies widely (Sunquist and Sunquist 2002), but averaged 248 km² for males (n = 5) and 133 km² for females (n = 5) in a radio telemetry study in Poland's Bialowieza forest (Schmidt *et al.*, 1997). Densities are typically 1-3 adults per 100 km², although higher densities of up to 5/100 km² have been reported from eastern Europe and parts of Russia (Sunquist and Sunquist 2002).

13. The Caracal (*Caracal caracal*)

The Caracal has long tail and slim body and is found in Central India. Recently it was located in district Panna of Madhya Pradesh. It is an ecological counterpart of the Cheetah and is specialized in hunting birds by jumping. Caracals feed on a

Figure 17.3: The Caracal (*Caracal caracal*).
Photo by Dr Razimuddin and Sathyaprakash, Hazaribagh.

variety of rodents, lizards and ground birds to small antelope. Caracals regularly kill even larger animals of moderate size such as Blackbuck.

Caracals are polyestrus, meaning they are not seasonal and can reproduce year round. They reach sexual maturity in less than 2 years. First litters have been recorded at 1 ½ years (18 months) of age. Their cyclicity lasts from 3-5 days and seems to peak in Feb., March and August. Gestation is 69-79 days. Litter sizes vary from 1-6 with an average of 2-3.

14. Indian Tiger (*Panthera tigris tigris*)

The Indian Tiger, also called Bengal Tiger came to India some 1,00,000 years ago from Siberia (Luo *et al.*, 2004).

They have big feet and long tail for arboreal life, webbed feet for swimming in wetland habitat, long legs and stout muscles for running down swift prey in open field. They have the highest leaping power (5 m high).The present population of tigers is 1703 (Recent Report by Ministry of Environment 2011). Average weight: Female 100 kg. Male 200 to 250 kg; Length: Female 75 to 100 cm Male: 155 to 225 cm, iii) Range: Female 10 to 39 km^2 Male: 30 to 105 km^2 iv) Distribution: India, Bangaladesh, Nepal, Bhutan and Myanmar.

A population of wild tigers comprises of animals belonging to different sex and age classes (Belsare 2008, Belsare 2011). The breeding females (also called resident or territorial) form the keystone of the tiger social structure. They hold on to fixed ranges of habitat containing high prey densities, monopolising the breeding within them. They mate with the large breeding males whose range may cover two to three female ranges. There are transient tigers, both males and females, who do not breed. After having dispersed away from their mother at the age of one and a half years to two years, the transient tigers wander back and forth through their natal range and

Figure 17.4: Tiger (*Panthera tigris tigris*).
Photo by Dr. D.K. Belsare, Van Vihar Bhopal.

surrounding areas, trying to establish a home range for themselves. As transients grow older and bigger, they compete aggressively with the breeders, and, sometimes, manage to kill breeders and appropriate their ranges. However, many transient tigers also perish in this process. Tiger cubs are born (two to four per litter) in a helpless condition after a short 105 day gestation. They depend on mother's milk for the first eight weeks or so, after which the female takes them to the kills. They gradually acquire hunting skills over the next one and a half years, often learning from their mother's ways, so as to feed for themselves after dispersal.

Although tiger walks on four legs, its walking style gives the impression of bipedal walk, because it uses two legs for one step (Belsare, 2011). If body weight is rested on fore and hind right limbs, the fore and hind limbs are placed forward to share the body weight. The right limbs are then placed forward after left limbs are placed on the ground. Thus these limbs share body weight by alternate steps. The bipedal walk is clearly seen when the tiger walks on soil in a straight direction.

The tiger's foot is much bigger than its impressions on the ground, because three-fourth of its upper part is covered with hair and one-fourth of its part is pad and fingers. The impressions on the ground, however, are those of pad and fingers. Although, the forepaw has five fingers and the hind paw has four fingers, the impressions on the ground are four fingers only. The fingers are egg-shaped and pad has three ridges. If the second finger is forward from the right side, it is the left paw and if it is from the left side, it is the right paw. Thus right and left legs are recognized from the forward placement of the second finger. If measurements of paw are carefully done, one can easily identify the individual tiger. It requires expert's handling of measurement.

The paw of male is bigger in size than that of the female. The paw of large sized tiger is 15 x 12 cm with an area of pad of 7.5 x 9 cm, whereas the paw of large tigress measures 13 x 11 cm with pad area of 6 x 8 cm. The paw of one-year-old tiger is slightly bigger than that of large sized tigress. The paw of young cubs of two to three month old is similar to that of the adult leopard, but it can be recognized from the characteristic shape of the pad. One can easily recognize the pugmarks of tiger, tigress and the cubs from such impressions, with some practice. The distance between the right and left fore limbs is called step, whereas the distance measured between the mark of right limb and the next mark of right limb is called the stride. If the impressions of paws of one or more tigers are seen on the ground, one can easily recognize the number of tigers on the basis of the, while walking. The pugmarks of tiger are used during census of tiger population.

The tiger's ungulate prey would include wild cattle (Gaur, Banteng, Kouprey and Water Buffalo), other bovids (Nilgai, Chousingha, Chinkara, Takin, Ox), wild goats and goat-antelopes (Tahir, Goral, Serow), several species of deer (Mouse Deer, Elk, Sika, Sambar, Barasingha, Thamin, Chital, Hog Deer, and Muntjacs), Tapir, Wild Pig and occasionally Rhino and Elephant calves. They cannot survive and reproduce if a habitat does not support ungulates at adequate densities (Jhala *et al.*, 2008).

Tiger needs at least 30 animals per annum, each weighing 30 kg (Novikov 1962). Schaller (1967) also estimated daily need of meat by an adult tiger as 5.5 to 7 kg. Tiger starts eating its prey from its hind portion, whereas the leopard starts eating from the belly side. Thus one can easily find out whether the prey is killed by tiger or by leopard. It can kill big as well as small animals, whereas the leopard can kill only small animals. The canine marks on neck of prey also indicate the type of cat that has killed the prey. If the distance between the marks is 7 to 8 cm, it is the tiger kill, whereas if it is 5 cm, it is the leopard kill. Tiger usually feeds twice or thrice a week and does not allow tigress to share the kill, but tigress allows the tiger to share its prey (personal experience at Bandhavgarh).

Tiger has ambush hunting strategy in which it gets closer to the prey, and rushes to capture it. A tiger's muscles can surge up the power needed to capture prey, but it gets tired quickly in comparison with ungulate muscles. Bonded around the cat's strong bones and flexible joints, these muscles are only capable of a short rush, with lots of twisting, turning and flexing

Management Interventions

- ☆ Ameliorative and Compensatory Management.
- ☆ Improvement of habitat that includes augmenting water sources, water regime development, eradication of weeds, and development and restoration of grasslands
- ☆ Development of communication and protection infrastructure,
- ☆ Patrolling and anti-poaching activities,
- ☆ Research and monitoring,
- ☆ Mitigation of man-animal conflicts,
- ☆ Inoculation of domestic cattle in and around PAs against contagious diseases.
- ☆ Visitor-use management (tourism) and interpretation.
- ☆ Maintenance of roads, check-barriers, patrolling camps, buildings, watchtowers, wireless network, water sources, vehicles etc are part of the day-to-day management activity.

Ecodevelopment

- ☆ Ecologically, socially and economically sustainable.
- ☆ People's active participation in protection of PA resources
- ☆ Central Government had been funding the eco-development programmes in the villages around Protected areas

Voluntary Village Relocation Activity

- ☆ Implemented in some Protected areas
- ☆ Each family relocated would get Rs. 1,00,000 rupees.

Compensation for damages Caused by wildlife

- ☆ Death of person due to wild cats....Rs. 20,000

☆ Injury by wild animals Rs. 7,500

☆ Cattle loss due to tiger or leopard Rs. 3,500

Monitoring

1. Population Estimation of wild animals
2. Training of personnel and members of Ecodevelopment Committes
3. Advisory Bodies for Managemnt and Conservation of Wildlife
4. Funding Mechanism

Initiatives taken by Government of Madhya Pradesh (Belsare, 2008)

1. Asiatic Lion Reintroduction project at Kuno-Palpur Sanctury
2. Establishment of Wildlife Health Monitoring, Disease Diagnostic and Research Cell, Jabalpur
3. Conservation of Biodiversity in Forests outside the PA network (i. biodiversity conservation concerns must be included in the management of territorial forest areas. iiPolice and the Forest departments jointly constituted a " Tiger Cell " in 1994 to curb illicit trade. iiiPreparation of database of wildlife related crimes, iv. Interact and co-ordinate with the authorities
4. Joint efforts of the police and forest personnel have yielded positive results.

Wildlife Intelligence Bureau/Antipoaching Squads

establishing Wildlife Intelligence Bureaus in 5 highly sensitive areas-namely Bhopal, Jabalpur, Satna, Itarsi, Sagar and Seoni. The mission of these bureaus is to collect and collate the intelligence regarding the wildlife crime and to help the field administration in preventing the wildlife offences from happening all together.

Constitution of Tiger Foundation Society

Aims

1. Provide protection to threatened and endangered species (with added emphasis on tiger conservation).
2. Provide help to the State Government and people in their effort to protect tiger habitats against fragmentation and destruction.
3. Provide support to the state government to control poaching of wild animals.
4. Provide help in controlling illicit trade in body parts of tigers and products made out of body parts.
5. Provide help in protection of tiger habitats and prey-base of tiger.
6. To make general public aware of the need for conserving biodiversity.
7. Reward those individuals (including forest department personnel) and institutions, organizations who have made special effort towards protection

of tiger. viii. Strengthen infrastructure required for protection of tigers in their habitats.

8. Pass an order enabling all National Parks and Sanctuaries to directly utilize the receipts from wildlife tourism for development of the Protected Areas.

Peoples Contribution towards Wildlife Conservation

1. Become member.
2. Inform Chief Warden Wildlife about illegal activities in national park or sanctuary.
3. Training and actual field census
4. Activities through a group through your Nature Club.

Wildlife Action Plan 2006-2016

1. Strengthening and enhancing the protected area network.
2. Effective Management of PA areas
3. Restoration of degraded habitats outside PA
4. Control of poaching and illegal trade in wild animal and plant species:
5. Speedy and effective amelioration of man-animal conflict
6. Monitoring and Research
7. Human Resource HRD and Personal Development.
8. Ensuring people's support and participation in wildlife conservation
9. Conservation Education and Interpretation of Protective Areas:
10. Tourism in Protected Areas
11. Domestic Legislation and International Conventions
12. Review combined effect of existing legislation on natural resource management. Prepare guidelines to enable conjunctive use of all laws and statutes by government departments, forest staff and NGOs
13. Enhancing financial allocations for ensuring sustained fund flow to the wildlife sector.
14. Integration of national wildlife action plan with other sectorial programme

References

Bashir, T., Bhattacharya, T., Poudyal, K. and Sathyakumar, S., 2011. Notable observations on the melanistic Asiatic Golden cat (*Pardofelis temminckii*) of Sikkim, India. *NeBIO* 2 (1): 2–4.

Belsare, D.K., 1994. Inventory and status of vanishing wetland wildlife of Southeast Asia and a operational management plan for their survival. In: *Global Wetlands: Old World and New*, (Ed.) W.J. Mitsch. Elsevier Sci., New York, pp 841–856.

Belsare, D.K., 1998. Survival of endangered mammals of Indian sub-continent: Facts and factors. *Proc 10th Congr. Zool. Bodhgaya*, (Ed.)B.N. Pandey. Daya Publ. Delhi.

Belsare, D.K., 2008. Strategy for conservation of tigers in tiger state. *Columbia J.*, 9: 1–14.

Belsare, D.K., 2011. *Vanishing Roar of Bengal Tigers.* RoseDogwood Pub. Pittsberg USA, pp. 1–156.

Bhattacharya, T., 1992. A brief note on some observations on the breeding biology of fishing cat (*Felis vivarrina*). *Tigerpaper*, 19: 18–20.

Chandawat, R.S. and Rawat, G.S., 1994. Food habits of snow leopard in Ladakh. *Proc. Int. Snow Leopard Symp.* 7: 127: 132.–986

Choudhury, A.U., 1992. The Clouded leopard in Assam. *Oryx*, 27(1): 51–53.

Choudhury, A.U., 1996. The clouded leopard. *Cheetal*, 35(1–2): 13–18.

Choudhury, A., 2007. Sighting of Asiatic golden cat in the grasslands of Assam's Manas National Park. *Cat News* 47: 29.

Dubey, Y., 1999. Sighting of rustyspotted Cat in Tadoba Andhari Tiger Reserve, Maharashtra. Journal of the Bombay *Natural History Society*, 96(2): 310.

Fox, J.L., Sinha, S.P., Chandawat, R.S. and Das, P.K., 1991. Status of the snow leopard, *Panthera uncial* in North-west India. *Biol. Cons.*, 55: 283–298.

Grassman, L.I. Jr. and Tewes, M.E., 2000. "Marbled cat in northeastern Thailand". *Cat News*, 33: 24.

Grassman Jr., L. I., Tewes, M. E., Silvy, N.J. and Kreetiyutanont, K., 2005. Ecology of three sympatric felids in a mixed evergreen forest in North-central Thailand. *J. Mammalogy*, 86: 29–38.

Haque, M.D. and Vijayan, V.S., 1993. Food habits of the fishing cat (*Felis viverrina*) in Keoladeo National Park, Bharatpur, Rajasthan. *J. Bomb. Nat. Hist. Soc.*, 90: 498–500.

Henschel, P., Hunter, L., Breitenmoser, U., Purchase, N., Packer, C., Khorozyan, I., Bauer, H., Marker, L., Sogbohossou, E. and Breitenmoser-Würsten, C., 2008. "*Panthera pardus*". IUCN Red List of Threatened Species.

IUCN Red List of Threatened Species, 2006. http://www.iucnredlist.org.

Jhala, Y.V., Gopal, R. and Qureshi, Q., 2008. Status of the Tigers, Co-predators, and Prey in India. National Tiger Conservation Authority, Govt. of India New Delhi, and Wildlife Institute of India Dehradun. TR 08/001 pp–151.

Luo, S.P. Kim, Johnson, J.H., van der Walt, W.E.J., Martenson, J., Yuhki, N., Miquelle, D.G., Uphyrkina, O., Goodrich, J.M., Quigley, H.B., Tilson, R., Brady, G., Martelli, P., Subramaniam, V., McDougal, C., Hean, S., Huang, S.Q., Pan, W., Karanth, U.K., Sunquist, M., Smith, J.L. and O-Brien, S.J., 2004. Phylogeography and genetic ancestry of tigers! Panthera tigris/DJE. *PLoS Biology*, 2-12/0e442.

Mallon, D., 1984. The snow leopard in Ladakh. *Int. Red Book, Snow Leopard*. 4: 23–37.

Mishra, C., Madhusudan, M. D. and Datta, A., 2006. Mammals of the high altitudes of western Arunachal Pradesh, eastern Himalaya: An assessment of threats and conservation needs. *Oryx* 40: 29–35.

Nowell, K., 2008. *Felis silvestris*. In IUCN 2008 Red List of Threatened Species.

Nowell and Jackson, 1996. *Wild Cats–Status Survey and Conservation Action Plan*. IUCN/SSC, Gland, Switzerland.

Novikov, G., 1962. Carnivorous mammals of the fauna of U.S.S.R. Israel Programme of Scientific Translation, Washington D.C.

Oli, M.K., Taylor, I.R., and Rogers, M.E., 1994. Snow leopard, *Panthera uncial* predation of live stock: An assessment of local perception in the Annapuram Conservation Area, Nepal. *Biol. Conserv.*, 68: 63–68.

Ravi, C. and Johnsingh, A.J.T., 1993. Management of asiatic lions in the Gir forest, India. *Symp. Zool. Soc. Lond.*, 65: 409–424.

Schaller, G.B., 1967. *The Deer and the Tiger*. Collins.

Sunquist, M. and Sunquist, F., 2002. *Wild cats of the World*. University of Chicago Press, Chicago, pp. 237–240.

Environment, Biodiversity and Traditional System, 2014 *Pages* 199–209
ISBN 978-81-7622-307-2
Editors: D.R. Khanna, G.S. Solanki and S.K. Pathak
Published by: BIOTECH BOOKS, NEW DELHI

Chapter 18

A Psychological Study of Environmental Awareness Correlated to Environmental Conservation Behaviour among School Students

Zokaitluangi

Department of Psychology,
Mizoram University, Aizwl – 796 004

ABSTRACT

The study investigated secondary school students' environmental awareness and environmental conservation behaviour by using the Environment Awareness Ability Measure (EAAM) and structure interview questionnaire to assess the conservation behavior of the participants. Two hundred students were selected through the purposive random sampling technique from secondary schools of Mizoram with due care of equal representation of gender (boys and girls) and ecological (rural and urban) representation. Results indicated that there were moderate scores on environmental awareness (depletion) across the samples, with positive significant differences in ecological difference and gender difference on environmental awareness and environmental conservation behavior. Results indicated the need of more environmental awareness and environmental conservation campaign in changing participants of behaviour for environmental conservation.

Keywords: Environment awareness, Environmental conservation, Ecology, Gender, Behaviour.

Introduction

A disaster is either a natural or man-made hazard which has come to fruition resulting in an event of substantial extent causing significant physical damage or destruction, loss of life, or drastic change to the natural environment. Some hazards that strike in areas with low vulnerability are not considered a disaster, as is the case in uninhabited regions (Quarantelli, 1998). All disasters are hence the result of human failure to introduce appropriate disaster management measures. Hazards are routinely divided into natural or human-made, although complex disasters, where there is no single root cause, are more common in developing countries.

A. **Natural Disaster** is a consequence when a natural hazard (*e.g.*, volcanic eruption or earthquake) affects humans and/or the built environment. Human vulnerability, caused by the lack of appropriate emergency management, leads to financial, environmental, or human impact. Natural disasters usually affect many people at a time in a number of ways. Its occurrence are always very sudden and alarming, usually unpredictable, highly stressful, limit of freedom, and harmful to behaviour. Natural disasters are landslide, earthquake, volcanic eruption, drought, flood, cyclone, etc.

B. **Man-made Disasters** are disasters resulting from man-made hazards (threats having an element of human intent, negligence, or error; or involving a failure of a man-made system), as opposed to natural disasters resulting from natural hazards. Man-made hazards or disasters are sometimes referred to as anthropogenic, that included Sociological hazards and technological hazards. Sociological hazards are crime, arson, civil disorders, terrorism, and war. Technological hazards are-Industrial disasters, Structural collapses, power outage, pollutions etc.

The 'Environmental Awareness' only implies knowledge about environment but also attitude, values and necessary skills to solve environment related problems. The environmental education is an inter-disciplinary and examines major environmental issues from local, national and international points of view. A host of studies found that significant differences on environmental awareness between male and female (Shobeiri, Omidvar, and Prahallada, 2006), and also Rou, Sabhlok (1995), Patel (1995), Szagun and Pavlov (1995) and Tripathi (2000) who reported sex has affect on level of environmental awareness among students. The environmental awareness difference on Socio-Economic status and also more environmental awareness among boys of better socio-economic status whereas among girl, lower socio-economic status had more environmental awareness as compared with boys.

Changing human behavior to save environment is exceedingly an important topic as environmental dilapidation and deterioration has been the frightful topic of environmentalists. Environmental specialists have repeatedly pointed out that a solution to environmental crises requires an environmental awareness which deeply rooted in human behavior as their value and behaviour causing changes leading to conservation of environment. Many may think right technology is the only solution even such advance technology required the awareness of prevalent of environmental deterioration to motivate scientist to make appropriate design, for its utmost maximum utilization, but all depends on behavioural change to effective solution of environment problems. Even high technology required particular behavior to ensure that people utilized available technology; the impact of any technology depends on human behavior. Finally, our behavior does have strong effects on the environment; anyone of us does either a positive of a negative impact on our environment.

The eclectic model (Bell *et al.*, 1996) applied to behavioural change for conservation of the environment can be explained as follow:

1. Build environment changed the availability of resources e.g natural environment such as air, water, and waste in the environment.
2. Perception of resources that can be influence by: 1) individual difference (sensitivity to litter, fuel consumption and saving) ii) Situational factors (affordance such as availability of recycling bin for litter collection) iii) Social conditions (social value and practices in preserving
3. Perception of personal responsibility for resources can be two ways (a) perceived need of environment preservation and personal responsibility b) perceived resources as unlimited and exploitable with waste materials effect as negligible).
4. Awareness of environment in danger can affect destructive behaviour to conserve resources.
5. A person has cope or adaptation or adjustment to the deteriorated environment. Several studies highlighted the impact of environmental deterioration such as Three Mile Island nuclear power station depleted 400 000 gallons of radioactive water lead to psychological and emotional distress among the residents. Buffalo Creek (Mining Dam) in West Virginia broke in 26 February, 1972, washed away houses, automobiles, killed 125 people, 5000 lost houses. Much more problems was aftereffect like anxiety of fear about disaster, change life style (Gleser, Cree, and and Winger, 1981), withdrawal symptom (blunted emotion, Erikson, 1976), depression (sad and subdue; Kilijanek and Drabek, 1979), stress related physical symptoms (somatic problems), unfocused anger (thinking about culprit or responsible person; Glaser, Green and Winger, 1981), regression (regresses to earlier stage of behavior; Tichner and Kapp, 1076), and nightmares (dream about dying or suffering self or relative; Glaser, Gree, and and Winger, 1981).
6. Possible after effect (depletion or exposure reduced quality of life for all induced more environmental conservation behavior or no repairable environment leading to psychological trauma with physical suffering including ulcer, adrenal enlargement, shrinkage of lymph, hypertension, cardiac problems, asthma, ulcer, cancer, etc.).

The problems of environment related behavior can be categorized as 1) problems of aesthetics (prevention and control of waste, protection of natural resources, control of man made environment with good plan 2) health related problems – control of water pollution, air pollution, noise pollution, soil pollution, toxic and radiation contamination. 3) resource problems – relating overconsumption of natural resources as water, forest, energy etc 4) reduction of consumption-over dependence on automobile.

The environmental awareness portrayed the needs of change attitude to destructive behavior causing environment deterioration from social trap, as single destructive behavior has minor impact but common use deteriorated environment. Individual use of fossil fuels seems harmless but collective use dangerously warms the entire planet(Kerr, 1988a), mass use of air conditioning (produced chlorofluorocarbons) creates on ozone hole over the planetary poles (Kerr, 1988b), collective use of fossils creates acid rains (Schindler, 1988), science advancement in lunching space vehicles left big amount of waste products endangering future lunches (Marshal, 1985), electrification of urban areas interferes astronomical studies of outer space (Waldrop, 1988).

Psychological interventions are aimed at changing the destructive behavior of individuals (Kempton *et al.*, 1992) through persuasive or informational massages e.g information about the energy of consumption of certain electrical appliances can improve the energy saving behavior of an individual help to adopt environmentally responsible behavior (Newhouse, 1990) but information alone may not sufficient enough to promote behavior change (Syme, *et al.*, 1987) unless global attitude change can work better. The degree of commitment of an individual affects the future behavior in saving environment (Lepper *et al.*, 1973) as high committer used less energy than low committers (Pallak *et al.*, 1980). Modeling can be effective if the model hold positive attitude for the perceiver (Bandura, 1977), having similarity with the perceiver (Newhouse, 1990) with the expectation of reward as the model has received (Winnett, *et al.*, 1981), Prompt can improve behavioral change that well time, well place, and the requested behavior can be easily enacted (Geller *et al.*, 1982). Consequent interventions such as appropriate reinforcement and feedback such as giving incentive that obliged the requested behavior, negative reinforcement for neglecting or minimizing requested behavior, and punishment for violating requested behavior (Bell *et al.*, 1996) are also effective. Feedback (providing information) about related behavior with the requested behavior (Winnett, neal, and Grier, 1979) is effective for attitudinal change toward save environmental behavior. There are two broad categories of conserving environment are recovery recycling and reclamation. Reduction of consumption is the easiest way of conservation of an environment by cutting down daily use and buying reusable materials for economic and environmental saving (De Young *et al.*, 1993). Mass transit system typically consumes far less fuel per passenger than private automobiles. Aluminum can be reused through recycle not losing its original purpose while recycling food container may contaminated harmful chemicals but can reclaim into diverse product (Bell, *et al.*, 1996).

Objectives

Based on the foregoing theoretical foundations, the aim of the present research study was to compare the boys and girls of high school students on the environmental awareness and conservative behaviour through psychological evaluation. The primary aim of this study was to examine the ecological differences (urban and rural), and gender difference (boys and girls) among high students with regard to environmental awareness and conservative behaviour. A secondary aim was to examine the effect of environmental awareness on conservative behaviour among samples.

Hypotheses

The following hypotheses were set forth for the conduct of the present study:

1. It was expected that there will be gender difference in environmental awareness among the selected samples of high school students.
2. It was expected that ecological difference will affect conservative behaviour differently among the groups.
3. It was expected that there will be gender and ecological interaction effects on environmental awareness and conservative behaviour among the samples.
4. It was expected that level of environmental awareness will predicted conservative behaviour.

Methodology

Sample

200 participants between 12 to 16 years of the high students were selected from Aizawl, capital of Mizoram state on purposive random sampling procedure with equal proportion of urban and rural, and boys and girls of under each cell (n=50) of the factorial designs '2x2' is 2 ecology (rural and urban) and x 2 gender (male and female) was employed.

Psychological Measures

The following psychological measures were employed such as

1. *Socio-demographic data sheet*: The socio-demographic data sheet is a semi-structured questionnaire, specially prepared by the researcher for the current study. It includes information like age, sex, education, name of the district and village, which class reading, etc. were recorded for necessary control of confounding variables.
2. *Environment Awareness Ability Measure (EAAM)* developed by Praveen Kumar Jha (1998). It contained 51 items that measures environmental awareness about pollutions (air, soil, water, and noise), environmental depletion (forest, natural resources, wild life etc). High score indicates high environmental awareness of the samples. The scale has reliability of.71 (split half) and.83 validity (Shobeiri, 2005).
3. *Conservative Behaviour scale:* The researchers developed the Likert type (3-point) scale consisting 18 items of negative and positive rating scale, having a reliability value of 0.80 Cronbach Alpha and.71 Split half reliability, high score means high conservative behaviour. The item validity was tested by Tetra choric correlation.

Results

Subject-wise scores on the specific items of the behavioural measure of the Environment Awareness Ability Measure (EAAM) developed by Praveen Kumar Jha (1998, Conservative *Behaviour scale:* (CBS: Zokaitluangi, 2011) were employed for the whole samples to ascertain the level of *environmental Awareness and conservative behaviour* of the samples.

The item- total coefficient of correlation (as an index of internal consistency and item validity) was ascertained for the scales/subscales of the behavioural measures as shown in Table 18.1 represent the reliabilities (Cronbach Alphas and Spearman Brown Coefficient) of specific scales/subscales of the behavioural measures. The reliability of the psychological scales showed alpha reliability Spearman Brown coefficients and inter-correlations among the two variables having positive significant relationships between EAAMI and CBS that pronounce trustworthiness of the selected psychological scales for the target population, and statistical assumptions were checked and fit for further analysis.

Table 18.1: Descriptive statistics, internal consistency, item validity and correlation matrix of the environmental awareness and conservation behaviour (S) for the whole samples.

Variables	*No. of Items*	*Cronbach Alpha (x)*	*Spearman-Brown Coefficient*	*Correlations*	
				EAAM	*CBS*
EAAM	51	0.77	0.68	1	86**
CBS	18	0.80	0.71	0.86**	1

Correlation is significant at the **= 0.01 level (2-tailed).

The bivariate relationships between the scales/subscales of the behavioural measures were computed (Table 18.1) and it indicated the positive significant relationships between the two scales (r= 0.86**) of the behavioural measures accounting for 'gender' and 'ecological' differences for the whole samples. In the urban areas, more or better consequent interventions will be there which includes appropriate reinforcement and feedback for those who obliged the requested behavior or negative reinforcement for neglecting or minimizing requested behavior, punishment for violating requested behavior (Bell *et al.*, 1996); that will reveal in higher environmental awareness with consequent behaviour of saving an environment.

The Table 18.2 reveals that higher scores by the girls students (rural and urban students) on environmental awareness and conservative behaviour (*i.e.* M=13.16; 10.15) than boy students (*i.e.* M= 13.04; 6.91) respectively, and Urban students (boys and girls) had high conservative bahaviour (*i.e.* M=21.83; 11.97.) than rural students (*i.e.* M=10.37; 5.09). The finding has confirmation with the earlier findings that environmental awareness has been found to be higher scores among girls than boys (Shobeiri, Omidvar, and Prahallada, 2006), and that high environmental awareness was highly associated with conservation behaviour. The urban students must have received better and more frequent awareness campaign about destruction of environment and associated harmful effects that may lead to higher commitment leading to higher conservation behaviour. Earlier finding supports that male litter more than female (Osborne and Powers, 1980).The degree of commitment of an individual to the issue likely affects the future behavior in saving environment (Lepper *et al.*, 1973), high commitment used less energy than low commitment (Pallak *et al.*, 1980), modeling can be effective as more model hold positive attitude for the perceiver were available in urban areas, and female are more prone to change (Bandura, 1977).

Table 18.2: Mean, standard deviation between independent variables of 'ecology' and 'gender' on environmental awareness and conservation behaviour (EAAM and CBS) for the whole samples.

Ecology	*Gender*		*EAAM*	*CBS*
Urban	Male	Mean	7.68	3.46
		N	50	50
		Std. Deviation	3.53	1.86
	Female	Mean	13.08	6.72
		N	50	50
		Std. Deviation	2.66	1.26
Rural	Male	Mean	18.42	10.36
		N	50	50
		Std. Deviation	2.46	1.32
	Female	Mean	25.24	13.58
		N	50	50
		Std. Deviation	3.61	1.77
Total Rural students		Mean	10.37	5.09
(Male + Female, N= 100)		Std. Deviation	4.10	2.24
Total Urban students		Mean	21.83	11.97
(Male + Female, N = 100)		Std. Deviation	4.60	2.24
Total Male students		Mean	13.04	6.91
(Cancer + other, N=100)		Std. Deviation	6.18	3.80
Total Female students		Mean	19.16	10.15
(Male + Female, N=100)		Std. Deviation	6.88	3.77

Results (Table 18.3) revealed significant independent effects of 'ecology' and 'gender' in all the analyses for test scores on EAAM and CBS respectively. The effect of ecology and gender interaction seems to be different for different behavioural measures. The results also indicated Significant (sig at.01) main effect on EAAM and CBS which indicated that there was interaction between the two main variables in the dependant measures. The F statistic for the test of between-subjects effects on dependent variables revealed significant difference among the score. This makes sense to test hypotheses. They are further look into in the following analyses.

Table 18.3: 2x2 ANOVA for 'ecology x gender' on environmental awareness and conservation behaviour (EA and CB) for the whole samples.

Dependent Variables	*Source of Variance*	*F -ratio*	*Sig.*	*Partial Eta Squared*
EA	Ecology	43.33	0.00	0.18
	Gender	345.29	0.00	0.64
	Ecology X Gender	288.74	0.00	0.82
CB	Ecology	107.49	0.00	16
	Gender	469.97	0.00	0.70
	Ecology X Gender	389.42	0.00.	0.86

There was a significant effects of ecology on EAAM (η^2=0.18), and intrinsic CBS (η^2=0.16); Gender effects on EAAM (η^2=0.64), and on CBS (η^2=0.70); and significant interaction effect of ecology and gender was on EAAM (η^2=0.82), and CBS (η^2=0.86). Results reveal that high effect size of ecology and together had a high impact on environmental conservation behaviour. That can be explain with the available finding that the importance of high income (Winkler and Winnett, 1982) that may be associated with opportunity to buy reusable material, stringent government regulation impact as mention by Stern (1992b).

Table 18.4: R, R^2, change statistics, durbin-watson, beta-values, t, and collinearity statistics in the prediction of scores on conservation behaviour by the environmental awareness for the whole samples.

Predictor	*R^2*	*Durbin Watson*	*Beta*	*t*	*sig*	*Collinearity Statistics*	
						Tolerance	*VIF*
EA	0.26	0.1.12	0.86	0.23	0.00	1.00	1.00

The regression model with CBS as criterion, and EAAM as predictors contributed 26 per cent emerged to be statistically significant. The Durbin-Watson statistics and the Collinearity statistics were robust in supporting the normality and the homogeneity. The results explained that higher environment awareness has given a reasons to saves environment for oneself life and for the whole community as environment is limited supply and has to share with fellow citizens which highly increasing every minutes. The awareness of the collective use dangerously warms the entire planet (Kerr, 1988a), mass use of air conditioning (produced chlorofluorocarbons) creates on ozone hole over the planetary poles (Kerr, 1988b), Collective use of fossils creates acid rains (Schindler, 1988), science advancement in lunching space vehicles left big amount of waste products endangering future lunches (Marshal, 1985), electrification of urban areas interferes astronomical studies of outer space (Waldrop, 1988).

The present studies highlighted the difference between urban and rural students, and boys and girls on their levels of environmental awareness, and relation to conservation behaviour among the selected samples. The present study was not free from limitations that non availability of earlier findings in the present context as most of the researched have been done in the western culture and even too concern with environmental destruction effects not the consequent behaviour of saving environment which was prime importance for future prevention. The samples were selected on purposive sampling is just to throw light on the need of scientific empirical research in this area, with more psychological scales with bigger samples size very much esteemed. The present finding was exploratory in nature as no research finding was available, not represent the different age group and strata to make generalization. With all those limitation it clearly reveals the relationships between EAAM and conservation behaviour, with ecological and gender differences on selected behavioural measures. It is very crucial to continue research in this growing area of study because of the importance of environmental destruction, needs of saving environment is the most alarming issues allover the globe, and to address the different aspects of environmental problems and environmental management to find a more comprehensive way to measure both. If environmental awareness can help predict or influence behavior to save environment then more future research will find prevention of environment through better management of environmental policy

Based on the finding, it was made some suggestions that more environmental awareness campaign reaching every individuals such as prevailing water, air, land, noise pollutions, toxic contamination, radiation contamination, global warming, limited natural resource with alarming rate of population increase, tragic climatic changes, alarming rate of technological wastes, human made ecological destruction, depletion of ozone layer, over outsourcing of natural resources and its harmful effect to physical and mental health. Environmental destructive behavior can be change with saving environment through awareness of social traps. Taking guides from psychological diagnostic and interventions for behavioural change to save environment is advisable as man is the centre of the environmental concern.

References

Bandura, A., 1977. *Social Learning Theory*. Prentice–Hall, Eaglewood Cliffs, NJ.

Bell, P.A., Greene, T.C., Fisher, J.D., and Baum, A., 1996. *Environmental Psychology*. Harcourt Brace College Publishers, New York.

De Young R, Duncan, A., Frank, J., Gill, N., Rotham, S., Shenor, J.L., Shorkin, A. and Zweizig, M., 1993. Promoting Sourcess reduction behavior. The role of motivational behavior. *Environment and Behaviour*, 25: 70–85.

Erikson, K.T., 1976. Loss of Communality at Buffalo creek. *American Journal of Psychiatrist*, 133: 302–305.

Geller, E.S., Winnett, R.A. and Everett, P.B., 1982. *Preserving the Environment: New Strategies for Behavioural Change*. Pergamon, New York.

Glaser, G., Green, B. and Winget, C., 1981. *Prolong Psychological Effect of Disaster: A Study of Buffalo Creek*. Academic Press, New York.

Kempton, W., Darley, J.M., and Stern, P.C., 1992. Psychological research for the new energy problems: Strategies and opportunities. *American Psychologist*, 47: 1213–1222.

Kerr, R.A., 1988a. Is the Greenhause here? *Science*, 239: 559–561.

Kerr, R.A., 1988b. Ozone hole bodes ill for the globe. *Science*, 241: 785–786.

Kilijanek, T.S. and Drabek, T.E., 1979. Assessing Long term Impact of a natural disaster: A focus on the elderly. *The Gerontologist*, 19: 555–566.

Lepper, M.R., Greene, D. and Nisbett, R.E., 1973. Undermining children's intrinsic intere4st with extrinsic reward: A test of the over justification principle. *Journal of Personality and Social Psychology*, 28: 129–137.

Marshal, E., 1985. Space junk grows with weapons test. *Science*, 239: 424–425.

Newhouse, N., 1990. Implications of attitude and behavior research for environmental conservation, *The Journal of Environmental Education*, 22, 1.

Pallak, M.S., Cook, D.A. and Sullivan, J.J., 1980. Commitment and energy conservation. In: *Applied Social Psychology Annual*, (Ed.) L. Bickman. 1: 235–253.

Patel, D.G. and Patel, N.A., 1995. An investigation into the environmental awareness and its enhancement in the secondary school teachers. *Prog. Edu*, 49(12): 256–259, 268.

Quarantelli, E.L., 1998. *What is a Disaster? Perspectives of Question*. Routledge, London.

Osborne, J.G. and Powers, R.B., 1980. Controlling the litter problem. In: Helping the Community: *Behavioural Applications*, (Eds.) G.L. Martin and J.G. Osborne, pp. 103–168.

Rou, S., 1995. A study of the awareness and attitude of teachers and students of high schools toward environmental education in Jabalpur district. *Ph.D. Thesis*, Ani Durgavati ishwavidyalaya, Indian Edu. Abstract, 1, Section 24, 62.

Schindler, D.W., 1988. Effects of acid rain on freshwater ecosystem. *Science*, 239: 149–157.

Shobeiri, S.M., 2005. A comparative study of environmental awareness and attitude of teachers and students of secondary schools in India and Iran. *PhD Thesis*, Department of Education, University of Mysore, India.

Shobeiri, S.M., Omidvar, B. and Prahallada, N.N., 2006. A comparative study of environmental awareness among secondary school students in Iran and India. *Int. J. Environ. Res.*, 1(1): 28–34.

Syme, G.L., Seligman, C., Kantola, S.J., and MacPherson, D.K., 1987. Evaluating a television campaign to promote petrol consumption. *Environment and Behaviour*, 19: 444–461.

Szagun, G. and Pavlov, V., 1995. German and Russian adolescents' environmental awareness. *Journal of Youth* Soc., 27(1): 93–112.

Tripathi, M.P., 2000. A comparative study of environmental awareness of students studying in central schools and other schools at 10+ level in Uttar Pradesh, Nation. *Journal of Education*, 6(1): 47–51.

Waldrop, J.M., 1988. Taking back the nigh. *Science*, 241: 1288–1289.

Winnett, R.A., Hatcher, J., Leckliter, I., Fort, T.R., Fishback, J.F., Riley, A/W., and Love, S., 1981. The effect of videotape modellingand feedback on residential comfort, the thermal environment and electricity consumption: Winter and summer studies. Unpublished document, Department of Psychology, VirginiaPolytechnic Institute and State University.

Winnett, R.A., Neal, M.S., and Grier, H.C., 1979. The effects of self monitoring and Feedback on residential electricity consumption. *Journal of Applied Behaviour Analysis*, 12: 173–184.

Environment, Biodiversity and Traditional System, 2014 *Pages* 211–215
ISBN 978-81-7622-307-2
Editors: D.R. Khanna, G.S. Solanki and S.K. Pathak
Published by: BIOTECH BOOKS, NEW DELHI

Chapter 19

Dynamics of N Mineralization in Soil Due to Addition of Solid Wastes

Angom Sarjubala Devi and K. Vanlalmalsawmkima*

Department of Environmental Science,
Mizoram University

ABSTRACT

Due to human activities different kinds of solid waste materials are discarded to soil. Among the wastes vegetable waste, pork waste, rice husk saw dust and cow dung are some wastes which can be easily degradable. The present work was carried out to study the changes in soil properties mainly N mineralization due to addition of these five types of wastes in a tropical lateritic soil in Tanhril, Mizoram University campus. The soil treated with the wastes were analyzed for organic C, total N and N mineralization monthly.

During the study it was found that the trend of N mineralization was found to be irregular in the initial seven months. However it attains a trend of increasing in rainy months and declining in cold months. In the last three months the mineralization rate as well as C and N content of the soil remain almost stagnant.

Keywords: N-mineralization, Pork waste, Saw dust, Solid waste, Vegetable waste.

Introduction

Various types of wastes from different sources *e.g.*: carpentry centers, markets and residential areas, meat processing and slaughter houses, rice mills and cattle farms have values for increasing soil fertility. However these wastes are often being

* *Corresponding Author:* E-mail: angom75@yahoo.com

disposed off. Therefore the main objective of the present study is to find out the changes in N mineralization in the soil due to treatment with different types of organic waste.

Materials and Methods

Study Site

The study was conducted in Tanhril 20 km. away from Aizawl city, within Mizoram University. The color of the soil is red and is lateritic type (Jha, 1997). The area receives an annual rainfall of 275cm. Average temperature ranges from20.6 to 30.2°C. Vegetation and dried organic matter were totally removed from the study site.

Treatment

For the study five types of wastes were collected. They are: Pork waste, husk of rice, vegetable waste, saw dust and dried cow dung. The vegetable waste and pork waste were chopped into small pieces, husk of rice, saw dust and cow dung was grinded in order to make the size very small. During the month of July, 08 pits having sizes of 1x2 m with a depth of 12.5cm were dug up in each site. The pits were filled with ten kilograms each of the wastes making up to 5cm and covered with 7.5cm of soil. Site one was filled with Pork waste, Site two was filled with husk of rice, Site three was filled with vegetable waste, Site four was filled with saw dust, Site five was filled with Cow dung and a control site (without any treatment) was also selected. Decomposition of the wastes was determined visually after every fifteen days.

Soil Analysis

Collection of soil for analysis was started monthly from the month of August, 08 by taking five replicates from each site. The fresh soil samples from the wastes treated sites were collected from 0-12.5 cm constituting the waste layer, they were mixed, grinded, sieved through 2mm mesh and analyzed for organic C (rapid dichromate oxidation method) and total N (Kjeldahl's digestion method). Organic C and N were estimated for the controlled site also. The N mineralization was estimated by following laboratory and field incubation methods outlined by Anderson and Ingram (1993). The net N mineralization of the treated soil was calculated from the difference between the mineral N(NH_3-N and NO_3-N) produced by the soil- waste mixture and that produced in the same period by the control soil, the same method was also adopted by various studies (Neve *et al.*, 2003; Christensen, 2004; Jin *et al.*, 2008 and Morvan and Nicolardot, 2009).

Results and Discussion

The pattern of decomposition can be written as cow dung>pork waste> vegetable waste>rice husk> saw dust. Overall soil organic C and N content (Figures 19.1 and 19.2).) were highest in soil treated with cow dung because the releasing of these nutrients from the waste took place at a faster rate due to its faster decomposition. Least organic C and N content were observed in soil treated with rice husk because of its low decomposition rate. The soil after treatment have higher C and N content compared to the initial C and N content in all the sites.

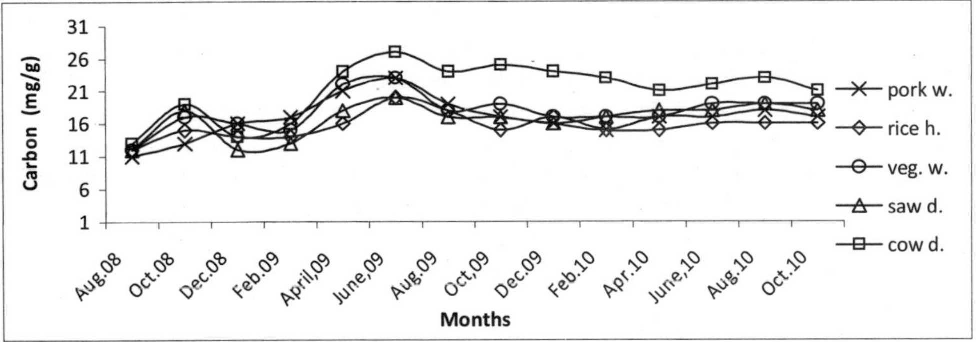

Figure 19.1: Carbon content in the treated soils in different months.

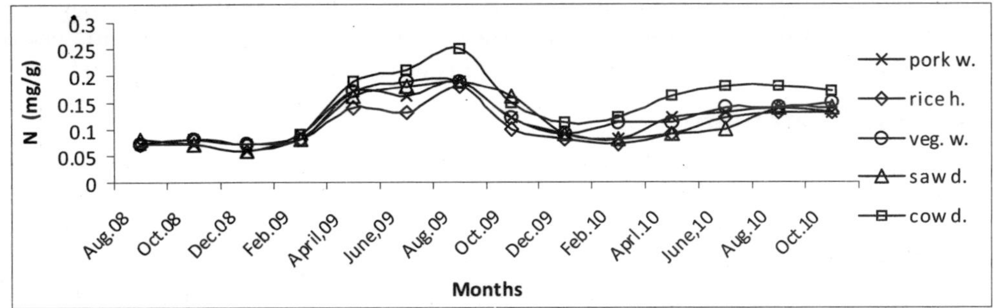

Figure 19.2: N content in the treated soils in different months.

The trend of N mineralization (Figure 19.3) was found to be irregular in the initial seven months. However it attains a trend of increasing in rainy months and declining in cold months in all the treatments. The increase in rainy months and declining in cold months may be due to increase in soil moisture during rainy months leading to more activities of the soil microorganisms and a decrease in their activities during dry months. Many studies have reported that the reduction in mineralized N with drying could be caused by reduced microbial activities and rewetting of the dry soil causes a flush of C and N due to mineralization indicating the microorganisms

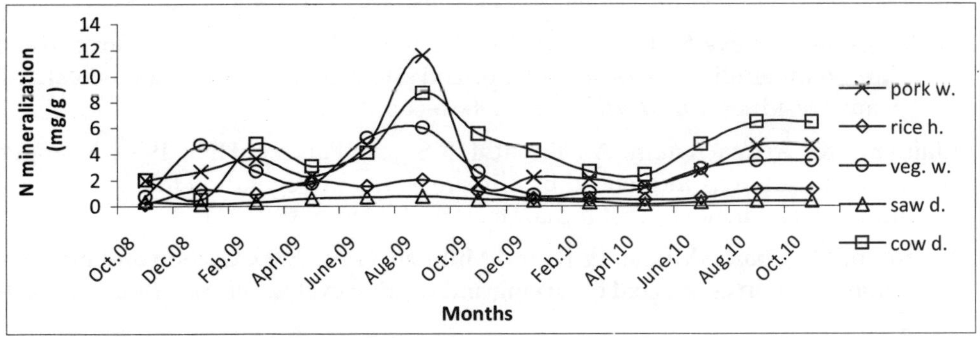

Figure 19.3: N-mineralization in the treated soil in different months.

regained activity upon rewetting (Griffin, 1981; Franzluebbers *et al.*, 1994; Maithani *et al.*, 1998; Pulleman and Tietma, 1999 and Maysoon *et al.*, 2005). In the last three months the rate of mineralization of N was stable as the wastes materials have completely decomposed. The rate of N mineralization was found to be higher in aal the treatments compared to the initial rate. The average N mineralization was found to be in the same order of decomposition pattern *i.e.*, cow dung> pork waste> vegetable waste > rice husk> saw dust. N mineralization rate of the pork waste on an average was high. The reason could be due to its faster decomposition rate. Same studies were reported by other researchers in their studies on animal wastes (Morvan and Parnaudeau, 2001).

From the study it was observed that due to addition of the solid wastes on soil, mineralization of N was increased as well as increasing the level of C and N. Therefore in this stage of development where there is increase in production of waste materials everyday, leading to shortage of dumping sites, usage of selected wastes materials can be taken up in order to enhance soil fertility as well as management of wasteland.

Acknowledgements

The authors would like to acknowledge the financial assistance given by CSIR, New Delhi, Scheme no. 24(0298)/08/EMR-II in conducting the present study

References

Anderson, J.M. and Ingram, J.S.I., 1993. *Tropical Soil Biology and Fertility: Handbook of Methods.* CAB International, 2: 82–102.

Christensen, B.T., 2004. Tightening the nitrogen cycle. In: *Managing Soil Quality: Challenges in Modern Agriculture,* (Eds.) P. Schjonming, S. Elmholt and B.T. Christensen. CABI Publishing, Cambridge, MA, USA, pp. 47–67.

Franzluebbers, K., Weaver, R.W., Juo, A.S.R. and Franzluebbers, A.J., 1994. C and N mineralization from cowpea plants part decomposing in moist and in repeatedly dried and wetted soil. *Soil Biol. Bioc.,.* 26: 1379–1387.

Griffin, D.M., 1981. Water potential as a selective factor in the microbial ecology of soil. In: *Water Potential Relations in Soil Microbiology,* (Eds.) J.F. Gardner Parr, W.R. Elliot L.F.. *Soil Sc. Soc. Am,.* Madison, WI, pp. 141–151.

Jha, L K., 1997. *Shifting Cultivation.* APH Publishing Corporation, pp. 194.

Jin, K., Sleutel, S., Neve, S.D., Gabriels, D.C., Jin, J. and Hofman, G., 2008. Nitrogen and Carbon mineralization of surface applied and incorporated winter wheat and peanut residues. *Biol. Fertl. Soils,* 44: 661–665.

Maithani, K., Arunachalam, A., Tripathi, R.S. and Pandey, HN., 1998. Nitrogen mineralization as influenced by climate, soil and vegetation in a subtropical humid forest in north-east India. *Forest Ecol. Man.,* 4373: 1–11.

Maysoon, M., CharlesMikha, W.R. and Milliken, G.A., 2005. Carbon and nitrogen mineralization as affected by drying and wetting cycles. *Soil Biol. Bioch.,* 37: 339–347.

Morvan, T., Dach, J. and Parnaudeau, V., 2001. Relationships between biochemical composition of manures and compost and their C and N transformations in soil. Book of abstracts of *11th Nitrogen Workshop*, INRA, Reims, pp. 153–154.

Morvan, T. and Nicolardot, B., 2009. Role of organic fractions on C decomposition and N mineralization of animal wastes in soil. *Biol. Fertil. Soils*, 45: 477–486.

Neve, D.S., Sleutal, S. and Hofman, G., 2003. Carbon mineralization from compost and food industry waste added to soil. *Nutrient Cycling in Agroecosystem*, 67: 13–20.

Pulleman, M. and Tietma, A., 1999. Microbial C and N transfer during drying and rewetting of coniferous forest floor material. *Soil Biol. and Bioch.*, 31: 275–285.

Environment, Biodiversity and Traditional System, 2014 *Pages* 217–225
ISBN 978-81-7622-307-2
Editors: D.R. Khanna, G.S. Solanki and S.K. Pathak
Published by: BIOTECH BOOKS, NEW DELHI

Chapter 20

Indigenous Knowledge and Zootherapeutic Analysis of Certain Wildlife Species and Conservation Practices in Arunachal Predesh

P. Chutia[1]* *and G.S. Solanki*[2]

[1]*Department of Zoology, D.R. College, Golaghat, Assam*
[2]*Department of Zoology, Mizoram University, Mizoram*

ABSTRACT

Present study deals with the use of 35 animal species as traditional medicines and cosmetics by indigenous people of the Arunachal Pradesh. Animal based medicine comes from insects, fishes, amphibians, reptiles. birds and mammals, but here it is emphasized on birds and mammals only. Medicinal use value of each animal was calculated by using standard formula. More medicinal use value, 0.2 has found in Himalayan Black Bear and Tiger which is followed by other large cats, rest of tiger and Horn bill with 0.09; lesser medicinal use value 0.04 has seen in ungulates. These results show how important this ethnozoological phenomenon is and indicate that traditional knowledge on zootherapy and impacts on biodiversity in the region.

Keywords: *Zootherapy, Wildlife, Biodiversity, Use value and Arunachal Pradesh.*

* *Corresponding Author:* E-mail: pabitra_sagar@yahoo.co.in

Introduction

Zootherapy is the treatment of diseases by the use of medicines that are obtained from the animals and their body parts. Marques (1994) explained zootherapy as "all human cultures that present a structured medical system utilize animals as medicines." Such a statement formed the basis of his 'zootherapeutic universality hypothesis.' Indeed, animals are therapeutic arsenals that have been playing significant roles in the healing processes, magic rituals, and religious practices of people's from different continents.

The consumptive interactions between human and animals has been shown both in indigenous and Western societies all over the world (Gudger 1925; Conconi and Pino 1988; Antonio 1994; Marques 1994; van Huis 1996; Jackson 1998; Costa-Neto 1996, 1999a, 1999b, 2000, 2003, 2004). Some authors have pointed out that animal-based medicines have been in use since antiquity (Angeletti *et al.*, 1992; Rosner 1992; Costa-Neto 2003).

Since uses of plants for therapeutic purposes was well known in ancient time and still use of plants for health giving is more popular. Studies on zootherapy have recently stimulated the interest in many researchers from different branches of science, who have recorded it a cultural practice and sought for compounds with pharmacological action (Werner 1970; But *et al.*, 1991;Lazarus and Atilla 1993). This interest increases when it is considered that the annual global trade in animal based medicinal products accounts for billions of dollars per year (Kunin and Lawton 1996). Marques (1997) in his publication mentioned that World Health Organization has selected 252 essential chemicals have been selected for therapeutic purposes, of them 11.1 per cent are plant origins, 8.7 per cent come from animals. Animal-derived compounds may bring considerable benefits to both producer and user in terms of monetary value and human welfare. Use of insects in the therapeutic activites is termed as entomotherapy (Costa-Neto 2003). Although entomotherapy is an ancient practice, it is still little known in the academic world. Only few authors have given the account of use of insects as medicines (Antonio 1994; Fasoranti 1997; Alexiades 1999; Zimian *et al.*, 1997; Green 1998; Namba *et al.*, 1998; Maya 2000; Costa-Neto 2003). They have been used live, cooked, ground, in infusions, in plasters, and in ointments as curative, preventive medicines as well as in magical-religious rituals. Compounds like terpenoids (triterpenoids and steroids, carotenoids, iridoids), sugars, polyols and mucilages, saponins, polyphenolic glycosides; quinines, anthraquinone glycosides, cyanogenic glycosides and alkaloids are the important compounds derived from the insects' body and used for ethno medicine. Kunin and Lawton (1996) have recorded that promising anticancer drugs, isoxanthropterin and dichostatin have been isolated from the wings of Asian sulphur butter flies (*Catopsilia crocale*, belongs to family pieridae) and from legs of Taiwanese stag beetles (*Allomyrina dichotomus* of Scarabaeidae family) respectively. Costa-Neto (2003) has reported that Cockroach (*Periplanata americana*) and other insects are good for Asthma.

However, very few studies are available on these aspects from Asia and South-east Asia (Solanki and Chutia 2004; Chutia *et al.*, 2008; Solanki and Chutia 2009). The use of traditional medicine derived from different categorizes animals is not uncommon in India. Tiger bone, horn of rhinoceros and wild buffalo are commonly used in traditional Chinese medicine (Jackson 1998; Barua 1998). Use of fruit bats (*Latidaens salimalii; Pteopus fanulus; P. hypomelanus; P. melanotus,*) to cure headache and female ailment is common in India and other South East Asian countries (Anon., 2003). Peoples of North Eastern Region also use animal and animal body parts in making traditional medicine (Lalramnghinglova 1999; Harit 2000, 2001, and 2002; Anon., 2003). The difference in killing of birds is due to the size of the population of the tribal groups and is also utilizes the animal and animal body parts for traditional medicine (Solanki *et al.*, 2004, 2005). Thus animals are the importance resources in traditional medicine vis-à-vis this tradition severely affecting the animal populations in their natural habitat. The impact of the zootherapy on natural animal populations is addressed here in this chapter. Information on animal losses due to these practices is not given in the introduction.

Materials and Methods

Questionnaire based information were collected about the animal based traditional medicine system and therapeutic activities prevailing among Apatani, Nyishi and Monpa tribal community.

The use-value of each animal was calculated on modifying the following equation:

$$UV = (\Sigma RM \times C)/N \text{ (Phillips and Gentry, 1991)}$$

where,

UV: Use value Index

RM: The total number of body parts use from individual animals,

C: The number of times which a particular animal has been cited, and

N: The total number informants questioned about Zootherapy.

The species utilizes in therapeutic purposes were identified by crosschecking of the vernacular names given by local folk with English name and reconfirmed by showing the photograph of the animals in the books (Mammals of India, Mammals of Arunachal Pradesh, Choudhery, 2002 and Birds of India, Ali, 1996).

Result and Discussion

The use of animal species by all tribal people in the northeastern region for the therapeutic purpose is one of the most consistent cultural practices. The animals and their therapeutic uses in different body ailments studied during three years (2002-2005). Mini-faunas includes annelids, arthropods mainly insects, mollusk, and some lower vertebrates like fishes and amphibians; larger faunas including reptiles, birds and mammals were recorded to be utilized for the therapeutic purposes. The faunal species extracted for the therapeutic activities have been quantified and expressed in Figure 20.1. The faunal categories utilized widely for the therapeutic purposes were mammals (48 per cent), birds (28 per cent), insects (17 per cent), and other categories were used in smaller quantity.

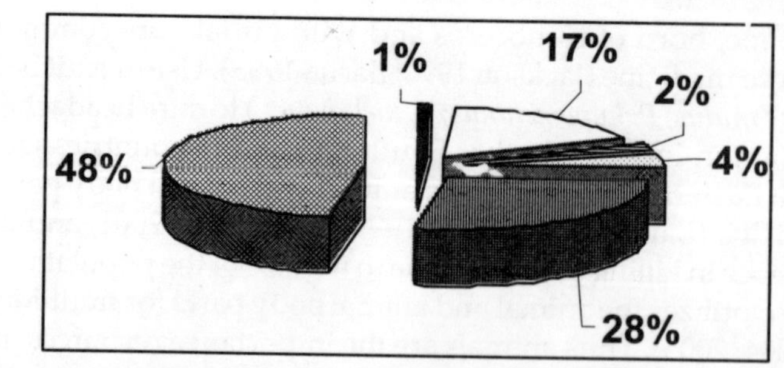

Fishes, Reptiles, Mollusks, Arthropods, Mammals, Birds

Figure 20.1: Therapeutic animal species utilize by tribal people.

Among the mammalian species harvested/killed, the maximum species harvested were the carnivorous, 38 per cent, followed by ungulates, 21 per cent, rodentia 15 per cent), primate 13 per cent), bear and chiroptera 5 per cent, and pholiodota was the least harvested (3 per cent)for therapeutic purposes. Among three tribal groups, Nyishi, Apatani and Monpa, the Apatani people extracted highest number (15) of carnivores species that constitute 40 per cent of extracted mammalian species, followed by Nyishi 36 per cent (n=13) and Monpa 35 per cent (n=10) for traditional medicine preparation. The ungulates are the next categories of mammalian species after carnivores used for traditional medicine. Amount of the extraction of this group of mammals by three tribes is 25 per cent by Monpa, 23 per cent by Nyishi and 19 per cent by Apatani people. The extraction of rodentia is nearly equal 17-16 per cent in all three tribal groups, whereas the primates were extracted maximum (14 per cent) by Apatani, the nyishi and monpa extracted this group of animal nearly equally, 9-10 per cent for traditional medicine. Bear extraction was 6 per cent by Nyishi 3 per cent by Apatani and Monpa people. The single species (Chinese pangolin, *Manis pentadactyla*) of pholiodota that was use equally 3 per cent (n=1) by all three groups of people. Chiropteran was extracted maximum by Monpa (7 per cent,), followed by Nyishi (6 per cent,) and by Apatani people (5 per cent), nearly equal.

Threatened Species as Zootherapeuticals

Out of 35 animal species utilized as source for the therapeutic activities by the indigenous people of the Arunachal Pradesh, 18 species are in list of threatened species as per the IUCN Red List 2000 and Wildlife (Protection) Act 1972. The threatened categories of mammalian species were regrouped and found that the 22 per cent of the total species belongs to endangered category, 50 per cent in Vulnerable, and 22 per cent in Lower Risked categories (Figure 20.2).

It is reflected from this study that the entire range of mammalian species extracted are not used in traditional medicine system except the Nyishi tribe. Nyishi people killed/extracted 35 mammalian species and utilized all of them for therapeutic purposes, Apatani tribe killed 39 mammalian species and utilized 35 species only where as Monpas utilized 29 mammalian species in therapeutic activities are of 38 killed.

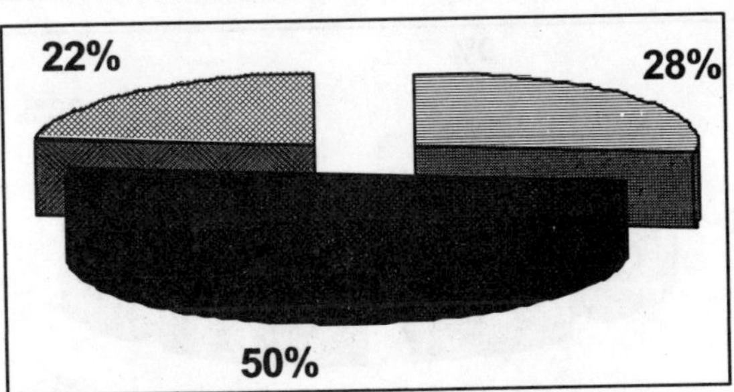

▩ Lower risked mammal, ▤ Endangered mammal, ■ Vulnerable mammal.

Figure 20.2: Composition of threatened categories of the mammals used in therapeutic activities.

Birds are also utilized in the traditional therapy. It indicates that the Passeriformes are used intensively by all the tribes; 31 per cent by Monpa, 30 per cent by Nyishi, and 29 per cent by Apatani followed by Ciconiformes, Strigiformes, Buceriformes, Piciformes, Psittaciformes and least number of Galliformes birds 5 per cent by Nyishi, 3 per cent by Apatani but no species in this utilized by Monpa tribe for therapeutic purposes (Figures 20.3–20.5).

Like mammals, all bird species extracted by tribal people are not used as therapeutic purposes. All three tribes extracted 53 bird species. Out of them, Nyishi people utilize 39 species, the Apatani people utilize 38 species and Monpa people utilize 29 species in their traditional medicine.

Mammal Body Parts in Therapeutic Activities

Generally whole body of mini-faunal species is used in therapeutic activities. The whole insect body (beetles, bug, bee and wasps) are used in traditional medicine of the indigenous people of the region. However, in large faunal species only specific body parts are used in particular body ailments. A total of 7 different body parts

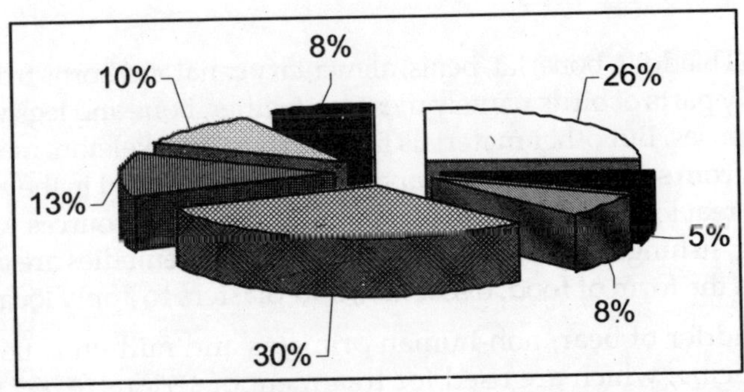

□ Ciconiforms, ▩ Passeriformes, ▤ Strigiformes, ■ Psittaciformes, ▦ Piciformes, ■ Galliformes, ▥ Buceriformes

Figure 20.3: Therapeutic use different categories of birds by Nyishi tribe.

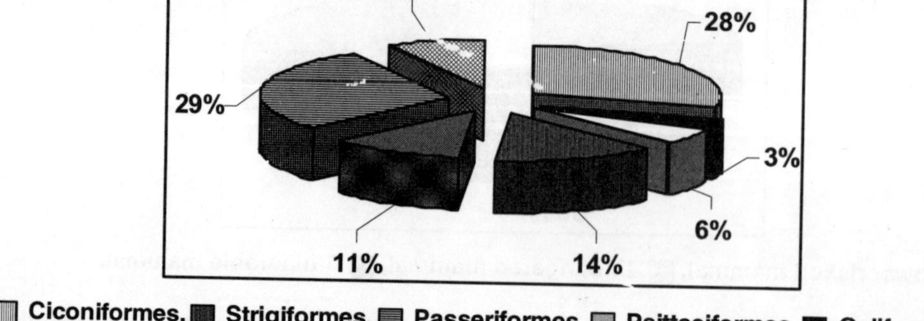

Figure 20.4: Therapeutic uses of different categories of birds by Apatani tribe.

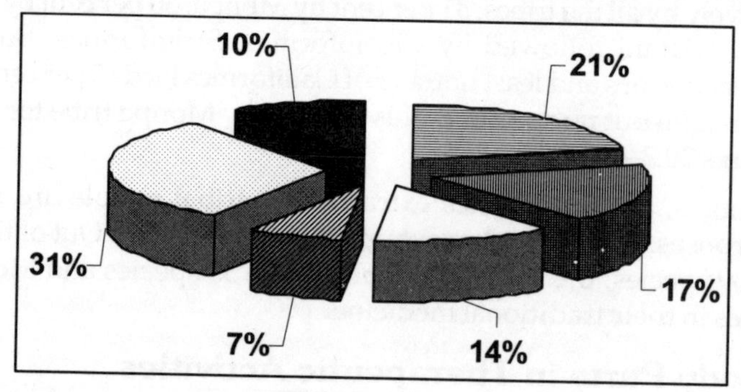

Figure 20.5: Therapeutic uses of different categories of birds by Monpa people.

namely meat, gall bladder, bone, fat, penis, alimentary canal and horns from mammals, 5 number of body parts of birds namely meat fat, feather, bone and legs were used for therapeutic purposes. But other materials like mucous, exoskeleton, mantle, tentacle and fresh blood from selective mini-therapeutic fauna were used in the elaboration of the remedies to treat locally diagnosed ailments. All of these resources were obtained through hunting, fishing and manual gathering. The folk remedies are administered to the patients in the form of food, decoction, and plasters to apply locally.

The gall bladder of bear, non-human primates and mithun is use to prepare traditional medicine, which are used for treatment of serious fever; like malaria, typhoid. Utilization pattern of the body parts is different in different tribes. Indigenous people of the region use meat of all the ethno-faunal species in their traditional medicine system. All three tribes, 54.0 per cent of mammals by Apatani, 51.4 per cent of ethno mammals for fat by Nyishi use fat and entire gall bladder of mammal species, and the Monpa used less mammalian species, 41.3 per cent. Mammalian bone is also used in traditional medicine system of all three tribes. Apatani people are utilized 40.5 per cent mammalian species, Nyishi people utilized 37.1 per cent of species and the Monpa people utilized 34.4 per cent of species for bone in their folk medicine. Penis and horns of mammals particularly ungulates are also used as raw materials of their traditional medicines. The Nyishi people used highest number of species, 20.0 per cent followed by Apatani people, 18.9 per cent and the Monpa people used minimum number of 17.2 per cent of animal species.

Meat, fat, feather, bone and legs of the different avian species are utilized as raw materials for traditional medicine. Number of avian species and their use by the tribal group is given in table 4.4. Meat is the commonly used from all of the extracted species. But the number of species (n) extracted for the purposes by three tribes are different. Among all of these three tribes, the Apatani people utilized 80 per cent (28 species), Nyishi people 76.9 per cent (30 species) and the Monpa people with lowest 68.9 per cent (20 species) for fat as raw materials of traditional medicine. Nyishi people utilized birds and their body parts for therapeutic activities. Highest 82 per cent of bird species which is followed by Apatani with 71.4 per cent and Monpa people, 62 per cent of species. Monpa people extracted 62 per cent of bird species for leg parts in their traditional medicine. But the Apatani people extracted 57.1 per cent and Nyishi people extracted 51.2 per cent of bird species in their traditional medicine. For bone, Nyishi people extracted 64.1 per cent of bird species, Apatani people extracted 62.8 per cent and the Monpa people extracted 51.7 per cent of bird species in traditional medicine.

About 90 animal species including both mini-faunas comprises 16 species of invertebrate and 3 number fish species. Large faunal species comprises 4 reptilian species, 39 bird species and 37, mammalian species were reported to be used by the tribal people for therapeutic in present study. Body parts and therapeutic use value of each of the animal species in the activities of therapeutics ailments, and cited number of ailment or disease (C, given in the formula) of each species. The larger cats namely *Panthera tigris*, and *P. paradas* and bears are the animal contribute to the traditional medicine through their body parts like bones, gall bladder and meat. It is found in this study that tiger, leopard, Himalayan black bear, sloth bear, primates, mithun and hornbills are the most variedly used animals. However, the therapeutic use value (UV) of tiger and Himalayan black bear is 0.20, hornbill, leopard and sloth bear was calculated to be, 0.1 and 0.09, respectively, indicates extensive use of the these animal categories. Some species with UV, ranging from 0.02–0.20 were found to be important resources for therapeutic activities. Non-human primates dominate this category.

References

Alexiades, M.N., 1999. Ethnobtany of Ese Eja: Plants, health and changes in an Amazonian society. *Ph. D. Thesis*. City University of New York.

Angelici, L.R., Agrimi, U., Curia, C., French, D. and Marfani, R., 1992. Healing rituals and sacred serpents. *The Lanset*, 340: 223–225.

Anon., 2003. *The Wildlife (Protection) Act, 1972 Amendment, 2003*. Nataraj Publishers, Dehradun, India.

Antonio, T.M.F., 1994. Insects as remedies for illness in Zaire. *The Food Insects News Letter*, 7(3): 4–5.

Barua, M., 1998. A Sanctuary Report (Kaziranga). *Sanctuary Asia*, 18(5): 56–64.

But, P.P.H., Tam, Y.K. and Lung, L.C., 1991. Ethnopharmacplogy of rhinoceros horn II: Antipyretic effects of prescriptions containing rhinoceros horn and water buffalo horn. *J. Eth. Pharmacol.*, 33: 45–50

Chutia, P., Solanki, G.S. and Singh, O.P., 2008. Extraction of mammals and its impacts on biodiversity of Arunachal Pradesh. In: *Proc. International Conference on Biodiversity Conservation and Management*, (Ed.) P. Natarajan. Cochin University of Sciences and Technology, Cochin, India, pp. 128–136.

Conconi, J.R. E. and Piano, M.J.M., 1998. The utilization of insects in the emperical medicine of ancient Mexicans. *J. Ethnobiol.*, 8(2): 195–202

Costa-Neto, E.M., 1996. Faunistic resources used as medicine by an Afro Brazilian Community from Chapada Diamantia National Park, State of Bahaia, Brazil. *Sitent*, 15: 211–219

Costa-Neto, E.M., 1999 a. Traditional use and sale of animals as medicines in Feria de Santana city. *J. Ethnopharmacol.*, 7(2): 6–9

Costa-Neto, E.M., 1999b. Healing with animals in Feria de Santana city. *J. Ethnopharmacol.*, 65: 225–230

Costa-Neto, E.M., 2000. Ethnotherapy in the country of Matina dos Pretos, State of Bahia. *The Food Insects News Letter*, 11(2): 1–3

Costa-Neto, E.M., 2003. The use of insects in folk medicine in the state of Bahia. *Hum. Ecol.*, 30(2): 245–262

Costa-Neto, E.M., 2004. Implication and application of Folk Zootherapy in the state of Bahia, Northeastern Brazil. *Sus. Dev.*, 12: 161–174.

Fasoranti, J.O., 1997. The place of insects in the traditional medicine of southwestern Nigeria. *The Food Insects News Letter*, 10(2): 1–5

Green, S.V., 1998. The bushman as an entomologist. *Antenna*, 22(1): 4–8

Gudger, E.W., 1925. Stitching wounds with the mandibles of ants and beetles. *J. American Medical Association*, 84: 1862

Harit, D.N., 2000. Non-human primates in Mizoram, India. *Him. J. Env. Zool.*, 14: 161–164

Harit, D.N., 2001. Zootheraputic use of animals in Health Care in Mizoram, India. *J. Nat. Con.*, 13(1): 117–122

Harit, D.N., 2002. Report on porcupine in Mizoram, India. *Him. J. Env. Zool.*, 16(1): 27–29.

Jackson, P., 1998. Information from the Cat Specialist Group of the World Conservation Union (IUCN). *Sanc. Asia*, 18(5): 88–89

Kunin, W.E. and Lawton, J.H., 1996. Does biodiversity matter? Evaluating the case for conserving species. In: *Biodiversity: A biology of Number and Difference*, (Ed.) K.J. Gaston, pp, 283–308, Oxford.

Lalramnghinglova, H., 1999. Ethnobiology in Mizoram state: Folklore medico-zoology. *Bull. Indian Hist. Med.*, 29(2): 123–148

Lazarus, L.H. and Attila, M., 1993. The toad, ugly and venomous, wears yet a precious jewel in his skin. *Neurobiology*, 41: 473–507

Marques, J.G.W., 1994. A fauna medicinal de San Blas (Panama) a hipotense da universalidate zootherapica, *Proceeding of 46th Annual Meeting*, Espirito Santo Federal University. 7: 687.

Marques, J.G.W., 1997. Fauna medicinal recuro do ambiente on ameaca a biodiversity? *Mutim*, 1(1): 4

Maya, E.M.A., 2000. Ethnoentomologia de la Comunidad, *Hnahunu EL–Dexthi–San Juanico Hidalgo*

Nama, T., Ma, Y.H. and Inagaki, K., 1998. Insect derived crude drugs in the Chines Song Dynesty, *J. Ethnopharmacology*, 24: 247–285.

Rosner, F., 1992. Pigions as a remedy for jaundice. *J. Medicine*, 92(5): 189–192.

Solanki, G.S. and Chutia, P., 2004. Ethno zoological and socio-cultural aspects of Monaps of Arunachal Pradesh. *Journal of Human Ecology*, 15: 251–254.

Solanki, G.S., Chutia, P. and Singh, O.P., 2004. Headgear: A cultural artifact and its impact on biodiversity in Arunachal Pradesh. *Rajiv Gandhi University Research Journal*, 7(1): 35–44.

Solanki, G.S., Chutia, P. and Singh, O.P., 2005. Ethnozoolgy of Nyishi tribe and its impact on biodiversity in Arunachal Pradesh. *Rajiv Gandhi University Research Journal*, 8(1): 89–100.

Solanki, G.S. and Chutia, P., 2009. Studies on ethno-medicinal aspects and zoo-therapy in tribal communities in Arunachal Pradesh, India. *International Journal of Ecology and Environmental Science*, 35(1): 67–76.

Van Huis, A., 1996. The traditional use of arthropods in Sub-Saharan Africa. *Procedings of Experimental and Aplied Entomology*, Ammstradam, 7: 3–20.

Weirner, D. 1970. Healing in Sierra Madre. *The Nat. Hist.*, 79(9): 61–66.

Zimian, D., Yonghua, Z. and Xiwn, G., 1997. Medicinal insects in China. *Ecolgy of Food and Nutrition*, 36: 209–220.

Environment, Biodiversity and Traditional System, 2014 Pages 227–233
ISBN 978-81-7622-307-2
Editors: D.R. Khanna, G.S. Solanki and S.K. Pathak
Published by: BIOTECH BOOKS, NEW DELHI

Chapter 21

The Qualitative Status of Tuikhur (Spring Water) and Hand Pump Water in the Eastern Flank of Aizawl City, Mizoram

Laltlankima[1], *Shiva Kumar*[1], *K.B. Singh*[2], *Vinod K. Bharati*[3] *and D. Chandrasekharam*[4]

[1]*Department of Geology, Mizoram University, Aizawl*
[2]*Department of Chemistry, Govt. Zirtiri Residential Science College, Aizawl*
[3]*Department of Chemistry, Govt. Kolasib College, Kolasib*
[4]*Department of Earth Science, IIT Mumbai, Pawai*

ABSTRACT

The main source of potable water in Aizawl city is through the supply of water maintained by Public Health Engineering Department PHED, Govt of Mizoram. The supply of water by PHED is irregular and inadequate. The people of Aizawl also depend on tuikhur (Artificially fabricated pit for accumulation of sub-surface water seepage) and hand pumps for the alternate source of potable water for their domestic purposes. The water samples were collected from different locations in the eastern flank of Aizawl for their physico-chemical analyses such as pH, electrical conductivity (EC), total dissolved solids(TDS), turbidity, phenolphthalein alkalinity, total alkalinity, total chloride, total hardness, iron and fluoride. Total coliform (MPN) and faecal coliform (MPN) were also determined for bacteriological characteristics in order to obtain the current status of potable water quality in the study area. The results show that the water quality of tuikhur and hand pumps which are used for domestic purposes are by and large within the acceptable limits.

Keywords: Alkalinity, Total dissolved solid, Tuikhur, Hand pump.

Introduction

Water is one of the most essential elements to good health. It is necessary for the digestion and absorption of food; helps maintain proper muscle tone; supplies oxygen and nutrients to the cells; rids the body of wastes; and serves as a natural air conditioning system. Water requirement for drinking and domestic uses depends on the physical and socio-economic status of the area. Water shortage problem may be regional or local in nature. Transporting water at large scale from one place to another is highly expensive, and hence, the demand of fresh water supply has to be fulfilled from indigenous resources. The growth of population and rapid urbanization enhances water requirement which exerts additional pressure on both quality and quantity of water. Therefore, the maintenance and preservation of existing good water source and quality should be another major component of any urban planning.

Aizawl is the capital city of Mizoram and it has been facing water problem due to urbanization and inadequate supply of water. The population of Aizawl is 291,822 as per 2011 census. The principal source of potable water in Aizawl city is by the supply through Public Health Engineering Department (PHED), Govt of Mizoram. The department procures water from Tlawng River only by vertical tube pumping with heavy machines. This supply is very limited not only during the non-monsoon seasons but also during the rainy season. The frequency of water supply varies from once a week to once a fortnight depending on the availability of water to the supply sources, which are fed by rain, weather and working condition of the machines. Therefore, a good chunk of population uses stream (tuikhur) and hand pump water for drinking purpose. Sometimes, water for household requirement is supplied by private tankers to meet the daily needs. The hand pump, tuikhur and tanker water are used without any treatment in the households including drinking purpose. The present paper discusses water quality of Eastern flank of Aizawl city in Mizoram state. There are several tuikhur and hand pumps in this part of the city. These tuikhur and hand pump are fed by sub-surface water and remain productive throughout the year. Water samples from 11 of tuikhurs and 6 hand pumps have been collected for the present study. The aim of this paper is to assess the physico-chemical and bacteriological characteristics of water from the selected tuikhurs and hand pumps located in the study area.

Material and Method

Tuikhur and hand pump samples were collected in dry and clean plastic containers using grab sampling method. pH, EC and TDS of all the samples were determined by the portable pH meter-335, EC meter and TDS meter repectively. Turbidity of the samples was measured by using digital Nephelo Turbidity meter-132 using formazin as standard. The most probable number (MPN) of coliforms was estimated by multiple tube dilution method. The multiple tube dilution technique was used to enumerate positive presumption and confirmed coliform test. Phenolphthalein alkalinity, total alkalinity, total chloride and total hardness were determined by titrimetric methods. Iron and fluoride were measured on Spectroquant NOVA-60.Total alkalinity and total hardness by titration methods.

Result and Discussion

The result of various physicochemical and bacteriological analyses is summarized in Tables 21.1 and 21.2. The pH in the samples ranges from 6.2 to 7.8 which are found well within prescribe limit for the quality of drinking water as specified by BIS (1991). The pH value of 7.0 is considered to be advisable and ideal one for most of the reaction as well as for the human being (Santhi, 2012).Electrical conductivity is the measure of salinity which greatly affects the taste and thus has a significant impact on the user acceptance of the water as potable. In the present study, the electrical conductivity varies from 88.2 to 1153µS/cm. EC values are higher in Tuikhur water in comparison with Hand pump water. Permissible limit for EC is 300µS/cm but in some of the samples the values are higher with the maximum of 1153µS/cm in Bawngkawn hmar Veng tuikhur.The total dissolved solid (TDS) ranged from 127 to 576 mg/l. According to the report of WHO(1996), TDS should be between 500-1500 mg/L. TDS value of tuikhur water is more in comparison with hand pump water. TDS values are found within desirable limit (500 mg/l)in most of the sample, but in one sample TDS is 576 which is under permissible limit (500- 2000 mg/l). It is reported that the higher concentration of TDS are aesthetically and may also produce distress in human and livestock (Yadav, 2010). Turbidity value ranged from 0.05 to 10.47. The WHO recommendation of turbidity is 5 NTU. All the samples except Bawngkawn hmar veng tuikhur are well within desirable limit. Bawngkawn hmar veng sample turbidity is more than permissible limit. The high turbidity may be due to local activity of nearby tuikhur at the time of collection of the sample. Only one sample has P-Alkalinity that is 10 mg/l whereas total alkalinity varied from 28 mg/l to 370 mg/l. Electric veng hand pump and Bawngkawn Hmar veng tuikhur samples' total alkalinity are higher than desirable limit of 200 mg/l (BIS, 10,500; 1983). All other samples total alkalinity are well within desirable limit i.e 200 mg/l. Total hardness ranges from 54mg/l to 392mg/l in the water samples. WHO (1996) recommended (100-500 mg/L) as safe permissible limit for hardness. Total iron is present in traces in majority of the samples, but in one hand pump sample from Ramhlun 'N' near Y.M.A V.Section godown has very high content of iron i.e 4.5mg/l. The maximum permissible concentration of iron in drinking water as per BIS 1983 and WHO guideline of 1993 is 0.3. Hand pump samples have more iron in comparison with tuikhur samples. This may be due to rusting in the hand pump tube well. Fluoride concentration in all samples is well within the desirable limit of 1mg/l (WHO, 2008). Total coliform and faecal coliform value were high in general but they were too high in four samples. The total coliform density varies from 0/100ml to 2400/100 ml. As per ISI drinking water standards, more than 10 MPN/100 ml should not be present in any sample but the values exceed this limit in all the samples except four water samples. These high values may be due to infiltration of sewage from drains as most of the tuikhur and hand pumps from where these samples were collected are situated near the sewage drains.

Conclusion

The results obtained from the physico-chemical and bacteriological studies revealed that water samples of selected areas are reasonably good for their use as

Table 21.1: Physico-chemical characteristics of tuikhur water.

Sample No.	Location	pH	EC µS/cm	TDS (mg/L)	Turbidity (NTU)	P-Alkalinity (mg/L)	Total Alkalinity (mg/L)	Total Hardness (mg/L)	Total Fe (mg/L)	Total F (mg/L)	Total Caliform	Faecal Caliform
1	Saron 1	**6.7**	450	240	1.5	Nil	38	54	Trace	0.6	0	0
2	Saron 2	6.5	439	340	3	Nil	58	158	Trace	0.3	23	0
3	Electric Veng 1	6.7	340	200	2	Nil	34	85	Trace	0.5	2400	0
4	Bawngkawn hmar veng	**7.8**	1153	576	10.47	10	370	392	0.1	0.5	1600	41
5	Ramthar	**6.5**	300	360	0.03	Nil	50	126	Trace	0.3	2400	1100
6	Armed Veng	**7.5**	210	380	2	Nil	38	124	Trace	0.5	93	15
7	Bethlehem 1	**6.2**	300	400	1	Nil	74	112	Trace	0.2	15	3.6
8	Dawrpui	**7.5**	88.2	500	0.14	Nil	60	44	Trace	0.4	540	7
9	Ramhlun 'N' 1	6.6	439	220	1	Nil	42	94	Trace	0.2	23	0
10	Ramhlun 'N' 2	6.8	798	239	0.53	Nil	28	126	Trace	0.3	130	49
11	Aizawl Venglai	6.5	344	375	3	Nil	98	248	Trace	0.2	2400	35

Table 21.2: Physico-chemical characteristics of Hand pump water.

Sample No.	Location	pH	EC µS/cm	TDS (mg/L)	Turbidity (NTU)	P-Alkalinity (mg/L)	Total Alkalinity (mg/L)	Total Hardness (mg/L)	Total Fe (mg/L)	Total F (mg/L)	Total Caliform	Faecal Caliform
1	Ramhlun 'N' 3	6.8	254	127	2.34	NIL	78	82	4.5	0.2	0	0
2	Bethlehem 2	6.6	300	200	2	NIL	170	152	0.1	0.3	0	0
3	Electric veng 2	6.7	280	136	1	NIL	240	240	0.2	0.7	0	0
4	Ramhlun 'E'	7.7	434	217	0.05	NIL	30	112	Trace	0.5	17	0
5	Ramhlun'N' 3	6.5	376	188	2.89	NIL	40	110	0.2	0.6	350	23
6	Ramhlun 'S'	6.6	400	200	0.39	NIL	54	106	0.2	0.4	13	2

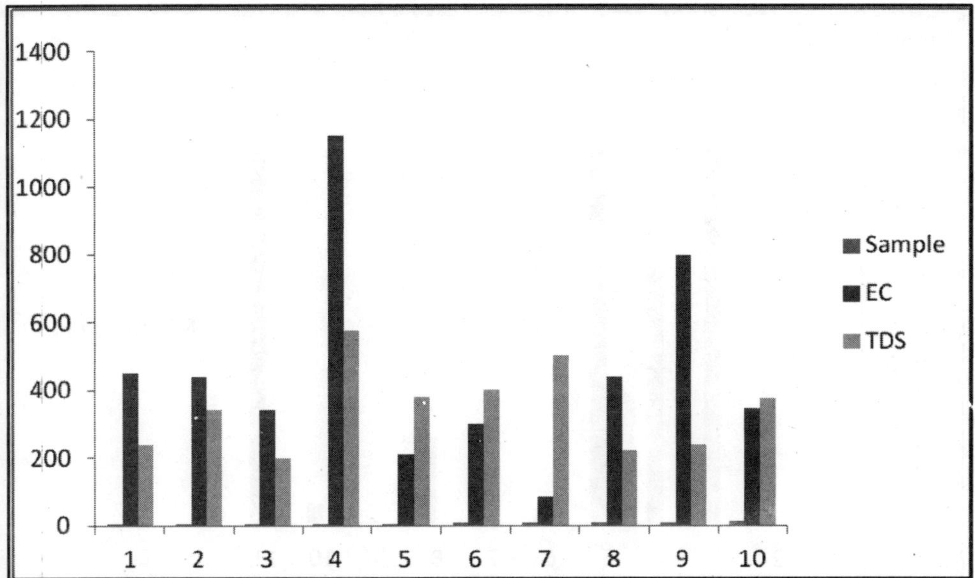

Figure 21.1: Variation of EC and TDS in tuikhur water.

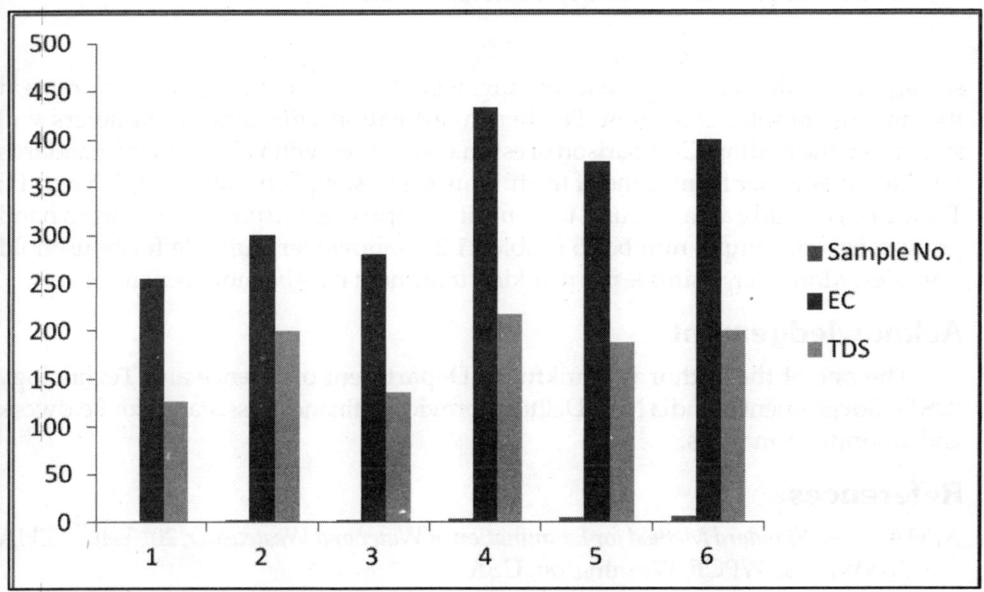

Figure 21.2: Variation of EC and TDS in hand pump water.

domestic and potable water as per WHO and ISI standards. Bacteriological parameter is in consistent among tuikhur samples spite of the fact that bacteriological contaminations are not found in some samples. But tuikhur samples from Bawngkawn Hmar Veng, Electric Veng, Aizawl Venglai and Ramthar are having high bacteriological contaminations. These highly contaminated tuikhur are located near

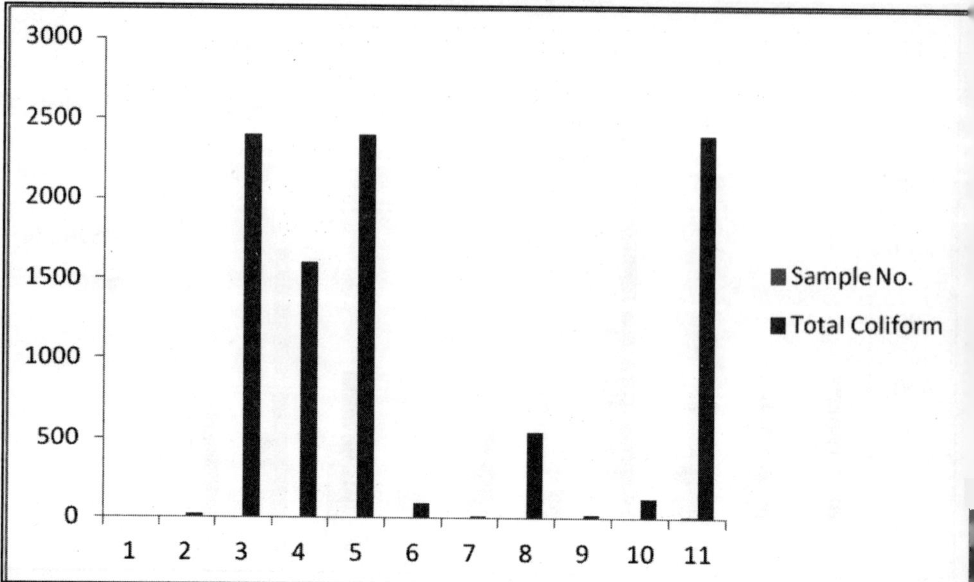

Figure 21.3: Variation of total coliform in tuikhur water.

sewage drains. It seems that these tuikhurs were fed by water from sewage drains at the time of samples collection. Further investigation with other parameters will strengthen the finding. Comparison of estimated values with WHO and ISI standards revealed that water from some of the tuikhurs (*viz.* sample number 3, 4, 5, 8 and 11; Table 21.1) of study area is found to be unfit for drinking purpose. Water from hand pumps, except sample number 5 (Table 21.2) is, however, suitable for household activities. Moreover, a simple pre drinking treatment may be more beneficial.

Acknowledgement

The one of the author is thankful to Department of Science and Technology (DST), Government of India New Delhi for providing financial assistance for fieldwork and laboratory analysis.

References

APHA, 1998. *Standard Method for Examination of Water and Wastewater*, 20th Edn. APHA AWWa and WPCF, Washington, USA.

BIS, 1983. *Specification for Drinking water, IS: 10500: 1983*. Bureau of Indian Standards, New Delhi.

BIS, 1991. *Specification for Drinking Water IS: 10500: 1991*. Bureau of Indian Standard, New Delhi.

Jain, P.K. Hydrogeology and quality of ground water around Hirapur, District Sagar (M.P.). *Poll. Res.*, 17(1): 91–94.

Kiran, A. and Jha, K., 2011. Assessment of drinking water quality of Madhubani town, Bihar, India. *Poll Res.*, 30(2): 201–204.

Kumar, T., Chandel, D., Singh, C.P., 2011. Drinking and irrigational feasibility of ground waters in Jaipur City, Rajasthan, India. *Poll Res.*, 30(2): 187–193.

Santhi, D., 2012. Assessment of water quality parameters in Kodumudiaru dam at Nanguneri Taluk of Tirunelveli District, Tamil Nadu, India. *Poll. Res.*, 31(1): 83–86.

Vyas, P.B., 2011. Assessment of drinking water Gandhinagar Town, Gujarat, India. *Poll. Res.*, 30(2): 161–163.

WHO, 1993. *Guidelines for Drinking Water.* CBS Publishers and Distributors, New Delhi.

Yadav, K. Kamalesh, 2010. Physico-chemical analysis of sewage drains at Darbhanga, Bihar, India. *Poll Res.*, 29(1): 31–32.

Environment, Biodiversity and Traditional System, 2014 ***Pages* 235–242**
ISBN 978-81-7622-307-2
***Editors:* D.R. Khanna, G.S. Solanki and S.K. Pathak**
***Published by:* BIOTECH BOOKS, NEW DELHI**

Chapter 22

Environmental Geomagnetism: Implications for Bio-magnetic Monitoring through Plant Leaves

Prabhat Kumar Rai, Lalita L.S. Panda, Biku Moni Chutia*

Department of Environmental Science, Mizoram University
Tanhril, Aizawl – 796 004, Mizoram

ABSTRACT

It is well known that air is an integral component of environment responsible for various life processes and hence persistence of life on earth. In recent Anthropocene era, rapid pace of industrialization and urbanization has given birth to dust/particulate matter (PM) pollution. Solid matter, which is composed of soil, anthropogenic metallic constituents, and natural biogenic materials, is termed as PM. Air quality is greatly being altered through the problem of particulate pollution which is inextricably linked with human health. Moreover, vehicle derived pollutants as well as industrial emissions simultaneously release deleterious fine-grained PM and magnetic particles into the atmosphere. These magnetic particles are derived from the presence of iron (as impurities in fuels, industrial emissions, street dust etc.), often a mix of strongly magnetic (magnetite-like) and weakly magnetic (haematite-like) iron oxides. Since conventional technologies for monitoring of PMs are not feasible in totality, they paved the way to bio-magnetic monitoring approach. Magnetic properties may act as a valuable tool in assessing the burning phenomenon of climate change, as demonstrated through the glacial and interglacial characterization of Chinese Loess Plateau on one hand, while on other hand it act as a proxy for particulate pollutants through roadside plant leaves. With the advent of environmental magnetism, magnetic measurement is becoming an important means in particulate pollution study. Present paper provides an overview on the problem of PM pollution, its environmental geomagnetic studies with

* *Corresponding Author:* E-mail: prabhatrai24@gmail.com

special reference to bio-magnetic monitoring through roadside plant leaves. Bio-magnetic monitoring of particulate pollution may biomonitor the extent of dust/particulate pollution in an effective way. More focused research works are required in the multifaceted environmental dimensions of magnetic monitoring particularly bio-magnetic monitoring of particulate pollution with roadside plant leaves and hence possess the potential to become a new frontier in the field of feasible pollution control technologies. Moreover, roadside plants may be linked with socio-economy and hence the livelihood of the tribal people.

Keywords: ***Livelihood, Human health, Environmental geomagnetism, Biomonitoring,. Particulates.***

Introduction

In the modern era of rapid industrialization and urbanization, air quality is greatly being altered through the release of huge amount of particulate matter (PM) which may be suspended (SPM) or respirable (RPM). It is well known that air pollution represents a threat to both the environment and human health. Impact of PM may be correlated with urban planning as well as topography of the particular region (Rai, 2011a, b).

PM/dust is actually solid matter, which is composed of soil, anthropogenic metallic constituents, and natural biogenic materials, is called dust/PM (Ferreira-Baptista and DeMiguel, 2005). Sources of PM may be natural (wildfires, volcano eruption and dust storms) or anthropogenic (energy intensive industries and traffic vehicles). Most importantly, PM also includes magnetic particles resembling magnetite or haematite. In environment, these magnetic particles with varying magnetic status may be associated with vehicular and industrial emissions (Figure 22.1a). This school of thought lead to researches which ultimately diverged into climate change studies as demonstrated through glacial and interglacial characterization of Chinese loess plateau and deep seas sediments (Figure 22.1a). In simple words, loess is actually an extensive deposit of windblown dusts. Also, it is well known that these Chinese loess plateaus are deposited over more than two million years without being perturbed, particularly in Mongolia and Tibet (Maher, 2009). Moreover, there exists a strong impact of East Asian summer monsoon on the deposition of loess. It comprises major components and minor fractions/traces of magnetic minerals *viz.* quartz, phyllosilicates etc. As far as seasonal variations of loess deposition are concerned, during glacial period of intense winter, there is more deposition as cold dry winds flow towards pacific (Figure 22.1b). However, less deposition takes place during interglacial phase of summer due to excessive weathering caused by development of soil and growth of vegetation. It is worth to mention that vertical sequences of loess or palaeosols may be analyzed for their magnetic properties *e.g.* magnetic susceptibility. Magnetic susceptibility is usually low for less weathered loess *i.e.* during winter and vice versa. Thus, magnetic signatures of loess assist in tracing of winter and summer monsoon and hence climate change that has been occurred from quaternary geological period (Maher, 2009). Another school of thought diverged into magnetic studies of particulates on roadside plant leaves leading to the concept of biomagnetic monitoring. Present paper was in the line of this school of thought and hence we confined our

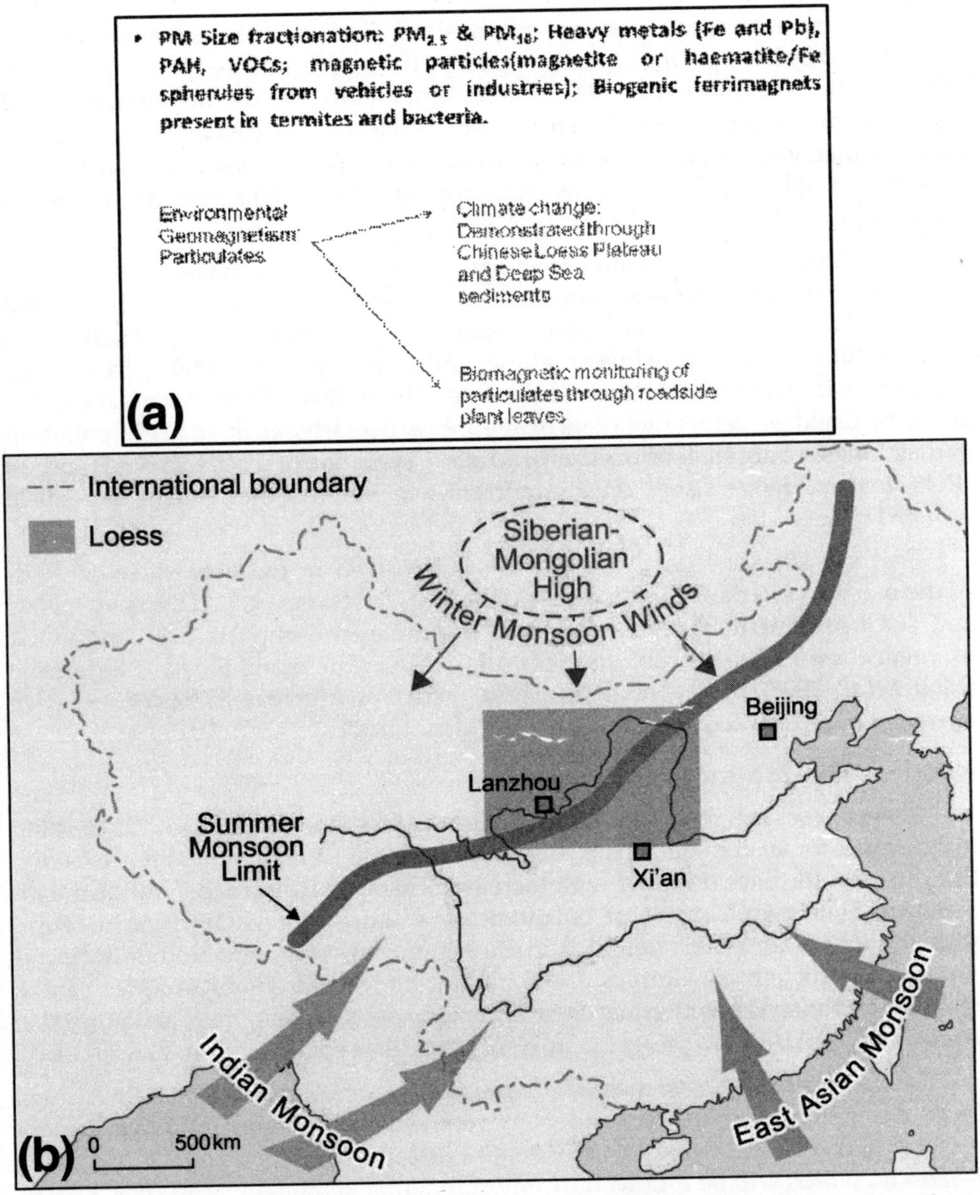

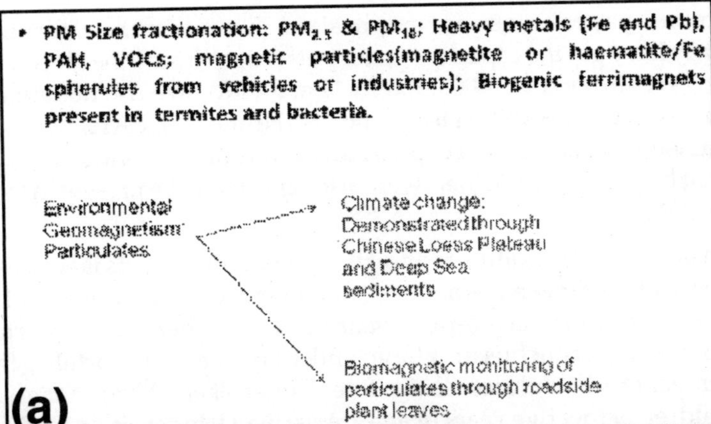

Figure 22.1: (a): Divergence of global researches on environmental magnetism; (b) Impact of East Asian summer monsoon on loess deposition.

quest towards bio-magnetic monitoring approach and its implications in air pollution science. Further, the plants used for magnetic biomonitoring may be linked with socio-economy or livelihood of the local people.

Measurements of the PM in ambient air are usually reported as the mass of particles with an aerodynamic diameter that is less than 2.5 µm ($PM_{2.5}$) or 10 µm

(PM_{10}). Aforesaid particle sizes are emphasized in view of their pertinent health impacts, as they have got the capacity to penetrated deep inside the alveoli or alveolar tissues of lungs (Pandey *et al.*, 2005; Rai, 2011 a, b). PM Size fractionation revealed that it comprises $PM_{2.5}$ and PM_{10}; Heavy metals (Fe and Pb), PAH, VOCs; magnetic particles (magnetite or haematite/Fe spherules from vehicles or industries). It is worth to mention that biogenic ferrimagnets were also reported to be present in termites and bacteria (Rai, 2011b).

Relation between PM pollution and Human health impacts is well established through a vast array of researches and well documented in literatures. $PM_{2.5}$ causes 3 per cent of mortality from cardio-pulmonary disease; 5 per cent of mortality from cancer of the trachea, bronchus and lung; and 10 per cent of mortality from acute respiratory infections in children under five (Cohen *et al.*, 2005; Maher, 2009; Rai, 2011a, b). Children below five years of age are particularly sensitive to PM pollution as their lungs are not as developed as of adults. Generation of reactive oxygen species (ROS) and oxidative stress is an important toxicological mechanism of particle induced lung cancer.

Further, Ultrafine particle exposure is involved in the very early onset of Alzheimer's and Parkinson's diseases (Maher, 2009). Moreover, PAH, Heavy metals and VOCs present on PM cause genotoxic and mutagenic impacts. PAH and VOCs in conjunction lead to the formation of bulky DNA adducts and blood cells damage (Moller *et al.*, 2008). In view of the aforesaid, there is an urgent need for cost-effective as well as non-destructive technology for PM monitoring.

Need of Bio-Magnetic Monitoring

There are several reasons to opt for magnetic biomonitoring approach. Existing technologies for abatement of particulates are neither cost-effective nor eco-friendly; PM_{10} values decrease not only with increased distance from roads, but also with increased height while conventional monitoring stations are distantly located from residential area and their height is in excess of 3 m; Magnetic biomonitoring is potentially efficient as samples are abundant and hundreds of samples can be collected and analyzed within few days; There is no power requirement and protection from vandalism; Act as a proxy for ambient particulate concentration (Rai, 2011a, b).

In present paper, I have mentioned the outcome of a pilot scale study in Aizawl, India, having implications for further environmental geomagnetic studies. Present research initiated a pilot study on dust deposition capacity of certain plants in Aizawl, Mizoram, which will be followed by environmental geomagnetic studies. Leaves from roadside trees form natural, widely distributed collection surfaces for pollutants. The recorded plants were *Hibiscus rosa-sinensis, Psidium guajava, Mikania micrantha, Mangifera indica* and *Bougainvillea spectabilis*.

Materials and Methods

Methods for dust deposition capacity are mentioned precisely in Table 22.1. Magnetic studies of particulates may be performed with wide range of magnetometers as shown in Figure 22.2. Roadside plant leaves are individually applied under magnetic field and then several parameters are assessed through hysteresis loop

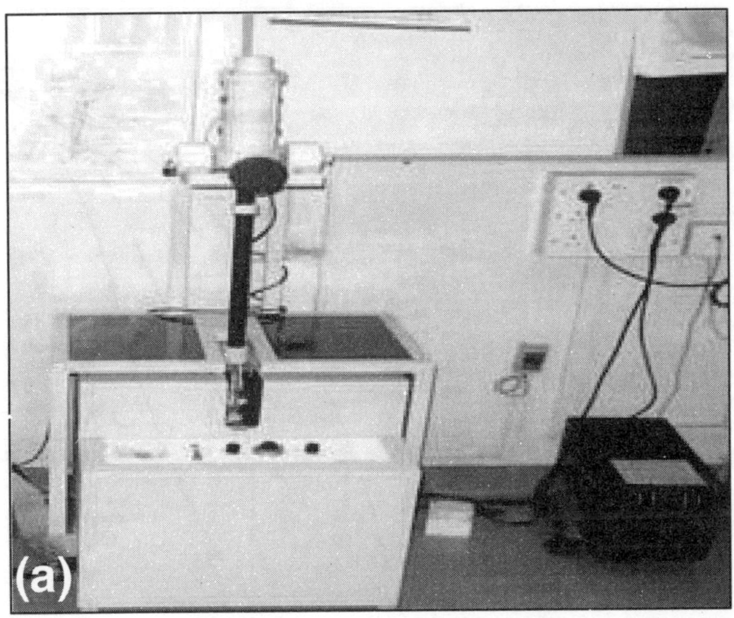

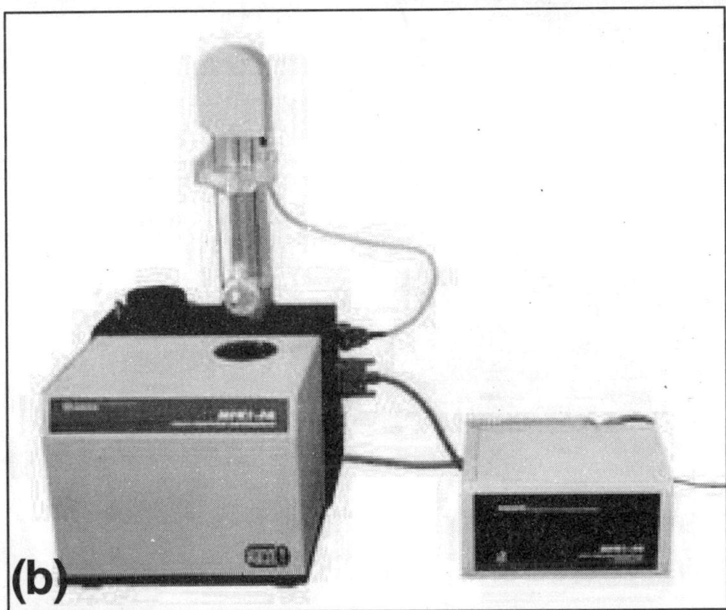

Figure 22.2: (a) Molspin vibrating sample magnetometer (VSM); (b)AGICO-MFK1-FA Multifunction Frequency Kappabridge for measuring magnetic susceptibility at IIGM, Mumbai.

(Figure 22.3). The resultant isothermal (*i.e.* room temperature) magnetic remanences (IRMs) are generally measured using a cryogenic/other applicable magnetometer. The magnetization values of plant leaves may be correlated with the dust content

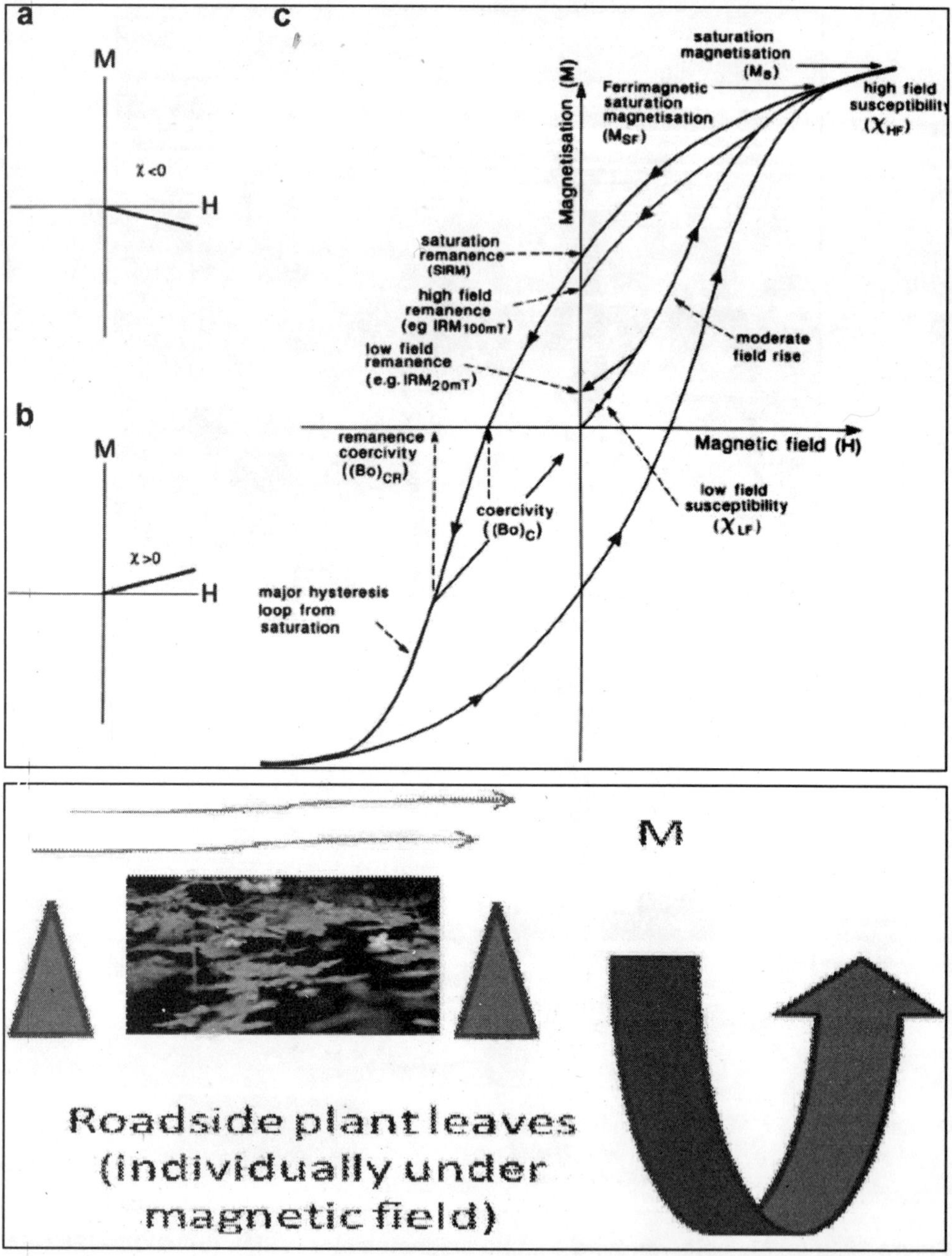

a

M

$\chi < 0$

H

b

M

$\chi > 0$

H

c

Magnetisation (M)

saturation magnetisation (M_S)

Ferrimagnetic saturation magnetisation (M_{SF})

high field susceptibility (χ_{HF})

saturation remanence (SIRM)

high field remanence (eg IRM_{100mT})

low field remanence (e.g. IRM_{20mT})

moderate field rise

remanence coercivity $((Bo)_{CR})$

coercivity $((Bo)_C)$

Magnetic field (H)

low field susceptibility (χ_{LF})

major hysteresis loop from saturation

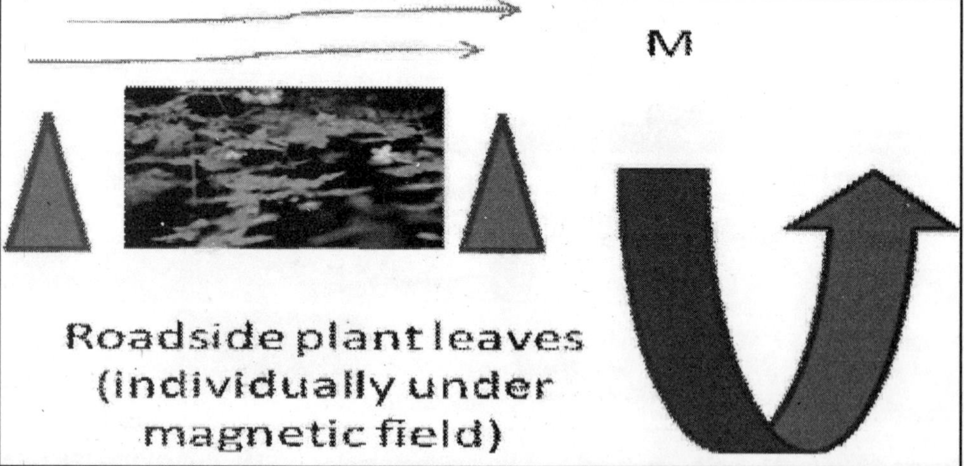

Figure 22.3: Diagrammatic representation of methodology used in studying the magnetic properties of dust loaded roadside plant leaves/typical magnetic hysteresis loop showing different magnetic properties (After Rai, 2011b).

and other constituents of dust (*e.g.* heavy metals), hence, act as an indicator/proxy of particulate pollution.

Results and Discussion

Season wise, dust content on plant leaves was highest during winter, followed by summer and least in rainy season. Among different pants investigated the decreasing order of dust deposition capacity was *Mangifera indica*> *Psidium guajava*> *Hibiscus rosa-sinensis*> *Bougainvillea spectabilis*> *Mikania micrantha,* which may be attributed to morphology, physiology and biochemistry of plant leaves tolerance to dust/particulates (Table 22.1). Since dust deposited on plant leaves may be derived from iron impurities in fuels or rock dust with magnetite, aforesaid results may be further studied by performing environmental geomagnetic studies of plant leaves and an overview of the concept is presented earlier in this paper. The 2D magnetization values of plant leaves may be correlated with the dust content and other constituents of dust (*e.g.* heavy metals). Further, detailed study on this aspect may find the suitable tolerant plants that can be effectively used in the air pollution remediation and hence Green Belt plantation.

Table 22.1: Seasonal variation in dust deposition capacity for certain roadside plants in Aizawl, Mizoram (Rai, 2011a, b).

Season	*Dust deposition capacity (mg/cm^2); METHODS -$W=w_2-w_1/w_2$ Where, W is dust content (mg/cm^2), w_1 is initial weight of tracing paper, w_2 is final weight of tracing paper with dust, and a is total area of the leaf (cm^2).*				
	Hibiscus rosa-sinensis	*Psidium guajava*	*Mikania micrantha*	*Mangifera indica*	*Bougainvillea spectabilis*
Rainy	0.0022	0.042	0.017	0.057	0.011
Winter	0.184	0.362	0.151	0.407	0.162
Summer	0.092	0.286	0.089	0.296	0.087

Acknowledgement

PKR wishes to acknowledge the Department of Biotechnology (DBT) vide research and Department of Science and Technology (DST) vide SR/FTP/ES-83/2009, for providing financial assistance in the form of research project (vide project no. BT/PR-11889/BCE/08/730/2009 and SR/FTP/ES-83/2009, respectively). Thanks are due to Dr. Onkar Nath Tiwari and Dr. Umesh Sharma for their useful discussion and cooperation in this work. Lalita L.S. Panda (DST-inspire fellow) is thankful to DST for The financial assistance received to one of the author LLSP is greatly acknowledged to the DST, New Delhi for awarding the D.S.T. inspire fellowship. We also wish to acknowledge and thanks to Professor Lalthantluanga, Vice Chancellor, Mizoram University, for his kind inspiration to accomplish the present work.

References

Cohen, A.J., Anderson, H.R., Ostro, B., Pandey, K.D., Krzyzanowski, M. and Kuenzli, N., 2005. The global burden of disease due to outdoor air pollution. *Journal of Toxicology and Environmental Health, Part A: Current Issues,* 68: 13–14, 1301–1307.

Ferreira-Baptista, L. and DeMiguel, E., 2005. Geochemistry and risk assessment of street dust in Luanda, Angola: A tropical urban environment. *Atmospheric Environment*, 39: 4501–4512.

Maher, B.A., 2009. Rain and dust: Magnetic records of climate and pollution. *Elements*, 5: 229–234.

Moller, P. *et al.*, 2008. Air pollution, oxidative damage to DNA, and carcinogenesis. *Cancer Letters*, 266: 84–97.

Pandey, S.K., Tripathi, B.D., Prajapati, S.K., Mishra, V.K., Upadhyay, A.R., Rai, P.K. and Sharma, A.P., 2005. Magnetic properties of vehicle derived particulates and amelioration by *Ficus infectoria*: A keystone species. *Ambio*, 34 (8): 645–647.

Rai, P.K., 2011a. Dust deposition capacity of certain roadside plants in Aizawl, Mizoram: Implications for environmental geomagnetic studies. In: *Recent Advances in Civil Engineering*, (Eds.) S.B. Dwivedi *et al.*, pp. 66–73.

Rai, P.K., 2011b. Biomonitoring of particulates through magnetic properties of road-side plant leaves. In: *Recent Advances in Environmental Chemistry*, (Ed.) D. Tiwari, pp. 34–37.

Environment, Biodiversity and Traditional System, 2014 *Pages* 243–249
ISBN 978-81-7622-307-2
Editors: D.R. Khanna, G.S. Solanki and S.K. Pathak
Published by: BIOTECH BOOKS, NEW DELHI

Chapter 23

Assessment of Air Pollution Tolerance Indices of Five Common Roadside Plant Species in Aizawl, Mizoram

Prabhat Kumar Rai, Lalita L.S. Panda, Biku Moni Chutia*

Department of Environmental Science, Mizoram University
Tanhril, Aizawl – 796 004, Mizoram

ABSTRACT

In urban areas, air pollution is more complex than most other environmental challenges. Monitoring of air pollutants is a prerequisite for air quality control. Their impact on the chemical composition of plants is often used as an indicator and a tool for monitoring environmental pollution. Monitoring with the help of biological indicators is simple, cheap and convenient method to ensure the state of local environment. In the present study, evaluation of air pollution tolerance index (APTI) of five selected plant species such as *Bougainvillea spectabilis, Psidium guajava, Lagerstroemia speciosa, Artocarpus heterophylla, Hibiscus rosasinensis*, growing along the road side of Aizawl, Mizoram, adjacent to the national highway, was carried out with a view to find out the air pollution tolerance as well as sensitivity of the plants species for abatement of air pollution, by selecting the tolerant and sensitive species among them. High values of APTI were recorded in *Artocarpus heterophylla* (9.34) and proved to be tolerant variety compared to other species while *Lagerstroemia*

* *Corresponding Author:* E-mail: prabhatrai24@gmail.com

speciosa showed least APTI value (6.53) and proved to be the most sensitive one among other species studied. *Bougainvillea spectabilis, Psidium guajava* and *Hibiscus rosasinensis* with APTI values of 7.75, 8.80, and 8.95 respectively can be considered as sensitive species.

Keywords: ***Air pollution tolerance index (APTI), Bio monitors, Chlorophyll, Ascorbic acid, Relative water content.***

Introduction

Both the developed and developing countries are facing the problem related to environmental pollution caused by anthropogenic activities of man. Air pollution has become a growing problem in cities throughout the globe and transportation is the major source of air pollution in many cities. In urban areas the mobile or vehicular population is predominant and significantly contribute to air quality problems. Air pollution is more complex than other environmental challenges. It is the human introduction of harmful substances into the atmosphere. Various strategies exist for controlling pollution but vegetation provides one of the natural ways of cleansing the atmosphere by absorption of gaseous and some particulate matter through leaves.No physical and chemical method is known to ameliorate air pollution. A suitable alternative may be to develop a biological method by growing plant in and around industrial and urban areas (Mondal *et al.*, 2011; Agarwal, 1988; Shannigrahi *et al.*, 2003; Sivasamy and Srinivasan, 1996; Fukuoka, 1997; Ghosh and majee, 2001). Plants act as scavengers for many airborne particulate in the atmosphere by providing an enormous leaf area for impingement, absorption and accumulation of air pollutants to reduce the pollutant level in the environment. Plant response to air pollution can be used to assess the quality of air that may provide early warning signals of air pollution trends. The identification and categorization of plants into sensitive and tolerant groups is important because former can serve as indicator and the latter as sink for the air pollutants in urban and industrial habitats.Plants sensitivity and tolerance to air pollutant varies with change in leaf extract pH, relative water content, Ascorbic acid content and total chlorophyll content. Study of single parameters may not provide a clear picture of the pollution induced change so Air pollution tolerance index(APTI), based on these parameters has been used to know tolerance levels of plant species.(Lui and Ding 2008;Singh and Verma, 2007, Chauhan, 2010). The usefulness of evaluating APTI for the determination of tolerance as well as sensitiveness of plant species were followed by several authors (Dwivedi and Tripathi, 2007; Liu and Ding, 2008). In the present study, the APTI of five selected plant species growing along the road side of Aizawl, Mizoram adjacent to the national highway were calculated.

Materials and Methods

Study Area

The research work was done in Aizawl, Mizoram (21p58′- 21p85′ N and 90p30′- 90p60′E), the capital of the state is 1132 meter ASL. The altitude in Aizawl district varies from 800 to 1200 meter asl. The climate of the area is typically monsoonal. The annual average rainfall is amounting to ca 2350 mm. The area experiences distinct seasons. The ambient airtemperature normally ranges from 20 to 30p C in summer and 11 to 21pC in winter (Laltlanchhuanga 2006).

Plants

Plants species studied were *Bougainvillea spectabilis, Psidium guajava, Lagerstroemia speciosa, Artocarpus heterophylla and Hibiscusrosasinensis*

Sample Collection

samples were collected in early morning and brought to laboratory in polythene bag and kept in fridge for further analysis. Leaves were collected in triplicate.

Procedure

The estimation of leaf extracts PH was assessed according to the method proposed by (Singh and Rao, 1983), relative water content by Sen. and Bhandari (1978), total chlorophyll (Arnon, 1949), ascorbic acid (Keller and Schwager, 1977).The APTI was determined by calculating these biochemical parameters, proposed by Singh and Rao, (1983).The air pollution tolerance index was calculated using the formula:

$$APTI = \frac{A(T+P)+R}{10}$$

where,

A: Ascorbic Acid (mg/g)

T: Total Chlorophyll (mg/g – f.w)

P: pH of the leaf extract.

R: Relative water content of leaf (per cent).

Results

APTI was calculated for five plants species growing along the road side of Aizawl, Mizoram and the data is shown in Table 23.1.

Table 23.1: APTI values of five selected plants in Aizawl, India.

Sl.No.	Plant Species	pH	Relative Waer Content (Per cent)	Total Chlorophyll mg/g	Ascorbic Acid mg/g	APTI
1.	*Bougainvillea spectabilis*	4.95	75.60	0.75	0.35	7.75
2.	*Psidium guajava*	5.56	84.95	1.23	0.45	8.80
3.	*Lagerstroemia speciosa*	4.35	65.06	0.21	0.07	6.53
4.	*Artocarpus heterophylla*	6.31	88.41	2.48	0.57	9.34
5.	*Hibiscus rosasinensis*	5.39	86.44	1.26	0.47	8.95

The biochemical parameters such as ascorbic acid, total chlorophyll, relative water content and pH play an important part in APTI determination. According to the Table 23.1, the result of determining the concentration of ascorbic acid for *Hibiscus rosasinensis, Bougainvillea spectabilis, Psidium guajava, Artocarpus heterophylla and Lagerstroemia speciosa* in Aizawl were 0.47, 0.35, 0.45, 0.57 and 0.07 mg/gm. The result of total chlorophyll concentration for *Hibiscus rosasinensis, Bougainville spectabilis, Psidium guajava, Artocarpus heterophylla and Lagerstroemia speciosa were* 1.26, 0.75, 1.23, 2.48 and 0.21 mg/gm. The third factor pH for the *Hibiscus rosasinensis, Bougainvillea spectabilis, Psidium guajava, Artocarpus heterophylla* and *Lagerstroemia speciosa* are 5.39, 4.95, 5.56, 6.31 and4.35 moreover forth factor RWC for *Hibiscus rosasinensis, Bougainvillea spectabilis, Psidium guajava, Artocarpus heterophylla* and *Lagerstroemia speciosa* were 86.44, 75.60, 84.95, 88.41 and 65.06 (per cent). Among the plant species studied, *Artocarpus heterophylla* possessed a highest APTI value (9.34) and *Lagerstroemia speciosa* got the lowest APTI value (6.53).

Discussion

The ability of each plant species to absorb and adsorb pollutants by their foliar surface varies greatly and depends on several biochemical, physiological and morphological characteristics. The variation in the biochemical parameters in the leaves can be used as indicator of air pollution for early diagnosis of stress or as a marker for physiological damage prior to the onset of visible injury symptoms (Tripathi *et al.*, 2009). Categorization of plants as sensitive or tolerant is determined by the level of these parameter in plants and thus plants show different susceptibility to different pollutants. Sensitive species are early indicator of pollution and the tolerant species help in reducing the overall pollution load (Nrusimha *et al.*, 2005). For this purpose, evaluation of plant with respect to their tolerance and sensitivity level to air pollution is essential for monitoring and abatement of air pollution. To evaluate the tolerance and sensitivity level of plant species to air pollution, Singh and Rao (1983) used air pollution tolerance index (APTI) which is a unique index, incorporating four different biochemical parameters *i.e.* Total chlorophyll, pH of leaf extract, ascorbic acid and relative water content.

Chlorophyll content of plant signifies its photosynthetic activity as well as the growth and development of biomass. Higher chlorophyll content in plant might favor tolerance to pollutants where as a considerable loss in total chlorophyll support the sensitivity nature of plant species. Studies (Tripathi and Gautam, 2007) also suggested that high level of automobile pollution decrease chlorophyll content in higher plants near road side. In this study, the highest chlorophyll content (in mg/g fresh wt.) was recorded in *Artocarpus heterophylla* (2.48) and lowest chlorophyll content in *Lagerstroemia speciosa* (0.21).

Ascorbic acid is regarded as an antioxidant, found in large amount in all growing plants and influences resistance to adverse environmental condition including air pollution (Keller and Schwager., 1977). The highest content of ascorbic acid was found in *Artocarpus heterophylla* (0.57 mg/g fresh weight) while the lowest value was recorded (0.07 mg/g fresh weight) in *Lagerstroemia speciosa.* The increased level of ascorbic acid reported may be due to the defense mechanism of the respective plant

(Tripathi and Gautam. 2007). Increased level of ascorbic acid in leaves will increase air pollution tolerance in plants (Chaudhury and Rao., 1977) whereas lower ascorbic acid content in the leaves of other plant species studied supports the sensitive nature of these plants towards pollutants particularly automobile exhaust.

The leaf extract pH ranged from that of *Lagerstroemia speciosa* recorded at 4.35 to that of Artocarpus heterophylla at around 6.31. A PH on the higher side gives tolerance to plant against air pollution (Agrawal, 1988). It was also reported that plants with lower pH are more susceptible while those with pH around 7 are more tolerant. The relative water content was lowest in *Lagerstroemia speciosa* (65.60) and highest in *Artocarpus heterophylla* (88.41). High water content within plant body helps to maintain its physiological balance under stress condition. It also serves as an indicator of drought resistance in plants. Therefore all these biochemical parameters together make up the APTI. Artocarpus heterophylla with high APTI value can be termed as tolerant among other species whereas plants species such as Lagerstroemia speciosus showing lowest APTI value is termed as sensitive plant and can be treated as bio indicator of pollution.

In nutshell, APTI determination for plants is important for future planting as plants have the ability to serve as excellent quantitative and qualitative indices of pollution (Jyothi and Jaya, 2010). Since bio monitoring of plant is an important tool to evaluate the impact of air pollution onplants, *Lagerstroemia speciosa* and *Bougainville spectabilis* can be used as bio monitors for air pollution stress whereas *Artocarpus heterophylla*, the tolerant among them can be effectively used in the air pollution amelioration purposes.

References

Agarwal, A.L., 1988. Air pollution control studies and impact assessment of stack and fugitive emissions from CCI Akaltara cement factory. *Report, NEERI, Nagpur, India.*

Agbaire, P.O. and Esiefarienrhe, E., 2009.Air pollution tolerance indices (apti) of some plants around Otorogun gas plant in Delta State, Nigeria. *J. Appl. Sci. Environ. Manage.*, 13(1): 11–14.

Arnon, D.S., 1949. Copper enzymes in isolated chloroplast. *Polyphenoxiase in Beta Vulgaris, Plant Physiology*, 24: 1–15.

Chauhan, A., 2010. Photosynthetic pigment changes in some selected trees induced by automobile exhaust in Dehradun. *Journal of New York Sciences*, 3(2): 45–51.

Das, S. and Prasad, P., 2010. Seasonal variation in air pollution tolerance indices and selection of plant species for industrial areas of Rourkela. *Ind. J. Env. Protec.*, 30(12): 978–988.

Dwivedi, A.K. and Tripathi, B.D., 2007. Pollution tolerance and distribution pattern of plants in surrounding area of coal-fired industries. *J. Environ. Biol.*, 28" 257–263.

Fukuoka, Y., 1997. Biometeorological studies on urban climate. *Int. J. Biometeorology*, 40(1): 54–57.

Ghose, M.K. and Majee, S.R., 2011. Air pollution caused by opencast mining and its abatement measures in India. *J. Environ. Management*, 63(2): 193–202.

Horaginamani, S.M. and Ravichandran, M., 2010. Ambient air quality in an urban area and its effects on plants and human beings: a case study of Tiruchirappalli, India. *Kathmandu Univ. J. Sci. Eng. Tech.*, 6: 13–19.

Jissy, J.S. and Jaya, D.S., 2010. Evaluation of air pollution tolerance index of selected plant species along roadsides in Thiruvanthapuram, Kerala. *J. Environ. Biol.*, pp. 379–386.

Joshi, P.C. and Swami, A., 2007. Physiological responses of some tree species under roadside automobile pollution stress around city of Haridwar, India. *Environmentalist*, 27: 365–374.

Keller, T. and Schwager, H., 1977. Air pollution and ascorbic acid. *Eur. J. For. Pathol.*, 7: 338–350.

Kulshreshtha, K., Rai, A., Mohanty, C.S., Roy, R.K. and Sharma, S.C., 2009. Particulate pollution mitigating ability of some plant species. *Int. J. Environ. Res.*, 3(1): 137 – 142.

Lakshmi, P.S., Sarvanti, K.L. and Srinivas, N., 2009. Air pollution tolerance index of various plants species growing in industrial areas. *An Int. Biannua J. of Environ. Sci.*, 2: 203–206.

Liu, Y.J. and Ding, H., 2008. Variation in air pollution tolerance index of plants near a steel factory: Implication for landscape-plant species selection for industrial areas. *WSEAS Transactions on Environment and Development*, 4: 24–32.

Mondal, D., Gupta, S. and Datta, J.K., 2011. Anticipated performance index of some tree species considered for green belt development in an urban area. *Int. J. Plant Science*, 2(4): 99–106.

Nrusimha, T., Kumar, K. Suresh and Srinivas, N., 2005.Air pollution tolerance index of tree species growing in industrial and traffic areas of Visakhapatnam. *Indian J. Env. Protection*, 25: 1057–1060.

Shannigrahi, A.S., Sharma, R.C. and Fukushima, T., 2003. Air pollution control by optimal green belt development for Victoria Memorial Monument, Kolkata (India). *Intern. J. Environ. Studies*, 60(3): 241–249.

Singh, S.K. and Rao, D.N., 1983. Evaluation of plants for their tolerance to air pollution. *Symp. on Air Pollution Control, New Delhi, Proceedings*, pp. 218–224.

Sivasamy, N. and Srinivasan, V., 1996. *Environmental Pollution and its Control by Trees.* Indian Agriculture Research Institute (IARI), Pusa, New Delhi, India.

Talreja, T., 2011. Biochemical estimation of three primary metabolites from medicinally important plant *Moringa oleifera*. *Int. Journal of Pharmaceutical Sciences Review and Research*, 7(2): 033.

Tripathi, A., Tiwari, P.B. and Mahima, Singh. 2009. Assessment of air pollution tolerance index of some trees in Moradabad city, India. *J. Environ. Biol.*, 30(4): 545–550.

Tripathi, A.K. and Gautam, M., 2007. Biochemical parameters of plants as indicators of air pollution. *Journal of Environmental Biology*, 28(1): 127–132.

Environment, Biodiversity and Traditional System, 2014 ***Pages* 251–255**
ISBN 978-81-7622-307-2
***Editors:* D.R. Khanna, G.S. Solanki and S.K. Pathak**
***Published by:* BIOTECH BOOKS, NEW DELHI**

Chapter 24

Phyto-monitoring of Particulate Pollution through Magnetic Properties for Certain Roadside Plants in Aizawl, Mizoram

Prabhat Kumar Rai*, Biku Moni Chutia, Lalita L.S. Panda

Department of Environmental Science, Mizoram University
Tanhril, Aizawl – 796 004, Mizoram

ABSTRACT

Air pollution is one of the serious problems due to the emission of deleterious fine-grained particulates and magnetic particles into the atmosphere. Particulate matter (PM) is one of the most important pollutants in view of their adverse impacts on human health and vegetation. Vegetation naturally cleanses the atmosphere by absorbing gases and some particulate matter through leaves, as they are continuously exposed to the surrounding atmosphere and is therefore the main receptor of particulate pollutants. It has been demonstrated that magnetic measurement is an important means in particulate pollution study through plant leaves. Leaves are effective biomonitors and may act as natural filter papers by trapping and retaining PM on their surfaces. Biomonitoring of particulate pollution through magnetic properties of plant leaves is a rapid and inexpensive alternative to conventional atmospheric pollution monitoring. Bio-magnetic monitoring as an application of environmental geomagnetism may act as proxy for PM pollution.

Keywords:* *Environmental geomagnetism, Biomonitoring, Particulates, Human health.

---

* *Corresponding Author:* E-mail: prabhatrai24@gmail.com

Introduction

Evaluation of health impacts arising from inhalation of pollutant particles < 10 µm (PM) is an active research area. Many studies highlighted the importance of particulates with an aerodynamic diameter of less than 10 µm (PM_{10}), which due to their small size, can penetrate deep into the human lung and cause most serious health effects including respiratory illness (Le Tertre *et al.*, 2002; Janssen *et al.*, 2005; Jerrett *et al.*, 2005).Alongside PM_{10}s are further grain size division of $PM_{2.5}$ and $PM_{0.1}$ (2.5 µm and 0.1 µm, respectively, again relative to their aerodynamic diameters).These fine and ultrafine particulates have higher burdens of toxicity as they become coated with heavy metals and chemicals, which, when inhaled, can absorbed into the body and may target specific organs (Morawska and Zhang 2002; Oberdorster 2002; Englert 2004).

Sources of particulate pollution may be natural or anthropogenic in nature. Emission sources may include natural processes such as wildfires, volcano eruption and dust storms. Man-made pollutions encompass combustion processes used for heating, power production, industry and traffic vehicles. One of the most major sources of environmental pollution in urban areas is road traffic, whereas other anthropogenic activities like power plants, metallurgy, mining, dust originating from fragile rocks are of minor importance.

Site Description

Mizoram (21p 56′- 24p31′N and 92p 16′- 93p26′E) is one of the eight states under northeast India, and it covers an area of 21,081 km^2. The tropic of cancer divides the state into two almost equal parts. The state is bordered by Myanmar to the east and south, Bangladesh to the west, and by the states of Assam, Manipur and Tripura to the north. In local Mizo language, Mizoram means land of highlanders, rise to just more than 2000 meter as near the Myanmar border. The forest vegetation of state falls under three major categories *i.e.*, tropical wet evergreen forest, tropical semi-evergreen forest and sub- tropical pine forest (Champion and Seth 1968).

Aizawl (21°58′- 21°85′ N and 90°30′-90°60′ E), the capital of the state is 1132 meter asl. The altitude in Aizawl district varies from 800 to 1200 meter asl. The climate of the area is typically monsoonic. The annual average rainfall is amounting to ca. 2350 mm. The area experiences distinct seasons. The ambient air temperature normally ranges from 20 to 30°C in summer and 11 to 21°C in winter (Laltlanchhuanga 2006).

The study is carried out in Aizawl district which is categorized in to five sub sites:

Site 1 Durtlang

Durtlang is a connecting road (NH) from Mizoram to Assam. So traffic density is high in that area.

Site 2 Zarkawt

Zarkawt is a commercial place in the city of Aizawl (Mizoram). Because of high traffic density the emission of dust particles is seen very high.

Site 3 Zonuam

This is a residential area with high vehicular movement.

Site 4 Ramrikawn

It is also a commercial area within market, bus-stand and food storage (food cooperation of India), so there is a heavy flow of traffics.As there is a bus stand vehicular movement is seen high.Stone quarrying activity leads to emission of dust particles.

Site 5. Mizoram University Campus (MZU)

MZU is a very young university, since it is under construction; it plays a very major role in the emission of dust particles. Vehicles including bus, taxi, truck etc. are also the source of pollution in MZU campus.

Materials and Methods

Collection of Leaves and Sample Preparation

Leaves were collected from five tree species. The recorded plants were *Mikania micrantha, Hibiscus rosa-sinensis, Bougainvillea spectabilis, Psidium guajava and Mangifera indica.* In order to assure a build-up of particulate matter, collection occurred after a minimum of 3 days of dry weather, as rain causes a wash off effect of particulates from leaves. Five Laves of similar size were collected from each tree at a height of approximately 2 m to avoid possible contamination from ground splash sessionly viz summer, winter and rainy season. Further, the leaves were packed into 10-cc plastic sample pots for their detailed analysis.

Magnetic Parameters Used in Environmental Geomagnetic Studies

Four magnetic measurements are carried out and an overview is provided in this section; these are magnetic susceptibility, Anhysteretic remanent magnetisation, Isothermal remanent magnetisation and Saturation Isothermal remanent magnetisation.

Magnetic Susceptibility (χ)

Magnetic susceptibility indicates the concentration of magnetic grains within a sample and has been used as a proxy of particulate pollution concentration in urban dusts and on tree leaves (Moreno *et al.*, 2003; Gautam *et al.*, 2005; Urbat *et al.*, 2004). A Bartington MS2B dual-frequency susceptibility sensor connected to an MS2 susceptibility meter according to Dearing (1999) and the measurements are taken at low frequency and high frequency.

Anhysteretic Remanent Magnetisation (ARM)

ARM is indicative of magnetic concentration and is also sensitive to the presence of fine grains ~0.04–1 lm (Thompson and Oldfield 1986), thus, falling within the respirable size range of $PM_{2.5}$ and having the potential to have a high burden of toxicity. ARM is induced in samples using a Molspin, A. F. Demagnetiser, whereby a DC biasing field is generated in the presence of an alternating field, which peaks at100 milli-Tesla (mT). The nature of this magnetic field magnetises the fine magnetic grains and the amount of magnetisation retained within the sample (remanence) when removed from the field is measured using a Molspin1A magnetometer. The samples are demagnetised to remove this induced field in preparation for the subsequent magnetic analysis (Walden 1999).

Isothermal Remanent Magnetisation (IRM) and Saturation Isothermal Remanent Magnetisation (SIRM)

Isothermal remanent magnetization reflects magnetic moment of plant leaves; each IRM value was determined by taking the mean of three acquisition-measurement cycles.

SIRM is actually the highest level of magnetic remanence that can be induced in a particular sample through application of high magnetic field. SIRM involves measuring the magnetic remanence of samples once removed from an induced field. Using a Molspin Pulse Magnetiser, a saturation isothermal remanent magnetisation (SIRM) of 800 mT in the forward field is induced in the samples. At this high magnetisation, all magnetic grains within the sample become magnetised and, therefore, SIRM indicates the total concentration of magnetic grains (Evans and Heller 2003) and, like Magnetic susceptibility, can be used as a proxy of PM concentration (Muxworthy *et al.*, 2003). All magnetic parameters are normalised by the dried weight of the sample to produce mass-specific values. Further explanations of these parameters are widely available (Evans and Heller 2003).

Results and Discussion

Results obtained for dust deposition capacity for five selected plants are mentioned in Table 24.1.

Table 24.1: Seasonal variation in dust deposition capacity for certain roadside plants in Aizawl, Mizoram.

Season	*Dust deposition capacity (mg/cm²); $W=w_2-w_1/w_2$ Where, W is dust content (mg/cm²), w_1 is initial weight of tracing paper, w_2 is final weight of tracing paper with dust, and a is total area of the leaf (cm²).*				
	Hibiscus rosa-sinensis	*Psidium guajava*	*Mikania micrantha*	*Mangifera indica*	*Bougainvillea spectabilis*
Rainy	0.0022	0.042	0.017	0.057	0.011
Winter	0.184	0.362	0.151	0.407	0.162
Summer	0.092	0.286	0.089	0.296	0.087

Winter season showed the height dust deposition capacity than summer and rainy season (Table 24.1). The dust deposition capacity was recorded among five plants were *Mangifera indica> Psidium guajava >Hibiscus rosa-sinensis> Bougainvillea spectabilis> Mikania micrantha*. Environmental geomagnetic studies and size fractionation of particulates will be the the part of future research area. As mentioned in the methodology, the magnetic parameters for these plants will be investigated with the methodology as developed and mentioned precisely in previous section.

This future research will enable us to correlate with the dust content and magnetic properties.

Acknowledgements

PKR wishes to acknowledge the Department of Biotechnology (DBT) vide research and Department of Science and Technology (DST) vide SR/FTP/ES-83/2009, for providing financial assistance in the form of research project (vide project no. BT/PR-11889/BCE/08/730/2009 and SR/FTP/ES-83/2009, respectively). Thanks are due to Dr. Onkar Nath Tiwari and Dr. Umesh Sharma for their useful discussion and cooperation in this work. Lalita L.S. Panda (DST-inspire fellow) is thankful to DST for The financial assistance received to one of the author LLSP is greatly acknowledged to the DST, New Delhi for awarding the D.S.T. inspire fellowship. We also wish to acknowledge and thanks to Professor Lalthantluanga, Vice Chancellor, Mizoram University, for his kind inspiration to accomplish the present work.

References

Booth, C.A., Shilton, V., Fullen, M.A., Walden, J., Worsley, A.T. and Power, A.L., 2006. Environmental magnetism: Measuring, monitoring and modelling urban street dust pollution. In: Air Pollution, (Eds.) J.W.S. Longhurst and C.A. Brebbia WIT Press, Southampton, 14: 333–341.

Dearing, J., 1999. Magnetic susceptibility. In: *Environmental Magnetism: A Practical Guide,* (Eds.) J. Walden, F. Oldfield and J. P. Smith. Technical Guide. Quaternary Research Association, Cambridge, England, 6: 35–62.

Dekkers, M.J., 1997. *Environmental Magnetism: An Introduction.* Geologie en Mijnbouw, 76: 163–182.

Evans, M.E. and Heller, F., 2003. *Environmental Magnetism: Principles and Applications of Enviromagnetics (International Geophysics).* Academic Press, London.

Gautam, P., Blaha, U. and Appel, E., 2005. Magnetic susceptibility of dust-loaded leaves as a proxy of traffic-related heavy metal pollution in Kathmandu City, Nepal. *Atmospheric Environment,* 39: 2201–2211.

Matzka, J. and Maher, B.A., 1999. Magnetic biomonitoring of roadside tree leaves: identification of spatial and temporal variations in vehicle-derived particulates. *Atmospheric Environment,* 33: 4565–4569.

Moreno, E., Sagnotti, L., Dinare's-Turell, J., Winkler, A. and Cascella, A., 2003. Biomonitoring of traffic air pollution in Rome using magnetic properties of tree leaves. *Atmospheric Environment,* 37: 2967–2977.

Shu, J., Dearing, J. A., Morse, A. P., Yu, L. and Li, C., 2000. Magnetic properties of daily sampled total suspended particulates in Shanghai. *Environmental Science and Technology,* 34: 2393–2400.

Urbat, M., Lehndroff, E. and Schwark, L., 2004. Biomonitoring of air quality in the Cologne conurbation using pine needles as a passive sampler–Part 1: Magnetic properties. *Atmospheric Environment,* 38: 3781–3792.

Environment, Biodiversity and Traditional System, 2014 ***Pages* 257–262**
ISBN 978-81-7622-307-2
***Editors:* D.R. Khanna, G.S. Solanki and S.K. Pathak**
***Published by:* BIOTECH BOOKS, NEW DELHI**

Chapter 25

Lead and Cadmium Contents in the Meat Tissues of Fish found in Water Ecosystems of Pumlen Lake of Manipur

Ksh. Birla Singh

Department of Zoology,
Pachhunga University College, Aizawl – 796 001
E-mail: birla.kshetri@gmail.com

ABSTRACT

Lead and cadmium levels were determined in the meat tissues of Common Carp (*Cyprious carpioroach*), Grass carp (*Clenopharyngodon idella) and* Silver carp (*Hypophthalmichthys molitrix*) netted in Pumlen Lake threatened by pollutions from major rivers from the valleys of Manipur. Concentrations in *Cyprious carpio*roach ranged from 0.122 to 1.04 mg/kg of Pb and from 0.024 to 0.242 mg/kg of Cd. In *Clenopharyngodon idella,* they were from 0.100 to 1.80 mg/kg of Pb and 0.056 to 0.345 mg/kg of Cd, while in *Hypophthalmichthys molitrix,* they were from 0.942 to 2.00 mg/kg of Pb and 0.095 to 0.405 mg/kg of Cd. The metal concentrations were low; no result exceeded admissible residue levels set forth by Regulatory authorities and no evident contamination of fish by pollution from the major rivers was detected.

Keywords: Lead, Cadmium, Pumlen lake, Manipur.

Introduction

Heavy metals like Lead (Pb) and Cadmium (Cd) are potential environmental contaminants with the capability of causing human health problems if their concentration exceeds beyond the permissible level in the food we eat. They are given special attention throughout the world due to their toxic effects even at very low concentrations [1]. Health safety is a very important aspect of food quality [2]. Hazardous substances enter the food chain through the soil, air and water and are the main sources of contamination for humans. Heavy metals are among the most dangerous contaminants. In spite of more and more rigorous norms concerning their presence in food, and despite the introduction of quality assurance systems, it is impossible to get rid of them from food. Their presence in food is a result not only of the contamination of raw materials, but they can also originate from machines and industrial plants, food additives, and materials used in cleaning and disinfection procedures [3]. Recently, various dumps in the rivers have become sources of serious environmental contamination. Very often they were built without due care which allows leaks to occur from the deposits [4]. Every kind of waste, including pesticides, contains some quantity of heavy metals that can escape into the river water, soil and ground water. Pumlen Lake is the second largest freshwater wetland situated in Thoubal District, a distance of 40km from Imphal city towards the southern lowlands of the valley. The associated lakes are Khoidum pat, Lamjao pat and Kakyaipat. The main sources of water are from the precipitation, surface runoff from its south-eastern denuded hills, the northern agricultural fields, and indirectly from the Sekmai River through Khoidum pat on the northern side as well as from the impounded Manipur Rivers. The aim of the study was to determine the concentration of lead and cadmium in meat tissues of three species of fishes collected in Pumlen Lake threatened over a long period of time by pollution from major rivers of the valley regions of Manipur.

Materials and Methods

Fish were netted in winter–spring in 2010 (Table 25.1): Common Carp (*cyprious carpioroach*), Grass carp (*clenopharyngodon idella*) and Silver carp (*Hypophthalmichthys molitrix*) was collected in Pumlen Lake of Manipur which has been influenced for over 10 years by pollution from the major rivers in the valley region of Manipur like Imphal River, Nambul River, Iril River and Thoubal River. The collected fishes were gutted and kept frozen until the analysis was performed. After defrosting, the wet solid tissue *viz.* ventral, dorsal and caudal tissue muscles were dry in the oven overnight and processed for analysis by dry- ashing method using Muffle furnace (up to 550°C) by following the method of Crosby, 1977[5]. Final analysis was done by using Atomic Absorption Spectrophotometer (Electronic Corporation of India Limited, Hyderabad- AAS 4139) installed at Panjab University, Chandigarh, India.

Results and Discussion

The concentrations of Pb and Cd analyzed as dry weight basis in the meat tissues, *i.e.* dorsal, caudal and ventral tissue of the Common Carp (*cyprious carpioroach*), Grass carp (*clenopharyngodon idella*), *Silver* carp (*Hypophthalmichthys molitrix*) are presented in Tables 25.1–25.3 respectively. The present study reveals no significant difference in the lead and cadmium content in the different tissues were observed in the three type of fishes studied. The concentration of the lead in the *Cyprious carpio*roach ranges from the minimum value of 0.122 mg/kg in the ventral tissue to maximum value of 1.04 mg/kg in the dorsal tissue. In the case of the cadmium, the concentration ranges from the 0.024 mg/kg in the ventral tissue to 0.242 in the caudal tissue (Table 25.1). In the *Clenopharyngodon idella*, the concentration of the lead ranges from the minimum value of 0.100 mg/kg in the ventral tissue the maximum value of 1.24 mg/kg in the caudal tissue. Moreover, the concentration of the cadmium recorded were ranges from the minimum value of 0.056 mg/kg in the ventral tissue to the maximum value of 1.24 mg/kg in the caudal tissue (Table 25.2). The concentration of the lead in the *Hypophthalmichthys molitrix* ranges from the 0.942 mg/kg in the ventral tissue to 2.00 mg/kg in the dorsal tissue. In the case of the cadmium, the concentration ranges from the minimum value of 0.095 mg/kg in the ventral tissue to the maximum value of 0.405 mg/kg in the dorsal tissue (Table 25.3). The study revealed that the concentration of the lead in the meat tissues of *Cyprious carpioroach, Clenopharyngodon idella, Hypophthalmichthys molitrix*) are less than those recorded by the Krystna *et al.*, [6], Kolman and Szczepkowski, [7], and reported by others [8,9]. As indicated the value of the lead recorded in the meat tissues of the 3 type of fishes are the within the permissible limit of the 2.5 mg/kg set by Meat and Food Product Order (MFPO), 1973[10] (Table 25.4). It has been reported that the half life of lead in blood, soft tissues, spongy bones (pelvis ribs and skull and cortical bones (mid tibia and mid femur) is 35, 40 days and 3-5 and 30 years respectively [11]. Tuorome in 1995 [12] reported that excessive lead accumulation in children is known to cause hyperactivity, reduced intelligence and antisocial behavior. In adult it is associated with heart

Table 25.1: Lead and cadmium concentrations (mg/kg) in different tissues of common carp (*cyprious carpio*roach). Value are mean ± SEM, n= 10 observations each.

Sample	*Pb*	*Cd*
Ventral tissues	0.122 ± 0.49	0.024 ± 0.02
Caudal tissues	0.156 ± 0.04	0.242 ± 0.04
Dorsal tissues	1.04 ± 0.05	0.067 ± 0.01

Table 25.2: Lead and cadmium concentrations (mg/kg) in different tissues of grass carp (*clenopharyngodon idella*, Value are mean ± SEM, n= 10 observations each.

Sample	*Pb*	*Cd*
Ventral tissues	0.100 ± 0.020	0.056 ± 0.02
Caudal tissues	1.24 ± 0.03	1.24 ± 0.06
Dorsal tissues	1.80 ± 0.03	0.345 ± 0.01

Table 25.3: Lead and cadmium concentrations (mg/kg) in different tissues of silver carp (*hypophthalmichthys molitrix*. Value are mean ± SEM, n= 10 observations each.

Sample	*Pb*	*Cd*
Ventral tissues	0.942 ± 0.02	0.095 ± 0.04
Caudal tissues	1.24 ± 0.02	0. 204 ± 0.04
Dorsal tissues	2.00 ± 0.01	0.405 ± 0.03

diseases, cancer and infertility. Lead could cause adverse effects on the renal and nervous systems and cross the placental barriers having the potential toxic effects on the fetus [12, 13]. However, in the present study, all of the samples had lead content below the MPL.

Table 25.4: Comparison of values of metals obtained in meat in this study and maximum permissible limit set by regulatory organizations.

	Present Study	*Maximum Permissible Limit (MPL) as per Regulatory Organization*	*Organization Name/Year*
Pb	0.122- 2.00 mg/kg	2.5 mg/Kg	Meat Food Product Order, 1973
Cd	0.024-0.405mg/kg	0.1 mg/Kg	FAO, 1980

The present study further revealed that the concentration of the cadmium recorded in the meat tissues of *Cyprious carpioroach, Clenopharyngodon idella, Hypophthalmichthys molitrix* are in agreement with those recorded by the Krystna *et al.*, [6], Luczynska *et al.*, [14] but less than the values recorded by the Michalak and Bulinski[16] and Marek [17]. However, the value of the cadmium in the meat tissues of the three types of fishes was below the permissible limit as proposed by FAO, 1980[18](Table 25.4). It is concluded that atmospheric deposition from urban and agricultural area may play an important role in the enrichment of meat produce from cadmium. Cadmium is amongst the most abundant heavy metal and is particularly toxic. The excessive content of this metal in food is associated with etiology of a number of diseases [19]. Cadmium exposure may cause kidney damage and skeletal damage [20]. Due to slow excretion of the metal, the concentration in the kidney increase, even at very low concentrations in the food. International Agency for research on Cancer (IARC) classified Cd as human carcinogen [20].

Conclusions

The results obtained from this study about the contents of lead and cadmium in the meat tissues of the three type of fishes netted from the Pumlen Lake of Manipur were compared favorably with the findings from other parts of the world and permissible limits stipulated by various regulatory agencies and organizations. The results of the study showed that the concentrations lead and cadmium in the meat of these fishes are generally below the safe limits and seems pose no threat to health. The results of this study supply valuable information about the heavy metals contents in the meat of the three types of commonly netted fishes from the Pumlen Lake of Manipur which has been threaten by the pollution from major rivers of the valley region of Manipur. These results can also be used to test the chemical quality of foodstuffs in order to evaluate the possible risk associated with their consumption by humans.

Acknowledgements

Thanks are due to Mr. V. P. Singh (Senior Technical Officer), Department of Zoology, Panjab University Chandigarh, India for helping me in analysis and Dr.

Tawnenga, Principal, Pachhunga University College, Aizawl for necessary corporations.

References

[1] Reilly, C., 1991. Metal contamination of food, 2nd ed. Elsevier Applied Science, London.

[2] Olêdzka R., 1999. Current problems of the health safety of food, *Farm. Pol.* 23: 1083–90.

[3] Skibniewska K.A., 2003. Diet monitoring for the assessment of human exposure to environmental pollutants, *Environ. Int.*, 28: 703–09.

[4] Skibniewska K.A., Guziur J., Grzybowski M., Szarek J(2003. DDT in the soil of a pesticide tomb and surrounding ecosystems, [in:] On Analytical and Environmental Problems, Galbacs, Z. (ed).), SZAB, Szeged, Hungary: 27–31.

[5] Crosby, N.T., 1977. Determination of metals in foods: A Review. *The Analyst*, 102: 223–268.

[6] Krystyna A. S., Janusz G., Zbigniew M., Stanisiaw Z., Mirosiaw G., Józef S.,(2008. Lead and cadmium content in the meat tissues of fish from water ecosystems in the vicinity of a pesticide tomb. *Int. Journal of Oceanography and Hydrobiology*, 37(3): 79–86.

[7] Kolman R., Szczepkowski M(2001. Heavy metals in the muscle tissues of fish from lakes in Mazury, Poland, *Komun. Ryb.* 4: 22–25.

[8] Hiyñczak A, J., Baranowska–Bosiacka I., Fokt M(1998. Content of some heavy metals in muscles and liver of bream (Abramis brama) netted in the Oder estuary, *Arch. Ochr. Œrod.*, 24(1): 69– 78.

[9] Michalak A, Buliñski R(1995. A study of some trace element content in Polish food products. Part XVIII. Content of lead, cadmium, nickel, zinc, copper and chromium in some frozen freshwater and marine fish, *Bromat. Chem. Toksykol.* 28: 33–38.

[10] Meat Food Products Order (MFPO), 1973. The Gazette of India, No.112, Dated 28.03.1973.

[11] Pueschel, S.M., Linakis J. G., Anderson A.C., 1996. Lead poisoning in childhood. Paul, H.(Ed.) Brooks Publishing.

[12] Tuormma, T.E., 1995. The role of chromium, selenium and copper in human and animal metabolism, *J. Orthomol. Med.*, 10: 149–164.

[13] WHO, 2003. Elemental mercury and inorganic mercury compounds: Human Health aspects. Concise International Chemical Assessment document.50.World Health Organization, Geneva

[14] Luczyñska I., Jaworski J., Markiewicz K., 2000. Some heavy metals in muscle tissue of fish from £añskie Lake, *Komun. Ryb.* 3: 22–25.

[15] Luczyñska I., Jaworski J., Markiewicz K(2001. Heavy metals in muscle tissue of fish from Mazurian lakes, *Komun. Ryb* 4: 22–25.

[16] Michalak A, Buliñski R., 1995. A study of some trace element content in Polish food products. Part XVIII. Content of lead, cadmium, nickel, zinc, copper and chromium in some frozen freshwater and marine fish, *Bromat. Chem. Toksykol.* 28: 33–38.

[17] Marek J., 1990. Heavy metals in the water environment of the River Barycza. Assessment of the threat to the fish industry, Zesz. Nauk. AR Wroc³aw (habilitation thesis no 82).

[18] FAO., 1980), Manuals of Food Quality Control 2. Additives, Contaminants, Techniques FAO Food and Nutrition paper, 14/2.

[19] Jarup, L., 2003. Hazards of heavy metal contamination. *British Medical Bulletin*, 68, 167–182.

[20] WHO., 1992. Cadmium. Environmental Health Criteria, Vol. 165, Geneva.

[21] IARC., 1993. Cadmium and cadmium compounds. In: Beryllium, cadmium, mercury and exposure in the glass manufacturing industry. IARC Monographs on the evaluation of carcinogenic risks to humans, Vol. 58: 119–237, International Agency for Research on Cancer, Lyon.

Environment, Biodiversity and Traditional System, 2014 ***Pages* 263–272**
ISBN 978-81-7622-307-2
***Editors:* D.R. Khanna, G.S. Solanki and S.K. Pathak**
***Published by:* BIOTECH BOOKS, NEW DELHI**

Chapter 26

Ethnomedicinal Usage of Plants by Mizo tribes in Skin Disease Treatment

Archana K. Verma[1], Vinod K. Bharati[2] and A.C. Shukla[1]*

[1]Department of HAMP, Mizoram University, Aizawl
[2]Department of Chemistry, Govt. Kolasib College, Kolasib

ABSTRACT

Traditional medicinal practice has been known for centuries in many parts of the world. Ayurveda, the science of life, prevention and longevity is the oldest and most holistic medical system available on the planet today. Herbal medicines are gaining growing interest because of their cost effective and eco-friendly attributes. Skin health is an important aspect of primary health care among many communities. India has a rich folklore of medicinal plant on healthcare. A large number of formulations have been used in India for treatment of cuts, wounds, burns and various skin diseases. Plants have evolved the ability to synthesize chemical compounds that help them defend against attack from a wide variety of predators such as insects, fungi and herbivorous mammals. By chance, some of these compounds, whilst being toxic to plant predators, turn out to have beneficial effects when used to treat human diseases.

The aim of the present paper is to explore the various ethnomedicinal uses in practice among the tribes inhabiting the villages around Aizawl city. Accordingly various field sessions were conducted to collect the samples and information from traditional healers on the use of the same in connection with the treatment of skin diseases. The indigenous knowledge of local traditional healers and the native plants used for medicinal purposes were collected

* *Corresponding Author:* E-mail: archanamzu@gmail.com

through questionnaire and personal interviews during field trips. Further, the scope of the work embraces to document 21 plant species, belonging to 20 genera and14 families being used traditionally in the treatment of skin ailments by the tribals of the study area.

Keywords: Healthcare, Folklore, Skin diseases.

Introduction

The socio-economic life of the Mizo people depends on vegetation from where they derive all their material requirements – timber, food, fuel wood, medicinal plants etc. About 95 per cent of the interior population depends on herbal medicine and nearly 98 per cent of raw materials are harvested from the wild plant resources without replenishing the growing stocks. The village herbal preparation includes uprooting of the plants which is detrimental to the individuals or sub-populations. This knowledge of using native plants and its associated medicinal practice have now become a part of the local tradition, culture, art, belief, and folklore (Pushpangadan and Kumar, 2005). During the last few decades, there has been an increasing interest in the study of medicinal plants and their traditional use in different parts of the world (Al-Quran, 2005; Hanazaki *et al.*, 2000; Rossato *et al.*, 1999). Documenting indigenous knowledge through ethnomedicinal studies is important for the conservation and utilization of biological resources. The World Health Organization (WHO) suggested that as many as 80 per cent people of the world depend on traditional medicine for their primary healthcare needs. Due to limited means of communication, poverty and unavailability of modern health facilities, many people, especially in rural areas, still rely on traditional medicines to treat common ailments and many of these people from the poorest link are involved in the trade of medicinal plants (Khan, 2002). The aim of the present paper is to explore the various ethnomedicinal uses in practice among the tribes inhabiting the villages around Aizawl city and to study the modes of application in different formulations commonly used.

Materials and Method

Mizoram is among the seven sisters state of India situated in the north eastern part of the country. The state of Mizoram lies between 21°58' and 24°35' north latitude and 92°15' and 93°29' east longitude. It shares its borders with Tripura, Assam, Manipur. With the capital city of Aizwal the state shares international borders with Bangladesh and Myanmar (Figure 26.1).The climatic condition of the state is favourable for the growth of a wide variety of fauna and flora making the place very rich in biodiversity. Several reconnaissance field trips were performed since December 2009-March 2010 in different village areas of Aizawl Mizoram namely Lungleng, Sairang, Tanhril, Riek, Lengpui etc. Information on the medicinal uses of plants was collected through questionnaires with tribals and other experienced elderly persons belonging to community and Herbal Practitioners. A specimen was collected with all plant parts *viz.* Flowers, leaves, stem and root. During collection of the specimen the names of locality, habitat, height, method of branching, colour of reproductive parts, common name, etc. were noted separately in a field book. Collection and processing of the voucher specimen were done as per routine Herbarium Method *viz.*

Figure 26.1: Location map of the study area.

identification of collected plants was made either in the field or in laboratory following the different flora. Identification and confirmation were done in BSI, Allahabad, NBRI, Lucknow and CIMAP Lucknow.

Results and Discussion

The present work emphasizes on the extensive collection, taxonomic distribution and use of plants by the Mizo tribes for the treatment of different kinds of skin diseases through field work. In the present paper 21 plants of such medicinal value were collected and documented during several field works. The photographs and taxonomic distribution of all the plants are presented in the the Plates 26.1 and 26.2

A. Albizia chinensis Osbek. Merr.

B. Alstonia scholaris R.Br.

C. Anogeissus acuminata Wall.

D. *Callicarpa arborea* Roxb.

E. *Cassia alata* Linn.

F. *Cassia tora* Linn.

G. *Centella asiatica* (Linn.) Urban.

H. *Clerodendrum infortunatum* Linn.

H. *Curcuma caesia* Roxb.D.

I. *Desmodium triflorum* (Linn.) DC.

J. *Eupatorium odoratum* Linn.

K. *Hedyotis scandens* Roxb.

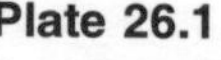

Plate 26.1

A. *Homalomena aromatica* Schott.

B. *Jatropha curcus* Linn

C. *Lindernia ruelloides* (Colon.) Pennel.

D. *Mussaenda macrophylla* Wall.

E. *Parkia roxburghii* G. Don

F. *Phyllanthus fraternus* Webster

G. *Pseudo-Drynaria coronans* (Wall. ex Mett.) Ching.

H. *J. Schima wallichii* Chois

I. *Securinega virosa* Roxb.

Plate 26.2

and Table 26.1 respectively. Among 21 plants 20 are angiosperms and 1 is pteridophyte belonging to 20 genera and 14 families. Further, 5 plants are from Fabaceae 3 from Euphorbiaceae; 2 each from Rubiaceae and Verbenaceae, and 1 each

Table 26.1: Ethnomedicinal plants of Mizoram and their traditional uses.

Sl.No.	Name	Local Name	Family	Parts Used	Mode of Administration
1.	*Albizia chinensis* Osbek. Merr.	Vang	Fabaceae	Bark	The bark infusion is used as lotion in scabies and other skin diseases.
2.	*Alstonia scholaris* R.Br.	Thuam-riat;	Apocynaceae	Latex	Latex is applied to wounds and boils.
3.	*Anogeissus acuminata* Wall.	Zairum	Combretaceae	Bark	Decoction of bark used externally on burn and ulcerated skin
4.	*Callicarpa arborea* Roxb.	Hnahkiah	Verbenaceae	Bark and young shoot	Juice of bark and white powder on young shoots applied externally on cuts and wounds.
5.	*Cassia alata* Linn.	Tuihlo/Dat plant	Fabaceae	Leaf	Leaf in the form of paste is applied on ringworm.
6.	*Cassia tora* Linn.	Kelbe-an	Fabaceae	Root, Leaf and Seed	Grind roots mix with lime juice and decoction of leaf/seed applied externally on various skin infections.
7.	*Centella asiatica* (Linn.) Urban.	Hnahbial, Darbengbur	Apiaceae	Whole plant	Juice of crushed plant is applied on itches, cuts and wounds.
8.	*Clerodendrum infortunatum* Linn.	Phuihnam-chhia	Verbenaceae	Root and Leaf	Decoction of the root is used externally in scabies. The leaf decoction is used topically as an antidandruff agent.
9.	*Curcuma caesia* Roxb.	Ailaidum	Zingiberaceae	Rhizome	The rhizome paste applied externally on bruises and leucoderma.
10.	*Desmodium triflorum* (Linn.) DC.	Bawngek-hlo	Fabaceae	Whole plant	Juice of fresh plant is applied externally on wounds, abscesses and other skin infections.
11.	*Eupatorium odoratum* Linn.	Tlangsam or pholeng or dettole plant	Asteraceae	Leaf	The leaf juice is utilized as an antiseptic. The juice is also applied externally to remove pinworm from the anus.
12.	*Hedyotis scandens* Roxb.	Kel-hnamtur	Rubiaceae	Leaf	Crushed leaves are rubbed to cure skin wart diseases

Contd...

Table 26.1–*Contd...*

Sl.No.	*Name*	*Local Name*	*Family*	*Parts Used*	*Mode of Administration*
13.	*Homalomena aromatica* Schott.	Anchiri	Araceae	Whole plant	Juice of whole plant is used as lotion in skin diseases.
14.	*Jatropha curcus* Linn.	Kangdamdawi	Euphorbiaceae	Stem	The juice of stem is applied topically on the burnt body part.
15.	*Lindernia ruelloides* (Colon.) Pennel.	Thasuih	Scrophulariaceae	Whole plant	Paste of the plant applied on infected skin.
16.	*Mussaenda macrophylla Wall.*	Vakep	Rubiaceae	Leaf and Root	The juice of leaves or roots is applied topically on ulcerated skin. It is also used topically at the site of snake bites.
17.	*Parkia roxburghii* G. Don	Zawng-tah	Fabaceae	Fruit	The green portion of the fruit is mixed with little amount of water and applied to wounds and scabies.
18.	*Phyllanthus fraternus* Webster.	Mitthi-sunhlu	Euphorbiaceae	Whole plant	The poultice/crushedleaves applied externally on scabies and cutaneous diseases.
19.	*Pseudo-Drynaria coronans* (J.) Sm.	Tuaibur or Awmvel	Polypodiaceae	Rhizome	Used to cure leucoderma and other skin problems.
20.	*Schima wallichii* Chois.	khiang	Ternstroemiaceae	Fruit	The fruit decoction is applied on wounds caused by insects like spider and scorpion. The bark juice is used externally as an antiseptic
21.	*Securinega virosa* Roxb.	Saisiak	Euphorbiaceae	Leaf	The leaves are boiled and the water is taken for bathing children suffering from scabies and measles.

from Asteraceae, Apocynaceae, Combretaceae, Apiaceae, Zingiberaceae, Araceae, Scrophulariaceae, Polypodiaceae and Theaceae family.

The plants under study are discussed with the Experts in the field for useful plant part are traditionally in practice with respective mode of administration vis-à-vis kinds of skin diseases along with their scientific and local names along with the family. The details collected in this regard are the recorded with corresponding tribes/ subtribes, local name, scientific name and photograph. From the plants under study, 24 per cent of medicines employed the leaves, 20 per cent the whole plant, 12 per cent root, 12 per cent bark, 8 per cent fruit, 8 per cent rhizome 4 per cent stem, 4 per cent young shoot, 4 per cent latex and 4 per cent seed.The distribution of useful plant parts are plotted and presented in Figure 262. Leaves are good store of alkaloids, tannins and inulins which are the active components of most herbal preparations (Okoegwale and Omefezi, 2001; Focho *et al.*, 2009).

Topical forms of administration of medicine are prevalent in the study area. The herbal medicines are prepared from different parts of the plants using fresh material and dried plant material. The ethnomedicines are prepared as a decoction, juice in adequate quantity of water, fresh paste in water and infusion for proper treatment of skin diseases (Figure 26.3). The source of majority of the plant species is forest and a very few are cultivated by farmers in an unplanned manner. This indicates that it may be difficult to propagate many indigenous plant species in farmlands. It is, therefore, required to train the herbal practitioners on the appropriate propagation techniques of these plant species for sustainable utilization (Ezekiel and Kisangau, 2012).

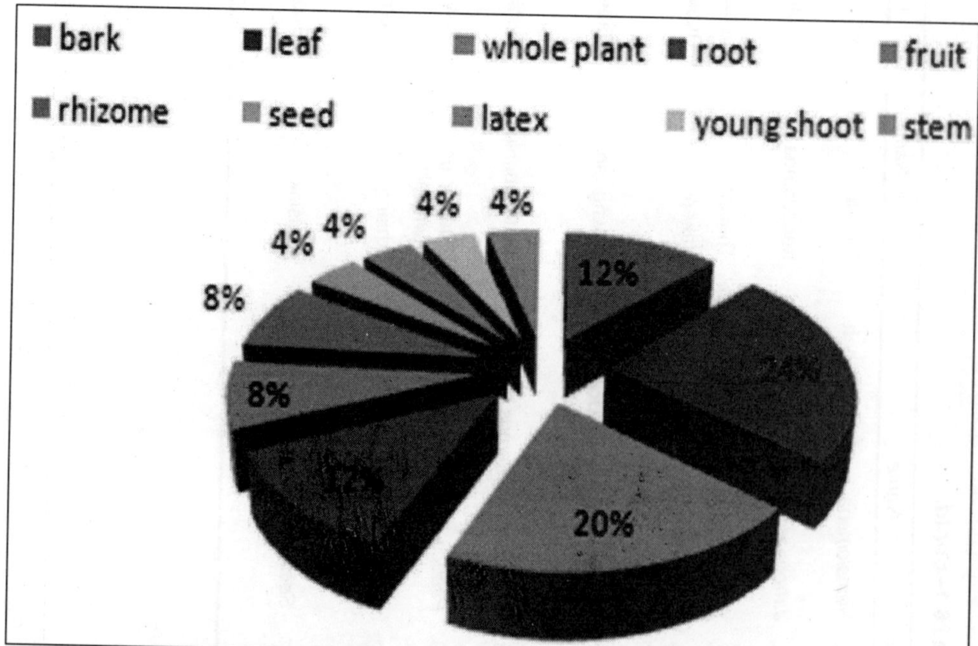

Figure 26.2: Parts of plants used in medicine.

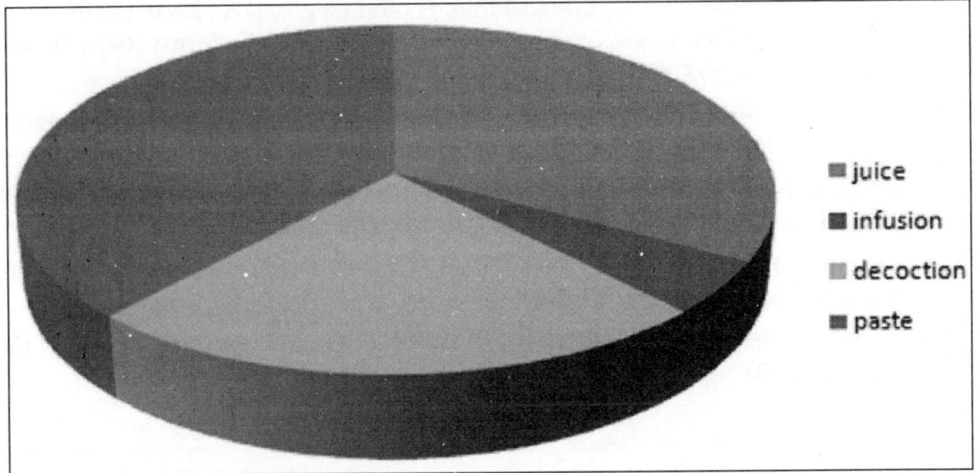

Figure 26.3: Modes of application.

Conclusions

The results of the study are suggestive of the fact that there is rich diversity of medicinal plants used to treat various skin ailments in the villages of Aizawl district of Mizoram. The local people and Herbal practitioners in the study area should be trained for scientific methods of harvesting medicinal plants many of which are on the verge of extinction. It is a demand of time to educate the tribal community of the study area to domesticate the valuable and threatened species of medicinal plants. The domestication of medicinal plants will create additional opportunities for the local people such as provision of an alternative income and could help reduce the pressure on the wild population.

Acknowledgement

The first author expresses her thankfulness as Principal Investigator to the Department of Science and Technology, New Delhi for providing financial assistance in form of DST sponsored project under Women Scientist Scheme – A. The author further is thankful to the Mizoram University authorities for providing necessary facilities for a smooth conduct of the research work.

References

Al-Quran, S., 2005. Ethnobotanical survey of folk toxic plants in southern part of Jordan. *Toxicon.*, 46: 119–126.

Ezakiel, A. and Kisangau, D.P., 2012. Ethnomedicinal study of plants used in villages around Kimboza forest reserve in Morogoro, Tanzania. *J. Ethnobiol. Ethnomed.*, 8: 1.

Focho, D.A., Newuh, M.C., Anjah, M.G., Nwana, F.A. and Ambo, F.B., 2009. Ethnobotanical survey of trees in Fundong, Northwest Region, Cameroon. *J. Ethnobiol. Ethnomed.*, 5: 17.

Hanazaki, N., Tamashiro, J.Y., Leitao-Filho, H. and Gegossi, A., 2000. Diversity of plant uses in two Caicaras communities from the Atlantic forest coast, Brazil. *Bio. Conserv.*, 9: 597–615.

Khan, A.V., 2002. Ethnobotanical studies of plants with medicinal and antibacterial properties. *Ph.D. Thesis*, Aligarh Muslim University, Aligarh, India.

Okoegwale, E.E. an Omefezi, J.U., 2001. Some herb preparation among the people of Isoko clan of Delta State, Nigeria. Niger. *J. Appl. Sci.*, 4: 2359–23571.

Pushpangadan, P. and Kumar, B., 2005. Ethnobotany, CBD, WTO and the Biodiversity Act of India. *Ethnobotany*, 17: 2–12.

Rossato, S.C., Leitao-Filho, H. and Gegossi, A., 1998. Ethnobotany of Caicaras of the Atlantic forest coast (Brazil). *Econ. Bot.*, 53: 387–395.

Environment, Biodiversity and Traditional System, 2014 *Pages* 273–281
ISBN 978-81-7622-307-2
Editors: D.R. Khanna, G.S. Solanki and S.K. Pathak
Published by: BIOTECH BOOKS, NEW DELHI

Chapter 27

Blue-Green Algal Biodiversity of Tripura

Rama Kant

Department of Botany, Ramkrishna Mahavidyalaya, Kailashahar, Unakoti, Tripura – 799 277
E-mail: ramakant.algae@gmail.com; rkojha_1@rediffmail.com

ABSTRACT

Blue-green algae (Cyanobacteria) are able to grow in all types of habitats including aerial, sub-aerial, aquatic, terrestrial and many other diverse habitats. North- East region of our country is well kwon for algal biodiversity including blue-green algae. India is known to have about 97 genera, 1100 species, 400 varieties and forma from various types of Indian habitats. Present communication deals with Blue-green algal biodiversity status of Tripura. Different habitats of Tripura have been explored during consecutive four years (June 2008-May 2012) for the occurrence and distribution pattern of Blue-green algae in agro-ecological conditions of Tripura. This study revealed the occurrence of 36 genera of blue green algae from all types of habitats of Tripura. Out of total 36 genera of blue-green algae, 12 genera belong to unicellular and colonial forms; 10 genera belong to non-heterocystous filamentous forms and 14 genera belong to heterocystous forms. Out of these total 36 genera of blue-green algae, about 15 genera are common in rice-field soils of Tripura. Present survey also revealed that *Aulosira* among heterocystous forms and *Oscillatoria* among non-heterocystous forms of blue-green algae are very common in rice-fields of Tripura.

Keywords: Biodiversity, Blue-green Algae, Rice-fields, Biofertilizer.

Introduction

Blue-green Algae (BGA)/Cyanobacteria are the most ancient group of organisms and evolved 3.5 billion years ago and became one of the most successful and ecologically significant organisms on earth with respect to longevity of lineage and impact on earth's early environment. In the geological time scale, the Precambrian era is considered as the age of Blue-green Algae. BGA is an interesting group of prokaryotic organisms endowed with remarkable genomic plasticity and are distributed in all biotypes of the world. Due to their occurrence in diverse habitat ranging from hot springs, where temperature reach upto 73°C to Antarctica, where temperature may go below the freezing points, oligotrophic to eutrophic, alkaline to acid, aquatic to terrestrial including the leaves and barks of trees, low to high light intensity and free living to symbiotic forms. They exhibit a wide range of thallus organisation ranging from unicellular to colonial, non-heterocystous to heterocystous forms, unbranched to branched filamentous micro-organisms. They are the unique organisms with a potential to manage oxygen evolving photosynthesis and anaerobic nitrogen fixation in a single thallus. BGA have tremendous potential in environmental managements as soil conditioners, biofertilizers, biomonitors of soil fertility, water quality, ameliorating agents-aiding in the reclamation of saline and usar soils and rehabilitation of degraded ecosystems through biosorption of metals and scavenging of toxic compounds besides their utilization, as feed for animals and protein supplements.

Blue-green algae have similarity with gliding bacteria in gross morphology and genetic organization to some photosynthetic bacteria in being able to fix both CO_2 and N_2 using light energy. Blue-green algae also have similarity to eukaryotic algae and higher plants in having chlorophyll-a and ability to use water as electron donor (Stewart, 1980). Some Blue-green algae show a variety of other nutritional modes including anoxygenic photosynthesis, photoheterotrophic growth, dark aerobic growth and dark anaerobic growth (Smith, 1982).

Ever since, it came to the knowledge that Blue-green Algae play an important role in nitrogen fertility of soils (De, 1939), numerous studies have been under taken by the scientist in India (Mitra, 1951, Singh, 1942; Pandey and Mitra 1965; Tiwari, 1972; Laloraya and Mitra, 1973; Tiwari and Pandey, 1976; Tiwari *et al.*, 2007, Venkataraman, 1972, 1975, 1981; Venkataraman and Kaushik, 1980; Subhashini and Kaushik, 1982; Mahajan and Patel, 1983; Maity and Santra, 1985; Kolte and Goyal, 1985; Pal and Santra, 1985; Singh, 1985; Anand and Revathy, 1987; Bongale, 1987; Bongale and Bharati, 1980; Goyal, 1987; Banerjee and Kumar, 1988; Deka and Bordoloi, 1991; Saika and Bordoloi, 1994; Nayak *et al.*, 1996; Kant *et al.*, 2003, 2004) as well as at global level (Junota, 1973; Roger and Kalasooria, 1981; Reynaud and Roger, 1978; Stewart *et al.*, 1969; Smith, 1982). But scientist, limited their work upto nitrogenase activity or floristic account of BGA of some region of India. Mostly they explored the plane and coastal regions and often they ignored remote localities of the country particularly Tripura. Anand (1989) reported the occurrence of total 207 forms blue-green algae from South India (Tamil Nadu and Kerala). Santra (1993) reported total 42 genera and 240 taxa from West Bengal. Although Tiwari *et al.* (2007) covered a wide part of the country, in this study they covered different localities of 11 states of India *viz.* Andhra Pradesh, Assam, Bihar, Karnataka, Kerala, Madhya Pradesh, Maharastra,, Manipur, Orissa, Tamil Nadu, and West Bengal and they made an exhaustive study on survey and distribution pattern of Blue-green algae of all most

.ll type of Indian habitats including rice field and they reported occurrence of total 530 taxa from all type of Indian habitats and 725 taxa from rice-fields of India.

The main aims of the present study were to explore the diversity of Blue-green Algae in different type of habitats including rice fields and to know their distribution pattern in different habitats in the totally ignored state of Tripura.

Material and Methods

In the present study, four districts *viz.* North Tripura, South Tripura, West Tripura and Dhalai of Tripura were surveyed during consecutive four years (June 2008 to May 2012) to study the occurrence and distribution of pattern of Blue-green algal diversity of Tripura. More than 5000 samples were collected from various localities of above mentioned districts of Tripura. All the collected samples were grown into the nitrogenous and nitrogen deficient BG-11 medium (Stanier *et al.*, 1971). After 5-7 days, BGA growth was observed under the stereoscopic binocular microscope and visually different colonies were transferred into the dishes containing liquid and solid BG-11 medium. All the strains were purified by repeated sub culturing and obtained pure cultures of morphologically different strains of Blue-green algae including 36 genera and 180 strains. Isolation of BGA was done either direct by collected material or by enrichment cultures. For enrichment cultures, 1 gm. air-dried well mixed soil was put in sterilized nitrogen deficient as well as nitrogenous BG-11 medium in petridishes (Kaushik, 1987). These petridishes were kept in 4-6 K lux light intensity for 14: 10 light and dark phase at 30 ± 2 °C temperature. The purity of the organisms was tested by incubating the cultures in glucose peptone media in the dark and by microscopic observation (Tiwari 1972). Morphological observations were taken with the help of research microscopes (Lieca DMLB and Radical RXLr-5) attached with digital cameras (DC300, Genoptic ProGres CT-5 and Nikon Coolpix 3400) and quine image analysis software. Isolated strains were identified upto the genus level with the help available literatures (Geitler, 1932; Desikachary, 1959; Drouet, 1968, 1973, 1978, 1981; Baker and Bold, 1970; Hindak, 1985, Kantz and Bold, 1969; Anagnostidis and Komarek, 1988, 1990; Komarek and Anagnostidis, 1986, 1989, 1998, 2001, 2005; Komarek and Komarkova, 2004).

Results and Discussion

In the present study, four districts *viz.* North Tripura, South Tripura, West Tripura and Dhalai district of Tripura were surveyed during June 2008 to May 2012 to study the occurrence and distribution of pattern of Blue-green algal diversity of Tripura. More than 5000 samples were collected from different habitats of various localities of above mentioned four districts of Tripura and total 180 strains of 36 genera of all groups of blue-green algae have been reecorded. Out of total 180 strains of 36 genera, maximum 70 strains are from heterocystous group of Blue-green Algae and 59 strains from unicellular and colonial; and 51 strains from non-heterocystous group. Analysis of data of distribution pattern also indicate that out of total 51 strains of non-heterocystous BGA, approximately 43.14 per cent strains were found growing rice-field of Tripura. As for as distribution of heterocystous forms is concerned, its percentage is approximately 28.57 per cent of total strains were found growing in rice-fields and non-heterocystous forms were 35.59 per cent of total unicellular and colonial strains. As for as percentage of total genera of different groups of blue-green algae is concerned, heterocystous BGA were 39 per cent and non-heterocystous BGA were 28 per cent and percentage of unicellular and colonial BGA was 33 per cent of total genera. The details of distribution pattern of strains in different habitats are given in Tables 27.1–27.3 and Figure 27.1. Analysis of data also indicate that share of total non-heterocystous BGA (non-heteroucystous filamentous and unicellular colonial BGA) was maximum (61 per cent) in comparison to heterocystous BGA (39 per cent).

Table 27.1: Showing genera and distribution pattern of unicellular and colonial BGA in all types of habitats and rice-fields of Triupra.

Sl.No.	*Genera*	*Strains of all Habitats*	*Strains of Rice Fields*
1.	*Aphanocapsa*	06	02
2.	*Aphanothece*	17	05
3.	*Chroococcus*	12	03
4.	*Coelosphaerium*	02	01
5.	*Gloeocapsa*	04	02
6.	*Gloeothece*	06	01
7.	*Merismopedia*	02	02
8.	*Microcystis*	02	01
9.	*Myxosarcina*	03	01
10.	*Stanieria*	02	01
11.	*Synechococcus*	02	01
12.	*Xenococcus*	01	01

Table 27.2: Showing genera and distribution pattern of non-heterocystous and filamentous BGA in all types of habitats and rice-fields of Triupra.

Sl.No.	*Genera*	*Strains of all Habitats*	*Strains of Rice Fields*
1.	*Arthrospira*	02	NF
2.	*Limnothrix*	02	01
3.	*Lyngbya*	12	05
4.	*Microcoleus*	06	02
5.	*Oscillatoria*	17	09
6.	*Phormidium*	05	02
7.	*Plectonema*	02	01
8.	*Pseudanabaena*	02	01
9.	*Schizothrix*	02	01
10.	*Spirulina*	01	NF

Table 27.3: Showing genera and distribution pattern of heterocystous BGA in all types of habitats and rice-fields of Triupra.

Sl.No.	Genera	Strains of all Habitats	Strains of Rice Fields
1	*Anabaena*	12	06
2	*Aulosira*	02	02
3	*Calothrix*	06	02
4	*Camptylonemopsis*	01	NF
5	*Cylindrospermum*	05	02
6	*Dichothrix*	02	NF
7	*Hapalosiphon*	02	01
8	*Nostoc*	23	02
9	*Scytonema*	07	02
10	*Stigonema*	01	NF
11	*Symphyonemopsis*	01	NF
12	*Tolypothrix*	05	01
13	*Westiellopsis*	02	01
14	*Wollea*	01	01

Singh (1961) described the role of blue-green algae in soil fertility of Indian rice fields. Anand (1989) reported the occurrence of total 207 forms blue-green algae from South India (Tamil Nadu and Kerala). Santra (1993) reported total 42 genera and 240 taxa from West Bengal. Although Tiwari *et al.* (2007) reported occurrence of total 1530 taxa from all type of Indian habitats and 725 taxa from rice-fields of India. They also revealed the occurrence of heterocystous and non-heterocystous BGA as 38.61 and 61.3 per cent respectively, which are slightly different from present study. Frequent occurrence of non-heterocystous forms than heterocystous BGA in rice-field of Tripura may be due to excessive application of agro-chemicals.

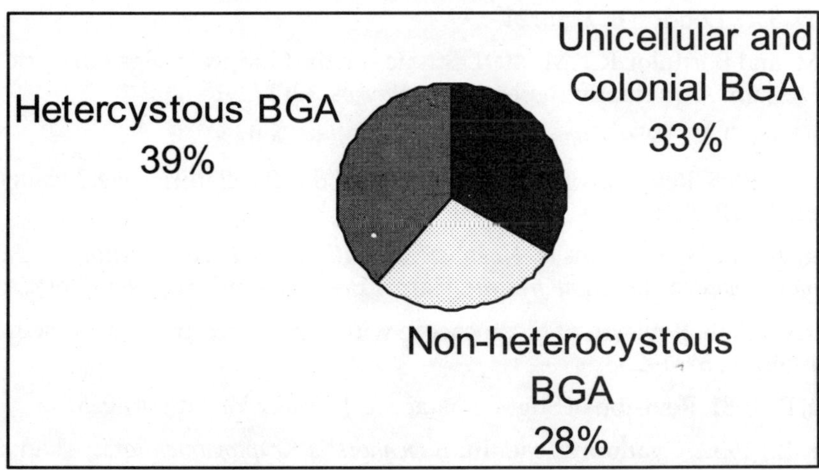

Figure 27.1: Showing blue-green Algae and percentage of different groups in Tripura.

Acknowledgement

Author is thankful to the Principal In-charge, Ramkrishna Mahavidyalaya, Kailashahar, Unakoti for providing laboratory facility. I am also thankful to Professor G.L. Tiwari for kind held in identification of the organism and for providing facilities for microscopic observations. I am also thankful to the University Grants Commission and Department of Biotechnology, Govt. of India, New Delhi, for providing financial assistance.

References

Anagnostidis, K. and Komarek, J., 1988. Modern approach to the classification system of cyanophytes. 3–Oscillatoriales. *Archiv. Hydrobio. Suppl.* 80,1–4 (Algological studies 50–53. 327–472.

Anagnostidis, K. and Komarek, J., 1990. Modern approach to the classification system of cyanophytes. 5–Stigonematales. *Algological studies* 59: 1–73.

Anand, N., 1989. *Handbook of Blue-green Algae of Rice Fields of South India*. Singh and Singh Publ. Dehradun: 80.

Anand, N. and Revathy, G., 1987. Blue-green Algae from rice fields of Tamil Nadu. *Phykos* **26**: 17–21.

Baker, A. and Bold, H.C., 1970. *Phycological Studies X. Taxonomic Studies in the Oscillatoriaceae*. The University of Taxas publication, pp 104–105.

Banerjee, M. and Kumar, H.D., 1988. Effect of carbon sources on growth and nitrogen–fixation of *Aulosira fertilissima* Ghose. *Biochem. Physiol. Pflazen.*, **183**: 51–58.

Bongale, U.D., 1987. Distribution of algal flora in the acidic paddy field soils of North Kannara district, Karnataka State, (India). *J. Indian Bot. Soc.*, **66**: 261–265.

Bongale, U.D. and Bharati, S.G., 1980. On the algal flora of cultivated soils of Karnataka State, India. *Phykos*, **19**: 95–109.

De, P.K., 1939. The role of the Blue-green Algae in nitrogen fixation in rice fields. *Proc. Roy. Soc.*, London B, **127**: 121–139.

Deka, M. and Bordoloi, R.P.M., 1991. Studies on the blue green algae from rice fields of Assam: A qualitative assessment. *Phykos*, **30:** 173–180.

Desikachary, T.V., 1959. *Cyanophyta*. I.C.A.R., New Delhi.

Drouet, F., 1968. Revision of the classification of the Oscillatoriaceae. *Monogr. Acad. Nat. Sci.*, Philadelphia., **15**: 1–370.

Drouet, F., 1973. *Revision of Nostocaceae with Cylindrical Trichomes. Formerly Scytonemataceae and Rivulariaceae*. Hafner Press (Macmillan), New York., pp. 292.

Drouet, F., 1978. Revision of Nostocaceae with constricted trichomes. Beih 2 *Nova Hedwigia.*, **57**: 1–258.

Drouet, F., 1981. Revision of Stigonemataceae. *J. cramer Vacluz.*, pp. 221.

Geitler, L., 1932. Cyanophyceae. In: *Rabenhorst's Kryptogamenflora*. Akademische Verlagsgesselschaft, Leipzig.

Goyal, S.K., 1987. Algal biofertilizer: An appraisal *J. Maharashtra Agri. Univ.*, **12**: 11–19.

Hindák, F., 1985. The cyanophycean genus *Lemmermanniella* Geitler. *Arch. Hydrobiol./ Algolog. Stud.* 40: 393–401.

Junota, 1973. Blue green algae in rice soils of logjakarat, Central Java. *Soil Biol. Biochem.*, **5**: 91–95.

Kant, R, Tiwari, O.N. and Tiwari, G.L., 2003. Distribution and Taxonomical description of Unicellular and Colonial cyanobacteria of rice fields in five districts of Uttar Pradesh. In: *Science and Ethics–Environmental care and sustainability*, Vol.2. (Ed.) Arvind Kumar pp. 107–122.

Kant, R., Tiwari, O.N., Tandon, R., and Tiwari, G.L., 2004. Data base Physiological characterization of Agricultural important unicellular Cyanobacteria of rice fields of U. P., India. *Bioinformatics* India, **2 (3)**: 11–16.

Kantz, T. and Bold, H.C., 1969. *Phycological Studies IX: Morphological and Taxonomic Investigations of Nostoc and Anabaena in Culture*. The University of Taxas Publication – 6924. Austin, U.S.A. pp. 67.

Kaushik, B. D., 1987. *Laboratory Methods for Blue-green Algae*. Associated Publ., Co., New Delhi: 171.

Kolte, S.O. and Goyal, S.K., 1985. Distributional pattern of blue green algae in rice field soils of Vidarbha region of Maharashtra State. *Phykos*, **24**: 156–262.

Komarek, J. and Anagnostidis, K., 1986. Modern approach to the classification system of cyanophytes. 2–Chroocacales. Arch. Hydrobio. Suppl. 73,2. 157–226.

Komarek, J., and Anagnostidis, K., 1989. Modern approaches to the classification of system of Cyanophytes 4, Nostocales. *Arch. Hydrobiol. Suppl.*, **82 (3)**: 247–345.

Komarek, J. and Anagnostidis, K., 1998. *Cyanoprokaryota* 1. Chroococcales. Jena Stuttgart Lübeck Ulm. G. Fischer.pp. 548.

Komarek, J. and Anagnostidis, K., 2001. *Cyanoprokaryota. Oscillatoriales. Susswasserflora von Mitteleuropa* Vol. 19. Part 2. 750 pp

Komarek, J. and Anagnostidis, K., 2005: *Cyanoprokaryota* 2. Teil/2nd Part: Oscillatoriales. *Süsswasserflora von Mitteleuropa* **19/2**, Elsevier/Spektrum, Heidelberg, 759 pp.

Komarek, J. and Komárková, J., 2004. Taxonomic review of the cyanoprokaryotic genera *Planktothrix* and *Planktothricoides*.-*Czech Phycology* **4**: 1–18.

Laloraya, V.K and Mitra, A.K., 1973. Studies on the Blue-green Algae of the paddy fields of India, Part–II. New members of Chroococcaceae and Oscillatoriaceae. *Nova Hedwigia*, **47**: 339–346.

Mahajan, A.D. and Patel, R.J., 1983. Nitrogen fixing Blue–green Algae from paddy fields of Kaira district, Gujrat. *Indian Botanical Contr.*, **1**: 133.

Maity, H. and Santra, S.C., 1985. Blue-green Algal flora of 24– Parganas, West Bengal (India). *Phykos*, **24**: 46–51.

Mitra, A.K., 1951. The algal flora of certain Indian soils Indian. *J. Agric. Sci.*, **21**: 357–373.

Nayak, H, Sahu, J.K., Adhikary, S.P., 1996. Blue-green Algae of rice-fields of Orissa state II. Growth and nitrogen fixing potential. *Phykos*. **35**: 115–118.

Pal, U.C. and Santra, S.C., 1985. Algal–flora of Midnapur, West Bengal (India)–I, *Cyanophyceae*. **24**: 12–17.

Pandey, D.C. and Mitra, A.K., 1965. Certain new Myxophyceae from the rice field soils of India. *Nova Hedwigia*, **10**: 85–96.

Reynaud, P.A. and Roger, P.A., 1978. N_2–fixing algal biomass in Senegal rice fields. *Ecol. Bull. Stockholm.*, **26**: 148–157.

Roger, P.A. and Kalasooria, S.A., 1981. *Blue Green Algae and Rice.* Publ. International Rice Research Institute, Manila, Philippines.

Saika, P. and Bordoloi, R.P.M., 1994. Blue green algal flora from rice fields of Assam. *Phykos*, **33**: 53–57.

Santra, S.C., 1993. *Biology of Rice-Fields Blue Green Algae*. Daya Publishing House, Delhi.

Singh, P.K., 1985. Nitrogen fixation by blue green algae in paddy fields. In: *Rice researches in India*. Indian Council of Agricultural Research, New Delhi, pp. 344–362.

Singh, R.N., 1942. *Wollea Bharadwajae* sp. nov. and its autoecology. *Ann. Bot.* London N.S., **6**: 593–606.

Singh, R.N., 1961. *Role of Blue–green Algae in nitrogen economy of Indian Agriculture.* Indian Council of Agricultural Research, New Delhi.

Smith, A.J., 1982. Modes of Cyanobacterial Carbon Metabolism. In: *The Biology of Cyanobacteria*. (Eds.) Carr, N.G. and Whitton, B.A., Blackwell Scientific Publications, Oxford, pp. 47–85.

Stanier, R.Y., Kunisava, R., Mandal, M., Cohen–Bazire, G., 1971. Purification and properties of unicellular blue-green algae (order Chroococcales). *Bact. Rev.* **35**: 171–205.

Stewart, W.D.P., 1980. Some aspects of structure and function in nitrogen fixing cyanobacteria. *Ann. Rev. Microbiol.*, **34**: 497–536.

Stewart, W.D.P., Haystead, A. and H.W. Pearson, 1969. Nitrogenase activity in heterocysts of blue green algae. *Nature*, **224**: 224–226.

Subhashini, D. and Kaushik, B.D., 1982. Nitrogen fixing potential of blue green algae from saline and alkali soils. *Acta Botanica Indica*, **10**: 321–322.

Tiwari, G.L., 1972. A study of the Blue–green Algae of paddy field soils of India. *Hydrobiologia*, **39 (3)**: 335–350.

Tiwari, G.L. and Pandey, R.S., 1976. A study of the Blue–green Algae of paddy field soils of India. *Nova Hedwigia*, **27**: 701–730.

Tiwari, G.L., Kant, R., Tiwari, O.N., Tandon, R. and Kushwaha, L.L., 2007. Distribution, diversity and characterization of cyanobacteria of rice fields. Proc. Nat. Acad. Sci. India. **77 (B), IV**: 387–402.

Venkataraman, G.S., 1972. Algal biofertilizer and rice cultivation. Pub. Today and Tomorrow's, New Delhi.

Venkataraman, G.S., 1975. The role of blue green algae in tropical rice cultivation. In: *Nitrogen fixation by free living microorganisms*. Ed. W.D.P. Stewart, Pub. Cambtidge Univ. Press, London, pp., 207–218.

Venkataraman, G.S., 1981. Blue green algae: A possible remedy to nitrogen scarcity. *Curr. Sci.*, **50**: 253–256.

Venkataraman, G.S., and Kaushik, B.D., 1980. Save on N – fertilizers by the use of algae on rice fields. *Ind. Fmg, Spl*. Issue no. 1, IARI, **30**: 27–30.

Environment, Biodiversity and Traditional System, 2014 ***Pages* 283–287**
ISBN 978-81-7622-307-2
***Editors:* D.R. Khanna, G.S. Solanki and S.K. Pathak**
***Published by:* BIOTECH BOOKS, NEW DELHI**

Chapter 28

Evaluation of Genetic Diversity and Approaches for Conservation of *Parkia timoriana* (DC.) Merr.

Robert Thangjam

Department of Biotechnology, School of Life Sciences,
Mizoram University, Aizawl – 796 004, Mizoram
E-mail: robertthangjam@gmail.com

ABSTRACT

Parkia timoriana (DC.) Merr., commonly known as tree bean, is a multipurpose leguminous tree commonly found throughout the north eastern Indian region. The pods are consumed as a special delicacy by the locals in the region. However, several reports on the infestation and mass death of this tree have been reported in recent years. For future breeding and conservation strategies, knowledge of the status of genetic diversity is essential as well as biotechnological approaches for its regeneration and genetic improvement. The paper discusses the status of the level of genetic diversity determined by using molecular makers and in vitro regeneration and genetic transformation techniques developed in this multipurpose tree legume.

Keywords: ***Parkia timoriana, Tree bean, Genetic diversity, Biotechnological tools, Conservation.***

Introduction

Parkia timoriana (DC.) Merr., Syn. *Parkia roxburghii* G. Don, commonly known as tree bean, is a multipurpose tree of the family Leguminosae and sub-family Mimosoidae. It is the most widespread species of *Parkia* in the Indo-Pacific region distributed from northeast India to Irian Jaya (Hopkins 1994). It is a highly branched and medium height (10-12 m) multipurpose tree having potential commercial and ecological significance in the region (Kanjilal *et al.*, 1982). The natives of north eastern India particularly in Manipur state, consumed the pod and the seeds as vegetable, salad and *chutney* in all its developmental stages or sun-dried for future use during off-season. The tree is well adapted to grow in different agro-climatic regions from colder hilly regions to the hotter plains. It is commonly grown wild in the *Jhums*, forests and backyard of houses throughout North-East India without any special care. It has high nutritional and medicinal value (Longvah and Deosthale 1998; Suvachittanont *et al.*, 1996). Trees bearing narrow and uniform pods having light green colour are superior in flavour than the others and accordingly thirteen cultivars were identified (Meitei and Singh 1990). Similarly nine varieties of tree beans were reported from Manipur based on their palatability and other eating qualities (Salam and Singh 1997). Tree bean is conventionally propagated through seeds and vegetative cuttings. However, vegetative cuttings are not feasible in view of low rooting percentage. Propagation using seeds is bestowed with severe fungal and pest problem during storage. During the last decade, many cases of mass-death of tree bean trees have been observed in the northeast Indian region due to the die-back symptoms seriously affecting socio-economic situation of the growers. Infestation of *Cadra cautella* on the pods collected from field and storage have been well documented (Thangjam *et al.*, 2003a). The problems of die-back symptoms in this tree have also been found to be associated with the infestation of *Anoplophora glabripennis* (Motchulsky) commonly known as Asian longhorned beetle. In recent years, the biotechnological interventions for the sustainable production, genetic improvement and conservation of plant species is gaining popularity because of its scientific advancement and reliability

Biotechnology can be defined as any technique that uses living organisms or parts thereof to make or modify a product, improve plants or animals or develop microorganisms for specific uses. It is invaluable in research on conservation of bioresources. Although modern biotechnology is a newly introduced science (less than 50 years old) its impact has greatly excited the imagination and provoked the concern of almost every part of the society worldwide. Developments such as the tomato that can be frozen and the cassava and other agricultural crops those has been genetically engineered for insect and virus resistance and which are in or near commercial release are simply spectacular by any standard. The richness of plant and animal diversity in developing countries is a major asset in agricultural development and, therefore, the conservation of such resources is fundamental to the progress and usefulness of biotechnology. The paper discusses the status of the level of genetic diversity of tree beans determined by using molecular makers and recent advances of *in vitro* regeneration and genetic transformation techniques developed.

Status of Genetic Diversity

Knowledge of the level, structure and origin of genetic variation within and between plant populations is a pre-requisite for the effective utilisation and conservation of the species. The factors determining the level and structure of genetic variation within plant species include evolutionary history characteristics, population density, mating system, and mechanisms of gene flow (Schaal and Learn

1988; Hamrick *et al.*, 1991). This intraspecific genetic diversity plays a critical role in the ability of populations to respond to specific adaptations such as resistance to disease or insect, tolerance to certain soil conditions or other attributes that may be of current or future value in forest tree breeding programmes (Rogers and Ledig 1996). Thus, the conservation of plant genetic resources involves not only preventing extinction but also ensuring the availability of resources for future use through adaptation to the changing environments. Although morphological characters and agronomic traits have been used traditionally to characterise levels and patterns of diversity, these traits alone represent only a small portion of the plant genome and are also influenced by environmental factors, thereby limiting their utility in describing the potentially complex genetic structures which may exist within and between taxa (Avise 1976). To overcome these constraints, various polymerase chain reaction (PCR) based DNA markers such as randomly amplified polymorphic DNA (RAPD), amplified fragment length polymorphism (AFLP), simple sequence repeats (SSRs) and inter-simple sequences repeats (ISSRs) have been widely used for the detection of genetic variation in many crops (Gupta and Varshney 2000). These 'markers' of genetic variation are independent of environmental factors and more numerous than phenotypic characters, thereby providing a clearer indication of the underlying variation in the genome of an organism (Avise *et al.*, 1989). Using randomly amplified polymorphic DNA (RAPD) markers genetic variation of eight genotypes of tree beans grown in Manipur, India was investigated at the DNA level. The results of the dendrogram and principal co-ordinate analysis were in considerable agreement with the overall representation of relationships among the genotypes studied. It is also observed that the similarities between genotypes from the different geographical locations revealed considerable discrimination (Thangjam *et al.*, 2003b).

In vitro Regeneration and Genetic Transformation

For a sustainable commercial cultivation of this tree, it would require large amount of superior quality and genetically improved planting materials that may be difficult to obtain by conventional methods of propagation. Thus, there is an urgent need for biotechnological interventions using tissue culture techniques for mass production of quality planting materials and genetic modification using gene(s) coding for insect resistance or proteinase inhibitors. The genetic improvement of any plant species requires the development of an efficient transformation and regeneration procedure for further transfer of gene of interest.

Micropropagation refers to *in vitro* mass production of plant propagules from any plant part or cell. Such propagules are used to raise whole plants. Through somatic embryogenesis individual cultured cells or small groups of cells undergo development resembling that of the zygotic embryo and the embryoids produced can be used to produce whole plants. Thangjam and Singh (2006) reported the potential for *in vitro* rapid regeneration of tree bean, using cotyledon explants cultured on MS and B5 basal media supplemented with various concentrations of 2,4-D, NAA and BAP. Successful callus induction was observed in all the treatments. Maximum percentage of callus induction was obtained in the 2,4-D supplemented basal media. However, the overall response of MS medium was found to be superior to that of B5 medium. Explants cultured on MS medium fortified with combinations of 2,4-D and BAP induced rapidly proliferating calli that turned more friable and nodular. These protuberances eventually developed into somatic embryos when transferred in basal MS suspension medium without growth regulators. *In vitro* regeneration and genetic transformation were also achieved using cotyledonary node explants (Thangjam and Sahoo, 2012). The ability to produce multiple shoots has been evaluated using semi-solid Murashige and Skoog (MS) basal medium and Gamborg's B-5 basal medium supplemented with various concentrations of α-naphthalene acetic acid (NAA) and 6-benzylaminopurine (BA) either in single or in combinations. The explants cultured in MS medium supplemented with combinations of 2.7 μM NAA and 11 μM BA showed the maximum frequency of multiple shoots (96.66 per cent) formation and number of shoots per explants (6.60) respectively. For rooting full and half strength MS medium supplemented with various concentrations of indole-3-butyric acid (IBA) and NAA were studied and the highest number of root formation was observed in full strength MS supplemented with 9.8 μM IBA. Using *Agrobacterium tumefaciens* strain EHA105 pCAMBIA2301 various optimum conditions for efficient transformation were determined by recording the percentage of GUS^+ explants. Following the optimized conditions, the co-cultured explants were cultured on semi-solid shoot regeneration medium (SRM) containing MS medium + 2.7 μM NAA + 11 μM BA + 100 mg/l kanamycin + 500 mg/l cefotaxime. After 8 weeks of culture, the regenerated shoots were rooted in rooting medium (RM) containing MS medium + 9.8 μM indole-3-butyric acid (IBA), 3 per cent sucrose, 7.5 mg/l kanamycin and 500 mg/l cefotaxime. Successful transformation was confirmed by histochemical GUS activity of the regenerated shoots, *npt*II gene PCR analyses of the regenerated kanamycin resistant plantlets and Southern analysis of putative transgenic PCR^+ plants.

Thus, the advances made on the application of biotechnological tools can be further applied to identify the superior genotypes, making strategies for conservation and genetic improvement of this multipurpose tree legume.

References

Avise, J.C., 1976. Genetic differentiation during speciation. In: *Molecular Evolution*, (Ed.) F.J. Ayala. Sinnauer, Sunderland, MA, pp. 106–122.

Avise, J.C., Bowen, B.W. and Lamb, T., 1989. DNA fingerprints from hypervariable mitochondrial genotypes. *Mol. Biol. Evol.*, 6: 258–269.

Gupta, P.K. and Varshney, R.K., 2000. The development and use of microsatellite markers for genetic analysis and plant breeding with emphasis on bread wheat. *Euphytica*, 113: 163–185.

Hamrick, J.L., Godt, M.J.W., Murauski, D.A. and Lordess, M.D., 1991. Correlations between species traits and allozyme diversity: Implications for conservation biology. In: *Genetics and Conservation of Rare Plants*, (Eds.) D.A. Falk and K.E. Holsinger. Oxford University Press, New York, pp. 75–86.

Hopkins, H.C.F., 1994. The Indo-Pacific Species of Parkia (Leguminosae: Mimosoideae). *Kew Bull.*, 49: 181–234.

Kanjilal, U.N., Kanjilal, P.C. and Das, A., 1982. *Flora of Assam*, Avon Delhi, 2: 151.

Longvah, T. and Deosthale, Y.G., 1998. Nutrient composition and food potential of *Parkia roxburghii*, a less known tree legume from northeast India. *Food Chem.* 62: 477–481.

Meitei, W.I. and Singh, A.I., 1990. Organoleptic test of treebean *Parkia roxburghii* G. Don. *Ind. Journ. Hill Farm.*, 3: 47–49.

Rogers, D.L. and Ledig, F.T., 1996. The status of temperate North American forest genetic resources. Report No. 16. University of California, Genetic Resources Conservation Program, Davis, Ca.

Salam, J.S. and Singh, S.B., 1997. Flavour, physical characters and pigment contents in the fruits of tree bean (*Parkia roxburghii*). *Ind. J. Hill Farm*, 10: 115–118

Schaal, B.A. and Learn, G.H., 1988. Ribosomal DNA variation within and among plant populations. *Ann. Missouri. Bot. Gard.*, 75: 1202–1216.

Suvachittanont, W., Kurashima, Y., Esumi, H. and Tsuda, M., 1996. Formation of thiazolidine-4-carboxylic acid (thioproline), an effective nitrite-trapping agent in human body, in *Parkia speciosa* seeds and other edible leguminous seeds in Thailand. *Food Chem*, 55: 359–363.

Thangjam, R., Damayanti, M. and Jitendra, G.S., 2003a. *Cadra cautella* Walker (Lepidoptera: Crambidae: Phycitinae): A pest on *Parkia timoriana* (DC.) Merr. in Manipur. *Curr. Sci.*, 85: 725–726.

Thangjam, R., Damayanti, M. and Jitendra, G.S., 2003b. Detection of genetic diversity in *Parkia timoriana* (DC.). Merr. using randomly amplified polymorphic DNA analysis. *J. Food Agric. Environ.*, 1: 46–49

Thangjam, R. and Maibam, R.S., 2006. Induction of callus and somatic embryogenesis of cotyledonary explants of *Parkia timoriana* (DC.) Merr.: A multipurpose tree legume. *J. Food Agric. Environ.*, 4: 335–339.

Thangjam, R. and Sahoo, L., 2012. *In vitro* regeneration and *Agrobacterium tumefaciens*–mediated genetic transformation of *Parkia timoriana* (DC.) Merr.: A multipurpose tree legume. *Acta Physiol. Plant*, 34: 1207–1215.

Environment, Biodiversity and Traditional System, 2014 ***Pages 289–295***
ISBN 978-81-7622-307-2
***Editors:* D.R. Khanna, G.S. Solanki and S.K. Pathak**
***Published by:* BIOTECH BOOKS, NEW DELHI**

Chapter 29

Water: The Elixir of Immortality Quotes from Vedas and Puranas

Rahul Verma

Department of Geology,
Pachhunga University College, Aizawl, Mizoram
E-mail: vrahul24@gmail.com

ABSTRACT

The Veds are the oldest (around 6500 BC) monumental scriptures and represent fountain of wisdom and knowledge in prehistoric India. Water has been regarded as the first and the foremost element in the Veds, which is the original source that gave rise to the evolution of the universe. Water, or the water element, has been vividly addressed in the Veds (*Rig-Ved, Sam-Ved, Yajur-Ved and Atharv-Ved*) and other ancient literature (*Brahmans, Upanishads, Smritis, Samhitas, Aranyaks,* and *Purans*) abundantly.

Every living and non-living body is made up five basic, gross elements of Nature, called '*Panchamahabhuta*'. Among the five basic elements (*Bhutas*)–Ether or Space or Firmament (*Akash*), Air (Vayu), Energy or Fire (*Tej or Agni*), Water (*Aapah*) and Earth (*Prithvi*), water occupies the highest place.

***Atharv-Ved* (3: 13: 6)**, prays the prime significance of water- *"The ability to see, hear and speak, are useless in the absence of adequate water. Water is the basis of life. Most life forms are born in water and live in it. O! Water stream come near me. You are the Elixir of immortality"*.

Indra, the god of water and heaven, is regarded as the most powerful of all gods, again, signifying the prime importance of water in Vedic period. *Maruts* are known as the god of winds and *Aryaman, Mitra, Varun* and *Som* are known as the facilitators of water.

The ***Rig-Ved hymn* RV6.70.4-5.2** describes *Dyava Prithvi* (Heaven and Earth) as decorated with ornaments of water, abundantly blessed with love of water, and as conservator of water.

According to the ***"Brahmanda Puran (Poorvardh 24.26-30),*** the sun is called *"Sahasrarashmi"* (having thousand rays).It is further divided into three significant groups:

"Vristisarjana" or *"Amrita"* (causing rain), *"Gharmasarjana"* (Causing heat) and *"Himsarjana"* (causing snow).

Atharv- Ved regards the earth (*Prithvi*) as lying hidden in the heart of a vast ocean. It probably refers to the *"Pangaea"* and *"Panthalasa"* relationship. In ***"Brihadaranyakopanishad* (II. 3.3),** the ratio of the land and water has been put as 1: 2 that is almost approximate to the modern calculation. ***Garun Puran* (54.4)** describes earth as floating on the water like a sailing boat, probably indicating the plate motion. ***Vayu Puran* (24.8, 26.7)** mentions *"Ekarnava"* when all the oceans intermingle in one another at the close of every *"Manavantra"* (probably, the reunion of all continents).

***Rig Ved Hymn* RV1.161.14** invites the *Maruts* (the god of wind) to create rain bearing conditions and showers on the earth. *"Sons of Strength, the Maruts, awaiting your arrival advances from the sky, Agni comes from the Earth, the winds traverses from the firmament, and Varun comes with undulating waters (clouds)"*. The hymn pictures the clear understanding of formation of clouds and rainfall through the hydrological cycle, during the Vedic period.

Rishi's in all the four *Veds* highly regard water as the form of god. The sage in ***Yajur - Ved*** pray *"O Water, thou art the reservoir of welfare and prosperity, sustain us to become strong. We look up to thee to be blessed by thy kind nectar on this earth. O Water, we approach thy to get rid of our sins"*

***The Rig-Ved Hymn* (RV 10.75.2, 4, 7),** signifies the importance of river. *"The straight-flowing, white coloured, bright-shining river moves along with her ample volumes through the realms; the inviolable river, most efficacious and speckled like a horse, is beautiful as a comely maiden"*.

***Atharv-Ved* (verse XIX.2.1-2.2)** classifies rivers into *"Haimavatih"* (originated from glaciers), *"Utsyah"* (originating from springs), *"Nishyandah"* (originating from small streams), *"Anupyah"* (originating from streamlets) and *"Dhanvanyoh"* (Desert rivers).

Ramayan and *Mahabharat* epics, also mention about the Ganga, Yamuna, Kalindi, Saryu and Narmada. Among the rivers of present North Eastern India, Ramayan period quotes reference of Lohiyta (Lohit/Brahmaputra) and Kalmahi (Kalni/Barak).

Water conservation for Social Welfare has also been highlighted in *Rig-Ved* and *Atharv-Ved*. ***Rig Ved* verses 1.12.18, 1.23.18, 5.32.2** refer to use of water from rains, rivers, wells, ponds etc. for agriculture, domestic and or other purposes with wisdom and efficiency. To meet the obligations towards the society, *Rig-Ved* stresses the need of construction of artificial canals to irrigate desert areas by deploying skilled persons (engineers).

***Atharv-Ved Hymn* 2.23.1, 6.100.2, 7.11.1** describes the objectives of water conservation through efficient use of available water resources and its conservation. It prescribes the learned men to provide water in the desert areas by wells, ponds, canals etc. and to take precautionary and preventive measures against the droughts, floods and natural calamities in advance.

It is a well documented and proven fact that the ancient philosophers and sages had identified water as the most important and inevitable element for the existence and survival of life in the universe. Apart from simple livelihood needs for the human beings, water is the nectar for performing religious rituals and social ceremonies, for purification of body and soul. The rivers are worshipped like goddesses, and considered sacred and the waters in rivers are treated as holy. River water is often described as life sustaining, medicinal, as ambrosia, cleanser of sins and are highly regarded as divine, protectors and blissful. This rich heritage of Indian culture is passing on since ancient Vedic times till today and shall continue to cherish till eternity....as long as the water flows on the planet earth.

Keywords: *Elixir of immortality, Vedas, Puranas.*

Introduction

Water has been regarded as the first and the foremost element in the Veds, which is the original source that gave rise to the evolution of the universe. Water, or the water element, has been vividly addressed in the Veds (*Rig-Ved, Sam-Ved, Yajur-Ved and Atharv-Ved*) and other ancient literature (*Brahmans, Upanishads, Smritis, Samhitas, Aranyaks, and Purans*) abundantly.

Every living and non-living body is made up five basic, gross elements of Nature, called '*Panchamahabhuta*'. Among the five basic elements (*Bhutas*)–Ether or Space or Firmament (*Akash*), Air (Vayu), Energy or Fire (*Tej or Agni*), Water (Aapah) and Earth (*Prithvi*), water occupies the highest place.

Water as the Basis of Creation: Quotes from Vedic Texts

Water: The Womb of Creation

Rig Ved gives an account of the beginning of the universe. In the beginning there was neither beginning, nor end, neither day nor night, no atmosphere, no sky. There existed only the "Self born".

Upanishads describe that in the beginning the "death" concealed all. The earth has originated from water that was produced after the worshipping of death (***Brihadaranyakopanishad.1.2.1-3***).

Prime Significance of Water

Atharv Ved signifies the prime importance of water and addresses water as the "Elixir of Immortality".

"The ability to see, hear and speak, are useless in the absence of adequate water. Water is the basis of life. Most life forms are born in water and live in it. O! Water stream come near me. You are the Elixir of immortality".

Atharv-Ved (3: 13: 6)

Ornamentation of Water

In the Rig Ved (*RV6.70.4-5.2*), *the impact of water on the Earth and Heaven, is expressed as -*

"Dyava Prithvi (Heaven and Earth) are decorated with ornaments of water, abundantly blessed with love of water, and as conservator of water."

Origin of Rain and Snow

According to "Brahmand Puran" (Poorvardh 24.26-30) Rain is considered as "Amrit" (Nectar). All the rain is controlled by the Sun. Sun is called as "Sahastra Rashmi" (having 1000 rays. These rays are further classified into:

1. Vristisarjana (Causing Rain)–400 rays
2. Gharmasarjana (Causing heat)–300 rays
3. Himsarjana (Causing snow)–300 rays

Pangaea and Panthalasa Relationship

The Plate Tectonic reconstruction has revealed that during the Precambrian Era, all continents were united as one mega landmass called "Pangaea", which was surrounded by a vast ocean known as "Panthalasa" Interestingly, the position of "Pangaea" on the Earth in the beginning, was compared with the position of heart in the human body.

Atharv Ved (*Atharv.XII.1.8*) regards the earth (Prithvi) as lying hidden in the heart of a vast ocean. It probably refers to the "Pangaea" and "Panthalasa relationship.

Concept of Plate Tectonics

Garun Puran (54.4) describes earth as floating on the water like a sailing boat. (Probably, indicating the plate tectonics). This concept is the nothing but the explanation of movements of lithospheric plates over the semi viscous "asthenosphere" under the influence of mantle convection current.

Reunion of all Aontinents

The modern theory of plate tectonics states that the plate motion is a uniform and continuous process in the Earth's cycle. Creation and destruction of plates is repetitive and cyclic process. It is very likely that in a future of Earth's process, all the continents may reunite and result in the reunion of "Pangaea".

The very concept of reunion of Pangaea is described in "Vayu Puran" (24.8, 26.7) as "Ekarnava", and this process is supposed to be repeated at close of every "Manavantra" (when all the seven oceans will intermingle into each other).

Knowledge of Land and Water Ratio

According to the "Brihadaranyakopanishad (II. 3.3), the ratio of the land and water has been put as 1: 2. As a matter of fact, it matches with the modern calculation.

Knowledge of Hydrological Cycle

The Vedic seers had concrete knowledge of 'hydrological Cycle". Agni (Evaporation), Varun (Clouds) and Maruts (Monsoon Winds) are the gods responsible for the hydrological cycle. This concept is well expressed in the following hymn-

"Invite the Maruts (the god of wind) to create rain bearing conditions and showers on the earth. "Sons of Strength, the Maruts, awaiting your arrival advances from the sky, Agni comes from the Earth, the winds traverses from the firmament, and Varun comes with undulating waters (clouds)".

Rig Ved.161.14

Cloud Formation

Conversance of cloud formation is well reflected in the work of Kalidas's Poorvameghdootam.

Brahmand Puran (22.36-40) gives an account of types of clouds. "Pakasya" are clouds originated on the windward sides, causing excessive rain."Pundra" clouds cause snow fall on the Himavats (Himalaya).Another cloud types are "Jimuta" and "Pushkarvarta" ("Cirrus clouds") and "Varshamegha" ("Nimbus clouds).

The Importance of River

"The straight-flowing, white coloured, bright-shining river moves along with her ample volumes through the realms; the inviolable river, most efficacious and speckled like a horse, is beautiful as a comely maiden".

Rig-Ved (RV 10.75.2, 4, 7)

Knowledge of Origin of Rivers

In Veds, Indra has been designated as the god of rivers and rains. Atharv Ved presents a classification of rivers on the basis of their origin. The rivers are classified as -

1. Haimavatih –Originating from Glaciers.
2. Utsyah – Originated from springs.
3. Dhanvanyoh – Originated from Desert.
4. Anupyah – Originated from Streamlets.

Such a detailed classification of rivers reflects the strong knowledge of geographical facts in the Vedic times.

References of Rivers

Ramayan and Mahabharat epics clearly mention about the Ganga, Yamuna, Kalindi, Saryu, Narmada, Godavari, Krishna and "Sapta Sindhu" (Fig.3). Rig Vedic texts explain the origin of Ganga River from the glacier of Himalaya, hence it is also known as "Haimavatih". Yamuna originates from Kalinda mountain and thus known as "Kalindi". The original habitat of Aryans is well known as "Sapta Sindhu" Pradesh (the region of seven rivers).

Rivers of peninsular India like Krishna, Godavari and Cauvery etc are referred to as "southern rivers" in the Vedic texts. Among the North Eastern rivers- there is reference of Lohiyta (Lohit/Brahmaputra) and Kalmahi (Kalni/Barak).

Concept of Water Conservation and Engineering Marvel

The concept of water conservation and optimum utilization has been beautifully explained and it reflects the farsightedness of the Vedic seers. Water conservation for Social Welfare has also been highlighted in *Rig-Ved* (*Rig Ved. 1.12.18, 1.23.18, 5.32.2*), *under the following heads:*

- ✰ To be used with wisdom and efficiency.
- ✰ To be utilized from rains, rivers, wells, ponds etc.
- ✰ To be used for agriculture, domestic and or other purposes.
- ✰ To be used to meet the obligations towards the society.
- ✰ To construction artificial canals to irrigate deserts.
- ✰ To deploy skilled persons (engineers) for such needs.

The Vedic people were skilled enough to construct wells and canals, as reflected from the above mentioned text. "Ramayan" mentions the excellence of engineering constructions. The reference of construction of "Ram Setu" (*Adam's Bridge*), between India and Lanka (Lanka Kand), throws the light upon the knowledge of engineering excellence in the "Ramayan Period". The satellite imagery obtained has proved the existence of the "Ram Setu".

Rivers ad Goddess

"O Water, thou art the reservoir of welfare and prosperity, sustains us to become strong. We look up to thee to be blessed by thy kind nectar on this earth. O Water, we approach thy to get rid of our sins"

Yajur -Ved prayer

This Yajur-Ved prayer truly highlights the high and heavenly position of rivers in the Vedic culture. The rivers are worshipped like goddesses, and considered sacred and the waters in rivers are treated as holy. The oldest living city of the World-Varanasi, situated on the bank of Ganga, and Rishikesh- Haridwar, are considered as the holy seats of Mother Ganga. The daily worshipping of 'Goddess Mother Ganga" is an age old tradition that is practiced for a long time.

Conclusions

It is a well documented and proven fact that the ancient philosophers and sages had identified water as the most important and inevitable element for the existence and survival of life in the universe. Apart from simple livelihood needs for the human beings, water is the nectar for performing religious rituals and social ceremonies, for purification of body and soul. The rivers are worshipped like goddesses, and considered sacred and the waters in rivers are treated as holy. River water is often described as life sustaining, medicinal, as ambrosia, cleanser of sins and are highly regarded as divine, protectors and blissful. This rich heritage of Indian culture tradition of worshiping water is passing on since ancient Vedic times till today and shall continue to cherish till eternity....as long as the water flows on the planet earth.

References

Atharv–Ved (*XII.1.8*)

Atharv–Ved (*2.23.1*)

Atharv–Ved (*3: 13: 6*)

Atharv–Ved (*6.100.2*)

Atharv–Ved (*7.11.1*)

Atharv–Ved (XIX.2.1)

Atharv–Ved (verse XIX.2.2)

Brihadaranyakopanishad (*1.2.1–3*)

Brahmanda Puran (Poorvardh 24.26–30)

Brihadaranyakopanishad (II. 3.3)

Garun Puran (54.4)

Matsya Puran (115.19)

Poorvameghdootam.5

Rig–Ved hymn (RV6.70.4–5.2)

Rig–Ved Hymn (RV 10.75.2,)

Rig–Ved Hymn (RV 4, 7)

Rig Ved (1.12.18)

Rig Ved (1.23.18)

Rig Ved (5.32.2)

Rig Ved (6.70.4–5.2)

Rig Ved (161.14)

Rig Ved. (1.12.18)

Rig Ved. (1.23.18)

Rig Ved. (5.32.2)

Vayu Puran (24.8, 26.7)

Vayu Puran (26.7)

Environment, Biodiversity and Traditional System, 2014 ***Pages* 297–311**
ISBN 978-81-7622-307-2
***Editors:* D.R. Khanna, G.S. Solanki and S.K. Pathak**
***Published by:* BIOTECH BOOKS, NEW DELHI**

Chapter 30

Ecology of some Lentic Systems in Barak Valley, Assam: A Preliminary Investigation

Tapati Das*, Mahabir Prasad Khandelia and Aftab Ahmed Shah

Department of Ecology and Environmental Science, Assam University, Silchar-788011

ABSTRACT

A preliminary limnological investigation was carried out in some lentic systems of Cachar district in Barak Valley, Assam during 2011. The study was carried out to gather information on the trophic status of the selected lentic systems especially in relation to its phyto- and zooplankton communities. The study revealed the change in species composition of plankton in relation to change in physico-chemical properties of water that again changed with the variation in general features and disturbances in the systems. Phytoplankton comprised a total of four classes belonging to Bacillariophyceae, Chlorophyceae, Euglenophyceae and Cyanophyceae and zooplankton comprised a total of three orders belonging to Cladocera, Calanoida and Cyclopoida. Dominance of Cholorophyceae amongst phytoplankton and Cladocera and Cyclopoida amongst zooplankton indicates the nutrient-enriched conditions in the sampling stations. Overall, the study reveals the presence of organic matters in the selected lentic systems and its greater accumulation with time unless there is initiation of some management options.

Keywords: Water properties, Phytoplankton, Zooplankton, Lentic systems, Barak valley.

* *Corresponding Author:* E-mail: das.tapati@gmail.com

Introduction

Deterioration of surface water quality because of pollution and its consequent effect on human health has become an important issue in management of aquatic resources. Lentic ecosystems perform many important environmental functions. For example, they recycle nutrients, purify water, attenuate floods, recharge ground water and provide habitats for several aquatic communities and wildlife (Loeb, 1994). Some of these ecosystems play a great role in human recreations like sporting and tourism and some provide a great economic benefit in terms of fishery and other aquatic resources. We know that most of the aquatic organisms form the most sensitive component of aquatic ecosystem and signal environmental disturbances (Carle, 1979).

The physico-chemical parameters and nutrient status of water body play an important role in governing the production of both phyto- and zooplankton communities which are the natural food of many fishes. Phytoplanktons are the primary producer and form the lowest trophic level in the food chain of any aquatic ecosystems, and they play an important role in production of autochthonous organic matter. Moreover, species composition and density of phytoplankton determine the quality of any water body (Bahura, 1991). Zooplanktons are the primary consumer, and they play a key role in the foundation of aquatic food web by being at the second trophic level in most aquatic ecosystems. Besides, zooplankton respond quickly to aquatic environmental changes (*e.g.*, water quality, such as pH, colour, odour and taste, etc.) because of their short life cycle, and are therefore used as indicators of overall health or condition of the aquatic habitats (Thorpe and Covich, 1991 and Carrick and Schelske, 1997). We also know that chemical measurements reflect water quality at a given time while biological assessment reflects condition that have existed in a given environment over a long period of time (Odiete *et al.*, 2003). In this regard, the study of planktonic communities can be very helpful in assessment of aquatic ecosystem potentials not only in terms of its biotic resources but also its trophic or health status. With this background information, the present study was undertaken to understand the ecology vis-à-vis the trophic status of some lentic systems, and hence their potentials in perspective of the well being of the local people as well as their immediate environment.

Study Area

Barak valley (24°8' N and 25° N latitude and 92°15' E and 93°15' E longitude) is situated in the southern part of Assam, north east India. It comprises of 3 districts namely Cachar, Karimganj and Hailakandi with a total area of 3,111 sq. miles. The present study was carried out around the campus of Assam University, Silchar. It is located at a distance of approximately 25 km from the Silchar town. During the last 5 years the present study area (in some patches) has changed from a village to a small sub-urban area due to increase in residences and small shops. Most of the people in this area are engaged in small business like grocery shops, photocopy shops, small cafeteria etc. while the rest are engaged in fisheries and agricultural activities.

Fisheries being one of the main occupations in this region for the local communities, the area is endowed with a large number of ponds of varied size. Some of the ponds are exclusively for household uses, some for fisheries and some are

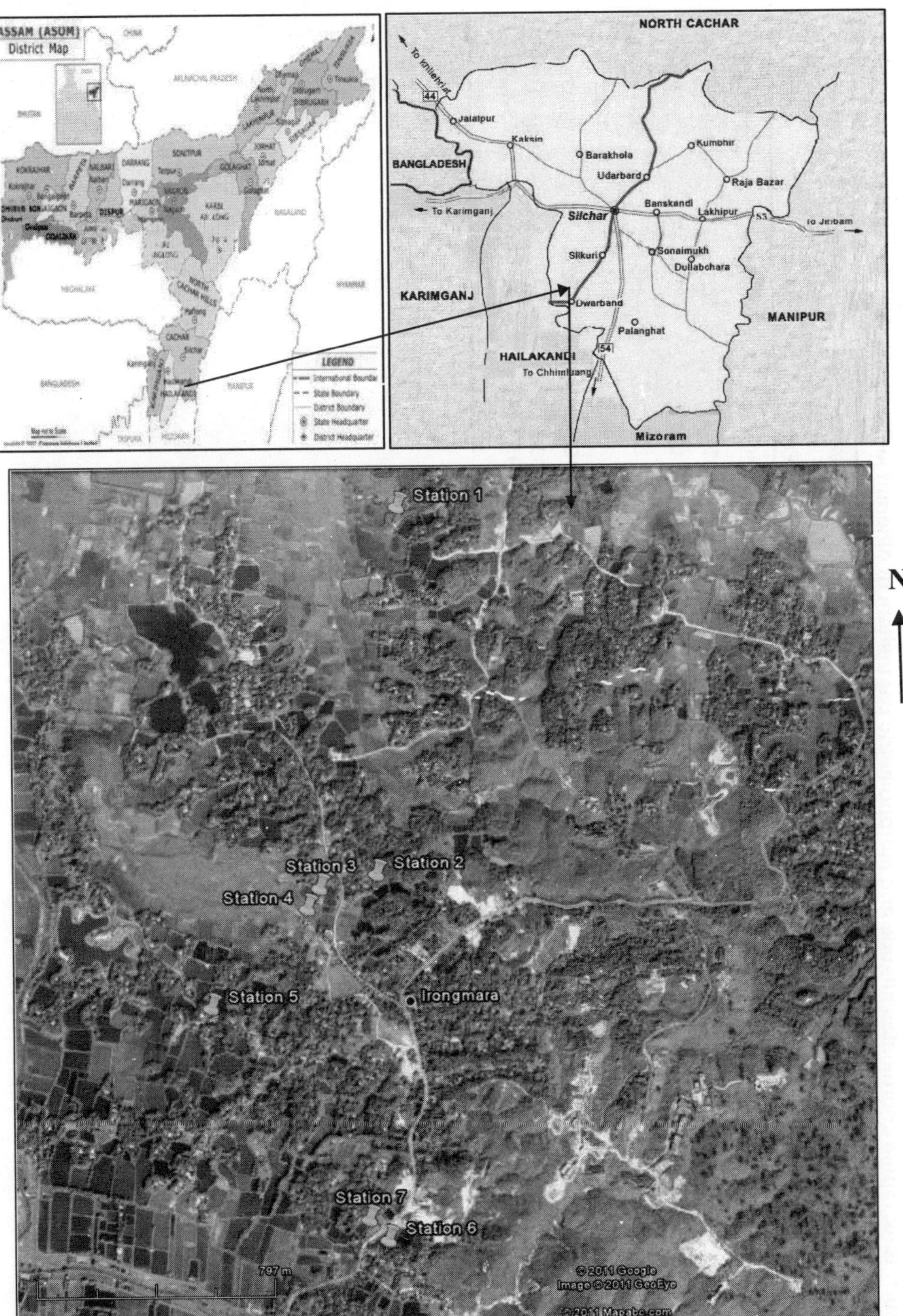

Figure 30.1: Map of the study area in Barak valley, Assam. The seven sampling stations have been marked on Google Earth imagery.

Table 30.1: Geographic coordinates altitude and morphometric parameters of different lentic systems in the study area.

Parameters	*Station 1 (Fishery-cum-paddyfield)*	*Station 2 (House Pond)*	*Station 3 (House Pond)*	*Station 4 (Fishery Pond)*	*Station 5 (Fishery Pond)*	*Station 6 (Roadside Pond)*	*Station 7 (Roadside Pond)*
Latitude	E092°44'	E092°44'	E092°44'	E092°44'	E092°44'	E092°44'	E092°44'
Longitude	N24°42'	N24°41'	N24°41'	N24°41'	N24°41'	N24°40'	N24°40'
Altitude (m asl)	5.6	32.4	14.78	13	16.2	31.4	38.2
Length (m)	115	129	150	254	225	145.7	153.8
Breadth (m)	24.5	98	145	174	75	135	128.2
Area (m^2)	2818	12642	21750	44196	16875	18225	19717
Water depth (m)	0.1±0.01 (0.74-1.3)	0.85±0.03 (0.78-0.99)	1.09±0.04 (0.83-1.27)	0.91±0.03 (0.64-1.02)	1.03±0.02 (0.97-1.14)	0.99±0.01 (0.91-1.04)	1.05±0.05 (0.65-1.22)

(Numbers in parenthesis designate range of mean values of the water depth)

combination of both while others are left abandoned on the sides of the road which ultimately turn into a habitat of aquatic weeds like *Eichhornia, Lemna, Pistia* etc. and also a breeding ground of mosquitoes, snails and other pathogenic organisms.

For the present study 7 different ponds of varying size and usages were selected (Figure 30.1). Geographic coordinates altitude and morphometric parameters of the different ponds are revealed in Table 30.1. On the other hand, the general features and disturbances in different ponds are shown in Table 30.2.

Table 30.2: General features and disturbances in the study area.

Common Features and Disturbances	*Stations*						
	Station 1 (Fishery-cum-paddy-field)	*Station 2 (House Pond)*	*Station 3 (House Pond)*	*Station 4 (Fishery Pond)*	*Station 5 (Fishery Pond)*	*Station 6 (Road-side Pond)*	*Station 7 (Road-side Pond)*
Riparian vegetation	**	**	*	*	**	*	*
Aquatic macrophytes	–	**	**	–	*	**	**
Human settlement	*	**	**	**	*	*	*
Agriculture and irrigation	***	–	–	–	–	–	–
Human bathing	–	**	*	–	*	–	**
Defecation on the bank	–	*	*	*	*	*	–
Washing clothes	–	**	**	–	–	**	**
Fishing	**	–	–	***	**	–	–
Influence of cattle	–	*	*	*	**	–	–
Influence of duck	–	**	*	*	*	*	*

–: Indicates absence; *: Less, **: Moderate, ***: High.

Materials and Methods

Water samples from surface water were collected from the sampling stations at monthly intervals from March to April, 2011. Samples for dissolved oxygen (DO) were collected directly in BOD bottles and were fixed in the field using alkaline iodide and manganous sulphate. Water temperature was measured with the help of a mercury bulb thermometer (0-50°C). For analyzing other chemical parameters of the water, the samples were kept in PVC bottles and were brought to laboratory.

At the same time, plankton samples were also collected by passing twenty five liters of the water through the plankton net of mesh size 50 μm and were preserved immediately in 4 per cent formalin which was later on brought to laboratory for identification.

In the laboratory, pH of the water sample was recorded using pH meter. Dissolved oxygen was estimated by Winkler's method. Total alkalinity and free carbon dioxide in the water samples were analyzed by titrating with strong acid and alkali, respectively. Other parameters like nitrate-N, and phosphate-P, were estimated following standard methods like phenol disulphonic acid method, and molybdenum blue method respectively (Michael, 1984; Eaton *et al.*, 1995; Tripathi and Govil, 2001).

Planktons were sorted out and identified to the lowest possible taxonomic level using a light microscope (Olympus; model- MLX). Phytoplanktons were identified at 40x magnification whereas zooplankton were identified at 10x magnification. Plankton identifications were done following Needham and Needham (1962), Twener (1978), Biswas (1980), Pentecost (1984), Anand (1998), and Battish (1999). The planktonic count was done by Lackey's Drop method (Lackey, 1938).

Trophic status of the water body using phytoplankton community was estimated by Compound Quotient (Tripathi and Govil, 2001). Besides community structure of zooplankton was also used to find out the Trophic index (Gannon and Stemberger, 1978) in the study area.

Different statistics and calculation of diversity indices were done using the statistical software PAST version 2.13. All the statistics were done following Zar (1974).

Results and Discussion

Physico-chemical Properties of Water

Station 1 had highest value of water temperatures and CO_2 accompanied by lowest value of DO (Table 30.5) as it had shallow water that warmed up very soon (Table 30.1). Besides, as this station was a fishery cum paddy field it also encountered higher input of organic matters that in presence of greater temperature might have lead to more decomposition thus paving the way for liberation of CO_2 to a greater extent. Station 2 which is a house pond, had greater value of total alkalinity and DO and lowest value of CO_2 (Table 30.5) thereby indicating relatively better water quality here which is suitable for various household use. On the other hand, Station 3 which is also a house pond had lowest water temperature and pH with greater phosphate-P. This might be due to the fact that this station has vegetation on its bank that had a shading effect which in turn lead to a low water temperature. Besides, the litter fall from the riparian vegetation and entrance of allochthonous matter from the settlement areas of human habitation lead to a hike in acidity of the pond water and thereby a drop in pH and a hike in the concentration of nutrients like Phosphate-P. Station 4 had lowest value of total alkalinity and BOD thereby indicating greater deposition of organic matter and less buffering capacity of the pond water. Station 5 had greater value of nitrate-N. This might be due to more quantities of nitrogenous matters in terms of the supplementary fish feed or perhaps due to deposition of fecal pellets of the fish stocked in this pond. Station 6 had greater value of pH perhaps due to washing of clothes by the local people there. Station 7 had lower value of nitrate-N and phosphate-P. This was perhaps due to the demand of these nutrients for aquatic macrophyte production which was abundant at that station to a considerable extent.

Phytoplankton and Zooplankton Communities

A total of 45 genera of phytoplankton (Table 20.4) and 13 genera of zooplankton were encountered (Table 20.5) in the study area during the sampling period. 4 classes

Table 30.3: Physico-chemical properties and One-way ANOVA of water in different stations of the study area.

Physico-chemical Properties	*Stations*							*F-ratio*
	1	*2*	*3*	*4*	*5*	*6*	*7*	
Transparency (m)	Transparent	0.23±0.003 (0.22-0.25)	0.16±0.004 (0.13-0.18)	0.09±0.002 (0.08-0.10)	0.16±0.003 (0.14-0.17)	0.13±0.004 (0.11-0.15)	0.19±0.004 (0.16-0.20)	–
Air temperature (°C)	21.7±1.10 (18-25)	18.4±0.50 (16.20)	16.9±0.64 (15-19)	17.5±0.27 (16-19)	18.35±0.37 (17-20)	19.2±0.60 (17-21)	19.85-0.72 (17.5-22)	–
Water temperature (°C)	22.41±0.47 (20.9-24)	20.06±0.46 (18-21.5)	19.8±0.61 (18-22)	20.11±0.38 (19-21.1)	20.61±0.48 (19-22.1)	21.88±0.70 (19-24)	22.14±0.94 (19-25)	0.38
pH	7.15±0.01 (7.11-7.20)	7.34±0.07 (7.15-7.63)	5.86±0.42 (4.75-6.87)	6.61±0.26 (5.51-7.32)	6.88±0.26 (5.73-7.39)	7.45±0.02 (7.38-7.52)	6.51±0.19 (5.65-6.88)	1.64
Total alkalinity (mgl^{-1})	33.6±0.89 (30-38)	67.4±2.05 (54-74)	24.2±0.87 (20-28)	19.8±0.96 (16-24)	38.2±1.78 (26-46)	61.6±1.73 (54-70)	31±1.77 (20-38)	54.71**
Dissolved oxygen (mgl^{-1})	7.9±0.09 (7.2-8.2)	13.42±0.12 (13-14.2)	12.74±0.09 (12.4-13.4)	8.98±0.11 (8.4-9.4)	9.52±0.16 (8.8-10.2)	10.12±0.11 (9.6-10.6)	10.16±0.11 (9.8-10.8)	71.33**
Biological oxygen demand (mgl^{-1})	2.22±0.09 (1.8-2.8)	2.54±0.07 (2.2-3)	2.18±0.13 (1.8-3)	2.4±0.08 (2.0-2.8)	2.46±0.08 (2.0-3.0)	2.52±0.14 (1.8-3.2)	3.46±0.12 (3-4.2)	24.98**
Free carbon dioxide (mgl^{-1})	3.99±0.16 (3.3-4.4)	2.97±0.52 (1.1-6.6)	3.33±0.29 (2.2-4.4)	3.78±0.23 (2.2-4.4)	3.08±0.27 (2.2-4.4)	3.3±0.37 (2.2-5.5)	3.41±0.26 (2.2-4.4)	0.77
Nitrate-N (mgl^{-1})	0.1±0.001 (0.097-0.108)	0.16±0.003 (0.154-0.171)	0.1±0.002 (0.098-0.111)	0.13±0.002 (0.122-0.137)	0.2±0.003 (0.185-0.206)	0.17±0.001 (0.166-0.18)	0.0026±0.001 (0.003-0.006)	126.15**
Phosphate-P (mgl^{-1})	0.013±0.0002 (0.012-0.013)	0.015±0.0002 (0.015-0.016)	0.017±0.00016 (0.016-0.017)	0.015±0.0002 (0.014-0.015)	0.014±0.0002 (0.013-0.014)	0.011±0.0002 (0.011-0.012)	0.005±0.0002 (0.004-0.006)	84.64**

Mean± SE, n=10; F-ratio: df= 6; **significant at $p < 0.01$.

(Numbers in parenthesis designate range of mean values of the physico-chemical properties of water in different stations of the study area).

Table 30.4: Distribution and mean density of phytoplankton in different stations of the study area (No. of individuals l^{-1}).

Systematic Positions				*Stations*						
Class	*Order*	*Family*	*Genera*	*1*	*2*	*3*	*4*	*5*	*6*	*7*
Bacillariophyceae	Naviculales	Naviculaceae	*Navicula*	2	5	19	6	1	3	6
	Pinnales	Diatomaceae	*Asterionella*	–	2	–	–	1	–	–
	Tabellariales	Tabellariaceae	*Tabellaria*	1	–	–	–	–	–	4
	Cymbellales	Cymbellaceae	*Cymbella*	–	3	–	14	1	–	–
			Encyonema	7	1	–	–	–	–	–
			Gryosigma	2	–	–	–	1	–	–
	Fragilariales	Fragilariaceae	*Synedra*	–	1	–	–	–	–	–
	Centrales	Melosiraceae	*Melosira*	3	1	–	–	–	–	–
		Rhizosoleniaceae	*Rhizosolania*	4	–	–	–	1	–	–
Chlorophyceae	Desmidiales	Desmidiaceae	*Arthrodesmus*	4	5	1	3	1	–	4
			Cosmarium	8	6	11	2	2	2	18
			Staurastrum	15	1-	3	2	–	–	2
			Pleurotaenium	2	–	–	–	–	–	–
			Euastrum	4	23	1	–	1	1	2
			Micrasterias	3	1	–	–	–	–	–
			Hyalotheca	–	–	–	–	–	2	–
			Sphaerozosma	2	1	14	24	7	2	24
		Closteriaceae	*Closterium*	28	6	1	–	2	1	16
		Zygnemataceae	*Spirogyra*	7	153	36	27	12	9	75
		Mesotaeniaceae	*Netrium*	–	–	1	–	–	–	–
	Chlorococcales	Chloroccaceae	*Chlorococcum*	1	10	–	7	2	2	3
		Selenastraceae	*Kirchnereila*	–	8	3	6	2	5	24

Contd...

Table 30.4–*Contd...*

Systematic Positions				*Stations*						
Class	*Order*	*Family*	*Genera*	*1*	*2*	*3*	*4*	*5*	*6*	*7*
		Oocystaceae	*Ankistrodesmus*	–	2	4	2	–	–	–
			Neprocytium	1	–	3	–	–	–	6
		Hydrodictyaceae	*Pediaastrum*	–	11	3	16	–	–	6
			Hydrodictyon	–	5	–	–	–	–	–
		Scenedesmaceae	*Scenedesmus*	6	74	40	70	42	24	272
			Coelastrum	–	2	–	–	–	3	1
			Crucigenia	1	–	–	–	–	–	1
	Chaetophorales	Chaetophoraceae	*Stigeoclonium*	–	1	–	–	1	–	–
			Uronima sp	–	–	1	–	–	–	–
	Microsporales	Microsporaceae	*Microspora*	3	25	1	–	–	–	–
	Ulotrichales	Ulotrichaceae	*Ulothrix*	2	–	–	–	–	–	–
	Volvocales	Volvocaceae	*Pandorina*	–	–	–	–	1	–	–
			Volvox	–	4	–	1	1		–
			Eudorina	–	–	–	1	–	–	–
Euglenophyceae	Euglenales	Euglenaceae	*Euglena*	3	49	12	14	2	4	5
			Phacus	–	2	–	–	6	–	1
Cyanophyceae	Oscillatoriales	Phycodnaviridae	*Spirulina*	–	1	2	–	–	–	1
	Chroococcales	Merismopediaceae	*Merismopedia*	–	3	–	–	–	–	–
			Aphanocapsa	–	2	–	1	–	–	–
		Chroococcaceae	*Chroococcus*	–	–	–	–	–	–	4
			Polycystis	–	1	–	–	–	–	–
		Microcystaceae	*Gloeocapsa*	3	–	–	–	–	–	–
	Nostocales	Nostocaceae	*Nostoc*	–	1	–	–	–	–	–
Total density:				112	419	156	196	87	58	475

–: Indicates absence of the genus concerned.

Table 30.5. Distribution and mean density of zooplankton in different stations of the study area (No. of individuals l^{-1}).

Systematic Positions					*Stations*						
Phylum	*Class*	*Order*	*Family*	*Genera*	*1*	*2*	*3*	*4*	*5*	*6*	*7*
Arthropoda	Branchiopoda	Cladocera	Daphniidae	*Ceriodaphnia*	1	–	12	4	–	–	–
				Moinadaphnia	–	–	1	1	–	–	–
			Sididae	*Sida*	1	–	2	–	–	–	2
			Chydoridae	*Eurycerus*	1	–	–	–	–	1	–
			Moinidae	*Moina*	1	5	10	4	–	–	1
		Anostraca	Chirocephalidae	*Eubranchipus*	–	1	–	–	–	–	–
	Maxillopoda	Calanoida	Diaptomidae	*Diaptomus*	2	1	–	2	2	4	3
			Centropagidae	*Limnocalanus*	–	–	2	–	–	–	–
		Cyclopoida	Cyclopidae	*Cyclops*	4	–	10	6	12	9	7
				Microcyclops	–	–	–	–	–	2	–
		Harpacticoida	Canthocamptidae	*Canthocamptus*	–	–	–	–	3	1	1
Rotifera	Eurotatoria	Ploima	Brachionidae	*Brachionus*	1	5	14	8	–	3	2
			Lepadellidae	*Lepadella*	–	–	–	1	–	–	–
Total density:					11	12	51	26	17	20	16

–: Indicates absence of the genus concerned.

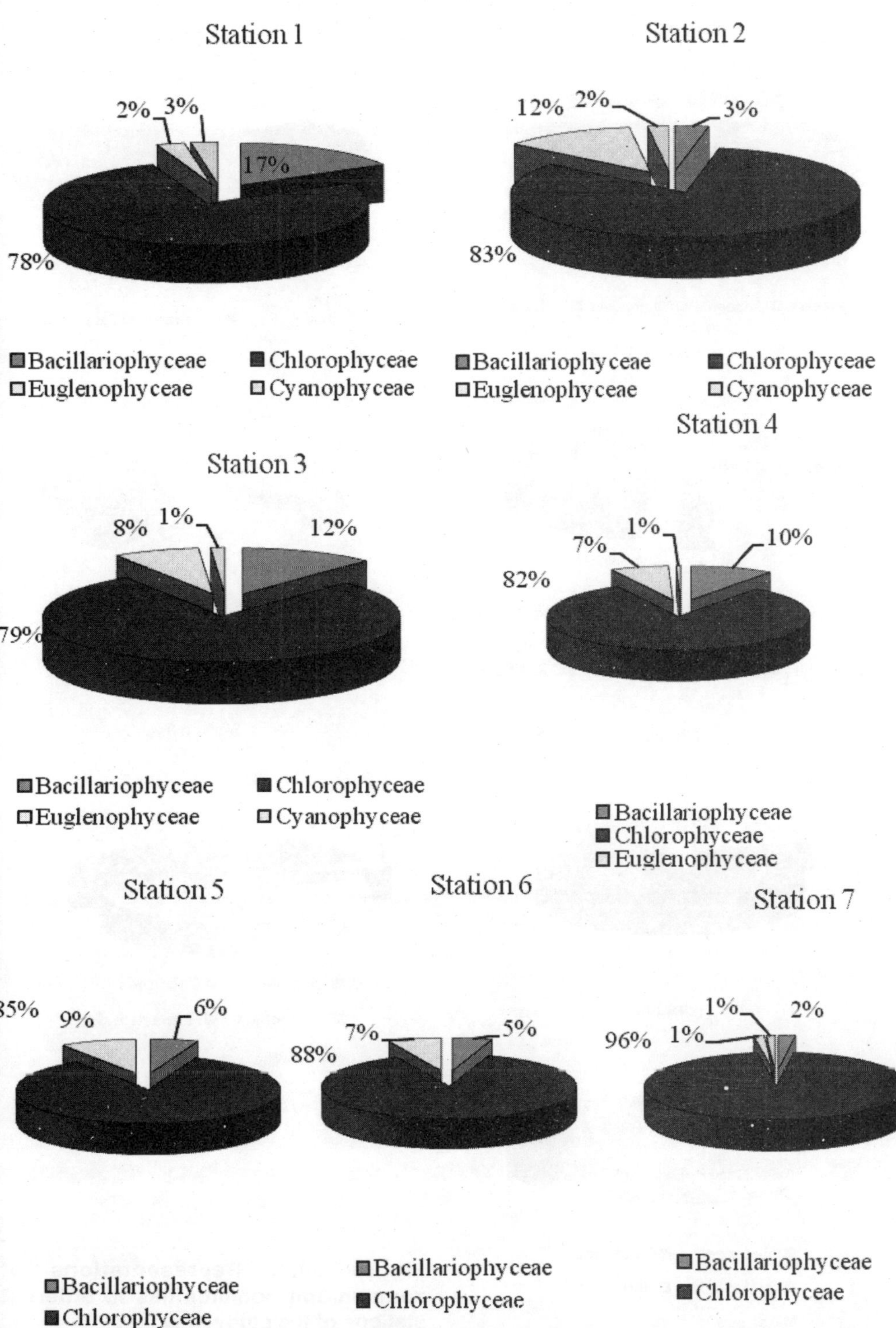

Figure 30.2: Representations of phytoplankton communities in different stations of the study area.

Station 1

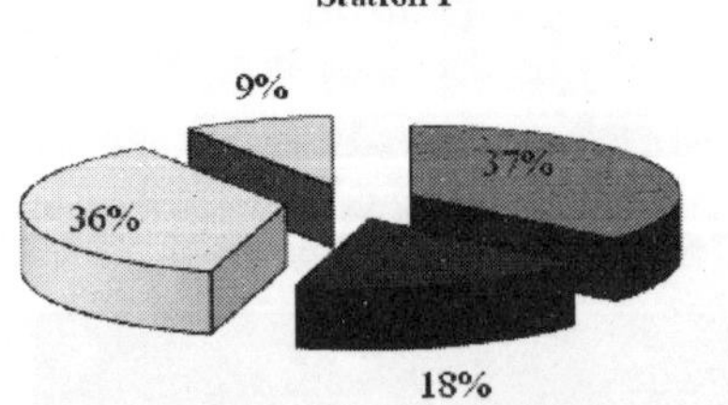

▪Cladocera ▪Calanoida ▫Cyclopoida ▫Rotifera

Station 2

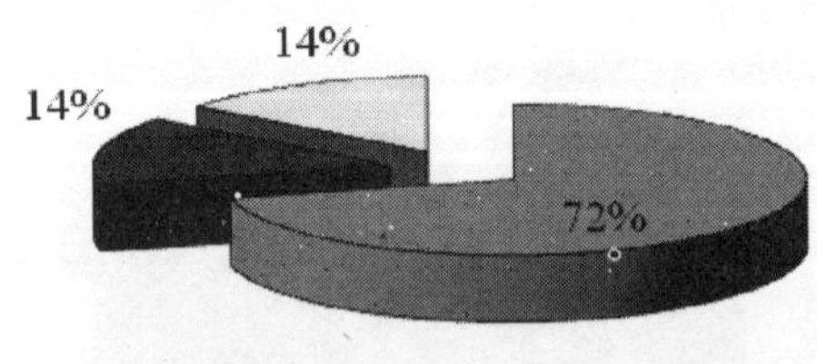

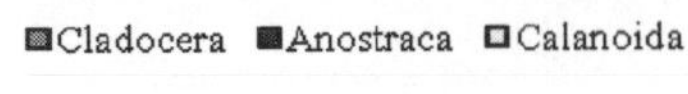

Station 3

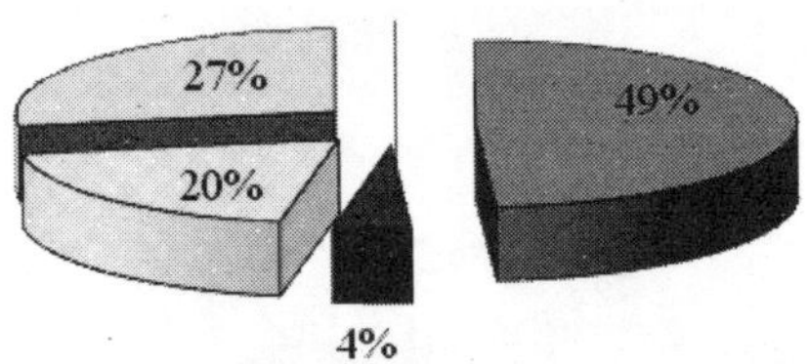

▪Cladocera ▪Calanoida ▫Cyclopoida ▫Rotifera ▪

Station 4

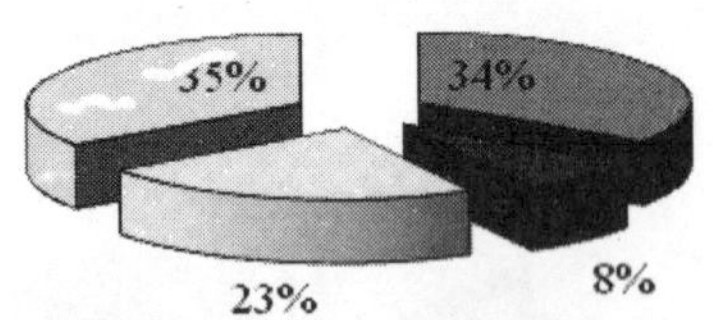

▪Cladocera ▪Calanoida
▫Cyclopoida▫Rotifera

Station 5

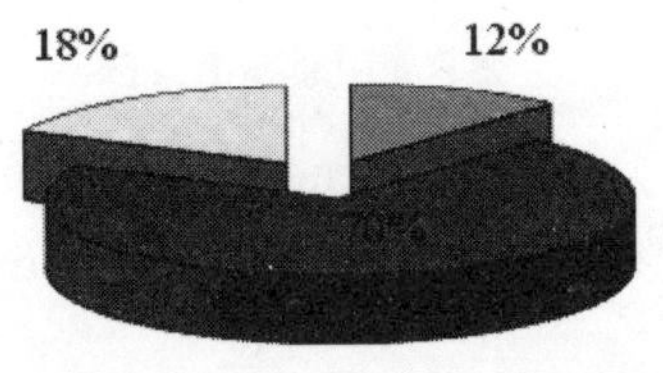

Station 6

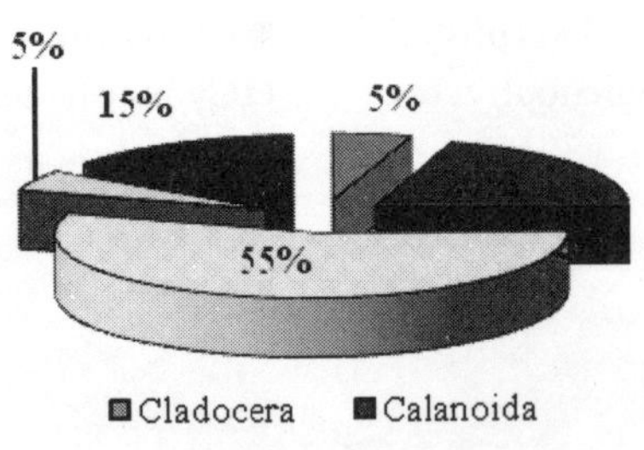

▪Cladocera ▪Calanoida
▫Cyclopoida ▫Harpacticoida
▪Rotifera

Station 7

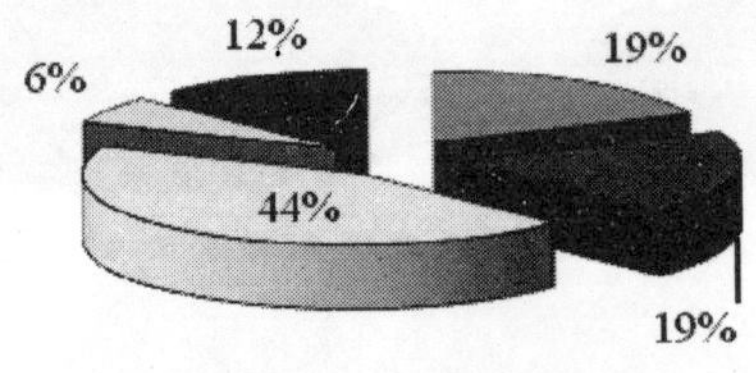

▪Cladocera ▪Calanoida
▫Cyclopoida ▫Harpacticoida
▪Rotifera

Figure 30.3: Representations of zooplankton communities in different stations of the study area.

of phytoplankton *viz.*, Bacillariophyceae, Chlorophyceae, Euglenophyceae and Cyanophyceae (Figure 20.2) and 3 classes of zooplankton *viz.*, Branchiopoda, Maxillopoda and Eurotatoria (Figure 20.3) were observed in the study area. Chlorophyceae was the most dominant phytoplankton group in all the stations. This might be due to the greater availability of nutrients in terms of nitrate, phosphate and sulphate in the study area (Mustapha, 2009). Amongst zooplankton, Cladocera was the most dominant group in stations 1, 2 and 3 (Figure 20.2). Station 4 was dominated by Rotifera followed by Cladocera. On the other hand, stations 5, 6 and 7 had greater density of Cyclopoida. It may be noted here that calanoid copepods are indicative of more oligotrophic conditions, while cyclopoid copepods and cladocerans are indicative of more eutrophic conditions (Gannon and Stemberger, 1978). Based on this information, it can be inferred that all the stations in the study area are eutrophic systems.

Diversity indices

Diversity indices are good indicators of pollution in aquatic ecosystem (Mason, 1998). In the present study diversity indices of phytoplankton ranged from 1.62 to 2.68 and that of zooplankton ranged from 0.80 to 1.77 (Table 30.6). According to Ali *et al.* (2003), diversity index greater than 3 indicates clean water; values in the range of 1 – 3 are characteristics of moderately polluted conditions and values less than 1 characterize heavily polluted condition. Based on all these observation it can be inferred that all the stations taken for the present study were moderately polluted.

Table 30.6: Diversity indices of plankton in different stations of the study area.

Parameters		*Stations*						
		1	*2*	*3*	*4*	*5*	*6*	*7*
Phytoplankton	Number of taxa	23	31	18	16	19	12	20
	Number of individuals	112	419	156	196	87	58	475
	Dominance index	0.11	0.19	0.16	0.18	0.27	0.22	0.36
	Shannon Diversity index	2.68	2.26	2.19	2.10	1.96	1.96	1.62
	Evenness index	0.63	0.31	0.50	0.51	0.37	0.59	0.25
Zooplankton	Number of taxa	7	4	7	7	3	6	6
	Number of individuals	11	12	51	26	17	20	16
	Dominance index	0.21	0.36	0.21	0.20	0.54	0.28	0.27
	Shannon Diversity index	1.77	1.14	1.67	1.73	0.80	1.50	1.54
	Evenness index	0.84	0.78	0.76	0.80	0.74	0.74	0.78

Trophic Status

Table 30.7 shows the trophic status based on the ratio of calanoid: cyclopoid plus cladoceran abundances in different stations of the study area. Gannon and Stemberger, (1978) revealed that greater values for the ratio (> 2) occurred in the more oligotrophic waters, whereas lower values (< 2) were found in more eutrophic systems. Based on this information it can be inferred that all the stations in the study area were eutrophic systems. Again CQ in different stations (Table 30.7) revealed that the water bodies in the study area were of saprotrophic status as the values of CQ was greater than 2 (Tripathi and Govil, 2001) in all the stations except station 1. However, when compared, CQ values were relatively less in stations 1, 2 and 3 but it increased in the remaining stations thereby indicating that the saprotrophy of the water bodies were greater in stations 4, 5, 6 and 7 thereby indicating relatively more organic matters in these systems.

Table 30.7: Trophic status of the lentic systems in different stations of the study area.

Parameters	*Stations*						
	1	*2*	*3*	*4*	*5*	*6*	*7*
Ratio of total number of Cyanophyceae, Chlorophyceae, Diatoms and Euglenophyceae: Total number of Desmids	1.53	2.04	2.29	3.38	3.48	3.53	3.37
Ratio of calanoids: cyclopoid plus cladoceran abundance	0.25	0.20	0.06	0.13	0.17	0.33	0.30

It may be mentioned here that stations 1, 4 and 5 are fishery ponds. If we compare the water quality of these ponds with relevant fishery standard under Indian condition, it seems that the water properties in these ponds are not within the standard as far as the concentration of total alkalinity and phosphate-P are concerned. This study therefore suggests initiation of some management strategies in these systems for better production of fish. As far as road side ponds (stations 6 and 7) are concerned some management options have to be initiated in terms of the aquatic weeds and other unwanted human activities around and in the ponds. In doing so, these may be converted to some productive systems that would give some returns in not only in terms of human well being but also the well being and sustainability of the local environment and the rural ecosystem.

References

Ali, M., Salam, A., Jamshaid, S. and Zahra, T., 2003. Studies on biodiversity in relation to seasonal variations in water of river Indus at Ghazi Ghatt, Punjab, Pakistan. *Pakistan Journal of Biological Sciences*, 6(2): 1840 – 1844.

Anand, N., 1998. *Indian Freshwater Microalgae*. Bishen Singh Mahendra Pal Singh, Dehra Dun.

Bahura, C. K., 1991. Diurnal cycle of certain Abiotic parameters of a fresh water lake the Ganjer lake (Bikaner) in the Thar desert of India. *J. Aqua. Biol.*, 16 (12): 45–48.

Battish, S. K., 1992. *Freshwater Zooplankton of India*. Mohan Primlani, New Delhi.

Biswas, K., 1980. *Common Fresh and Brackishwater Algae Flora of India and Burma*. International Book Distributors, Dehradun.

Carle, F.L., 1979. Environmental monitoring potential of the Odonata with a list of rare and endangered Anisoptera of Virginia, United States. *Odontologia*, (8): 319–323

Eaton, A.D., Clesceri, L.S. and Greenberg, A.E., 1995. *Standard Method for the Examination of Water and Wastewater*, 19th Edn. American Public Health Association, Water Works Association and Water Environment Federation.

Carrick, H.L. and Schelske, C.L., 1997. Have we overlooked the importance of small phytoplankton in productive waters? *Limnol. Oceanogr.*, 42: 1613–1621.

Gannon, J.E. and Stemberger, R.S., 1978. Zooplankton (especially crustaceans and rotifers) as indicators of water quality. *Transactions of the American Microscopical Society*, 97: 16–35.

Lackey, J.B., 1938. The manipulation and counting of river plankton and changes in some organisms due to formalin preservation. *Public Health Reports*, 53: 2080–2093.

Loeb, S.L., 1994. *Biological Monitoring of Aquatic Systems*. CRC Press, USA.

Michael, P., 1984. *Ecological Methods for Field and Laboratory Investigations*. Tata McGraw-Hill Publishing Company Limited, New Delhi.

Needham, J.G. and Needham, P.R., 1962. *A Guide to the Study of Freshwater Biology*. Holden-Day Inc., San Francisco.

Odiete, W.O., Nwokoro, R.C. and Daramola, T., 2003. Biological assessment of four courses in Logos metropolis receiving industrial and domestic waste discharge. *Nig. Environment Soc.*, 1(1): 1–14.

Pentecost, A., 1984. *Introduction to Freshwater Algae*. Richmond Publishing Corporation, England.

Thorp, H.J. and Covich, P.A., 1991. *Ecology and Classification of North American Freshwater Invertebrates*. Academic Press, San Diego, California.

Tripathi, B.D. and Govil, S.R., 2001. *Water Pollution: An Experimental Approach*. CBS Publishers and Distributors, New Delhi.

Twener, W.B., 1978. *The Freshwater Algae (Principally Desmidieae) of East India*. Bishen Singh Mahendra Pal Singh, Dehra Dun.

Zar, J.H., 1974. *Biostatistical Analysis*. Prentice-Hall Inc., New Jersey.

Environment, Biodiversity and Traditional System, 2014 *Pages* 313–322
ISBN 978-81-7622-307-2
Editors: D.R. Khanna, G.S. Solanki and S.K. Pathak
Published by: BIOTECH BOOKS, NEW DELHI

Chapter 31

Host Plant-Based Biological and Genetic Variations in *Aphis gossypii* Glover (Homoptera: Insecta) Populations from North-East India

Parichita Ray Choudhury[1], Samir K Sil[1] and B.K. Agarwala[2]*

[1]Genetics and Molecular Biology Laboratory,
Department of Human Physiology,
[2]Ecology and Biodiversity Laboratory, Department of Zoology,
Tripura University, Suryamaninagar. West Tripura – 799 022

ABSTRACT

The aphid *Aphis gossypii* Glover is a highly polyphagous species with mainly parthenogenetic reproduction. This study shows that genotypes living on wild plants of taro and brinjal behave as genetically distinct host races. Colonization success rate after reciprocal host transfer is very poor. *A. gossypii* clones from taro survived when transferred to brinjal but larval mortality was as high as hundred percent in the second generation. Brinjal clones when transferred to taro could not survive even in the first generation. Further, taro clones of *A. gossypii* showed significant differences in their biological performances from the brinjal

* *Corresponding Author:* E-mail: parichitarc@gmail.com

clones. Preliminary data on RAPD patterns also indicated host-specific genetic variation. Results indicate that *A. gossypii* populations from taro and brinjal show host specialization which may be a consequence of the contrasting feeding environment. These results point towards greater potential of sympatry in this aphid.

Keywords: Aphis gossypii, Reciprocal-host transfer, Host race.

Introduction

Asexual populations of the cotton aphid, *Aphis gossypii* Glover (Homoptera: Aphididae: Aphidini), a worldwide pest in agriculture, horticulture and greenhouse crops (Agarwala and Ghosh, 1985; Blackman and Eastop, 1984). *A. gossypii* Gover (Homoptera: Aphididae: Aphidini) is extremely polyphagous and show clonal diversity in relation to host plants in several parts of the world (Inaizumi, 1981; Guldemond *et al.*, 1994). This aphid species is now considered to consist of distinct genotypes that vary with respect to their ability to reproduce and host preferences (Takada and Murakami, 1988; Zhang and Zhong, 1993). Recent morphological, biochemical, and host plant preference studies have shown that a number of aphid species, notably polyphagous species, consist of genetically different forms, *i.e.* host races, or even appear to represent complexes of several separate species (Muller, 1985; Guldemond, 1990). Available literatures suggest that *A. gossypii* consists of several genotypes that are adapted to different host plants and might be considered as host races (Jaenike, 1981; Diehl and Bush, 1984; Agarwala and Das, 2007). Precise identification and population differentiation of *A.gossypii* populations is essential proper control of this pest species and to predict its occurrence on different crops as the various strains of the aphid show different host preference, insecticide resistance, and life cycle pattern (Saito, 1989; Furk and Hines, 1993; Han *et al.*, 1998; Toda *et al.*, 2004; Komazaki and Toda, 2008).

Wild plants of taro, *Colocasia esculenta* (L.) *esculenta* Schott, (Family Araceae L., and brinjal, *Solanum torvum* Swartz (Family Solanaceae), occur commonly in the moist and hot climate of Tripura (22°56′ to 24°32′ N and 91°10′ to 91°21′E) and elsewhere in north- east India (Deb, 1981). These plants attract *A. gossypii* which form small to large colonies in undersides of leaves and tender shoots of these hosts. In field, asexual wingless viviparous morph (apterae) of *A. gossypii* showed differences in life colour and colonization behaviour on the two host plants which provided different food environments for colonization by aphids. It is predicted that *A. gossypii* of wild plants of taro and brinjal might be different races due to influence of host environment. Environment mediated host races were earlier described in *A. gossypii* populations from chrysanthemum and cucumber (Guldemond *et al.*, 1994).

Previous studies involving host plant-transfer experiments, suggest that the species *A. gossypii* probably consists of several genetically differentiated host races, but this has not been genetically demonstrated (Blackman and Eastop 1985; Guldemond *et al.*, 1994; Wool *et al.*, 1995). Recently, high resolution techniques such as the random amplified polymorphism DNA are in use to access the genetic structure of *A. gossypii* populations (Peng *et al.*, 2000; Komazaki *et al.*, 2011). The RAPD had successfully been used to detect DNA polymorphism in several aphid species (Black *et al.*, 1992; Cenis *et al.*, 1993; Puterka *et al.*, 1993).

In the present study, developmental and reproductive characters of *A. gossypii* populations from wild plants of taro and brinjal were investigated to assess any variation and two RAPD primers were also used to detect variation at genetic level.

Materials and Methods

Aphids

Apterous parthenogenetic viviparous aphids of *Aphis gossypii* were collected from taro and brinjal plants, respectively, found in the wild at five different locations, separated by about 2000 m distance from each other, in and Agartala, north-east India (23.50°N latitude, 91.25°E longitude). These aphids were used to raise twenty stock cultures, ten each of *A. gossypii* on taro and on brinjal plants, respectively, under greenhouse conditions (24±1°C temperature and 16: 8 L: D photoperiod). Aphid cultures on individual potted plants were confined in nylon net cages in segregated locations. Aphids from these clones representing two different genotypes from the two host plant species were used to measure differences in their biological performances. For determining the mean relative growth rate, generation time, reproductive duration and fecundity parthenogenetic females of both the aphid species from their clones were placed individually in leaf cages and monitored (Blackman 1987) to obtain parthenogenetic descendents.

Performance Measurements

Mean relative growth rate (MRGR) is correlated with the intrinsic rate of increase (Dixon 1985) and thus, a good measure of performance. MRGR was calculated following Watt and Hales (1996):

$$\text{MRGR} = \frac{\log_{10} \text{adult weight} - \log_{10} \text{birth weight}}{\text{Developmental time}} mgmg^{-1}d^{-1} \backslash$$

and expressed as mg increase in weight of aphids born per mg of the mother aphid per day.

The intrinsic rate of increase (R_{max}), a measure of rate of increase of a population under controlled conditions, was calculated using the formula:

$$R_{max} = \frac{\log_e (R_0)}{G}$$

where, R_0 is the net reproductive rate and **G** is the mean length of a generation, determined as under (Dublin and Lotka, 1925):

$$G = \frac{\Sigma\ x\ .lx\ .bx}{R_0}$$

where, ***lx*** is the proportion of females surviving, ***bx*** is the number of female offspring produced per female during its reproductive time and ***x*** is the age of female adults.

Generation time (number of days from birth to the production of the first offspring) presented here as GT and weight at the day of first reproduction (AW) were assessed as a measure of performance of the aphid clones on the two host plants.

Host Transfer Experiments

Aphids from both the clones were subjected to reciprocal host transfer to record the colonization success in a new food environment. Individual nymphs, 0-12 h old, were released at the apical most part of potted plants of similar age of field hosts (control) and laboratory hosts (treatments). These aphids were allowed to and produce nymphs for the second generation. If successful, a third generation was produced. Two experiments were set up using parental clones of *A. gossypii* from the two host plant species. In the first treatment, *A. gossypii* aphids were transferred individually from the wild taro field host to the laboratory host, wild brinjal, and in the second treatment *A. gossypii* aphids were transferred from the wild brinjal field host to wild taro as the laboratory host. In both the treatments, performances of aphids on their respective field host plants were considered as the controls. In each case of host transfer, ten replicates were used to record the success rate of survival and reproduction by apterous viviparous aphids of a host plant leading to the establishment of colony. Aphids that either failed to develop to the adult stage in the first generation or failed to produce second generation or third generation were considered to be unsuccessful.

DNA Extraction

Extraction of genomic DNA of single aphid was carried out as described by Yang (1999). In brief, single aphid was crushed in 0.5 ml Eppendorf tube, filled with 200 µl extraction buffer (1 per cent SDS, 50 mmole/L Tris-Cl pH 8.0, 0.25 mmole/L NaCl, 25 mmol/L EDTA), incubated for 45 min at 65°C, added 100µl of 3 mol/L sodium acetate (pH 7.0), then placed the tube 1 h at -20°C. Centrifuged for 10 min at 12000 r/min, and transferred the supernatant to a new tube, added two times volume of pure ethanol and placed the tube for more than 1 h at -20 °C. The precipitated DNA was pelleted by centrifugation for 20 min at 12000 r/min. After a wash with 70 per cent ethanol, the pellet was dried and re -suspended in 100 µl pH 8.0 TE buffer (10 mmol/L Tris-HCl, 1 mmol/L EDTA pH 8.0).

RAPD-PCR Reaction

PCR were performed in 25µl of reaction mixture containing 3 µl of template DNA, 1.5 unit of Taq polymerase, 0.1 mmole/L dNTPs, 20 ng of primer and 10X buffer supplied by the enzyme manufacturer. 25 µl of mineral oil was added on top of the reaction mixture.

The two primers used in this study were decamers of arbitrary sequence supplied by Genei/Merck Specialities PVT. LTD (A07: 5′-GAAACGGGTG-3′and A09: 5′-GGGTAACGCC-3). Genetic distance (D) was calculated based on the results amplified by these two primers. Amplifications were carried out in Applied Biosystems 2720 thermal cycler. The program is as follows: 3 min at 94°C; 40 cycles of 30 s at 94°C, 1 min at 36° C, 2 min at 72 C, then 10 min at 72° C. The PCR products were electrophoresed in a 1.4 per cent agarose gel containing 0.5 µg/ml ethidium bromide in 1X TBE buffer. DNA bands were visualized and photographed unse UV lights.

Data Analysis

Data of the third generation aphids, wherever available, were used to compare results of performance measurements. This was done to allow the aphids sufficient time for acclimatization to the laboratory rearing conditions. All weights in this study were taken in a Mettler microbalance (www.met.com) sensitive to 2 µg. Each of the developmental and reproduction parameters that were measured from the wingless aphids from different *A. gossypii* clones were compared using the Student's *t* test. Origin 7 (www.originlab.com) was used for the analysis of data.

Genetic distance (D) was calculated as follows:

$$D = 1 - \frac{Nxy}{(Nx + Ny - Nxy)}$$

D: The genetic distance between aphid clones of taro 'a' and aphid clones of brinjal 'b'

Nxy: The number of bands shared by a and b

Nx: The number of bands in a, *Ny*= the number of bands in b

Results and Disucssion

Apterous aphids of *A. gossypii* clones from wild plants of taro and brinjal respectively, showed significant differences in mean relative growth rate, generation time and in their weights at the day of first reproduction.

MRGR from twenty replicates each of the *A. gossypii* clones from brinjal and taro host plants also showed significant difference (t-test: $t\ value = 8.83, df = 18, P < 0.000$). MRGR was higher (>1.27 times) in *A. gossypii* clones that were reared on wild taro in comparison to that of aphids from wild brinjal plants. Aphids from wild taro plants achieved higher rate of increase than the aphids from wild brinjal plants. Small size aphids of taro clones took shorter time to complete the development and this may be responsible for higher R_{max} in these aphids.

Aphid weight at the day of first reproduction of taro clones were 1.30 times and 1.13 times bigger, respectively, in comparison to aphids reared on brinjal host. Generation time of brinjal clones were longer by 19 per cent in comparison to clones reared on taro plants.

Aphids of *A. gossypii* clones from wild species of brinjal plants perished all together when transferred to wild taro plants (Figure 31.2A). Likewise, aphid clones from wild taro hosts could not survive when transferred to wild brinjal plants (Figure 31.2B).

No change in the RAPD pattern was detected over ten parthenogenetic generations of the *A. gossypii* clones, one of taro and the other of brinjal, indicating the reliability of these markers. A comparison of the *A. gossypii* clones of taro and brinjal host plants revealed few shared bands and many polymorphic bands. Out of these polymorphic bands, only well amplified and reproducible ones were considered for further analysis. Figure 31.1 showed that the number and position of PCR chart

Table 31.1: Mean values of biological characters studied in *A. gossypii* clones from wild species of taro and brinjal host plants.

Performance Measurements	*Mean ±SEM* Values (range: minimum-maximum)*		*t Test*	
	Taro (n=20)	*Brinjal (n=20)*	*t Value*	*P Value*
Adult Weight (mg)	0.106±0.004a (0.098-0.118)	0. 093±0. 003b (0.072-0.104)	3.153	0.005
GT^2 (days)	5. 90±0. 100a (5.00-6.00)	7. 40±0.163b (7.000-8.000)	7.317	0.000
$MRGR^3$ (mg/mg/day)	0.18±0.003a (0.174-0.20)	0.14±0.004b (0.131-0.163)	8.83	0.000
$R_{max}{}^4$ (aphid no./female/day))	8.49± 0.49a (8.076-9.30)	5.02± 0.12b (4.802-5.12)	24.11	0.000

*: Different letters with mean values in a row indicate significant differences between the treatments by Student's t test at P<0.05.n=20.

2: Generation time.

3: Mean relative growth rate.

4: Intrinsic rate of increase.

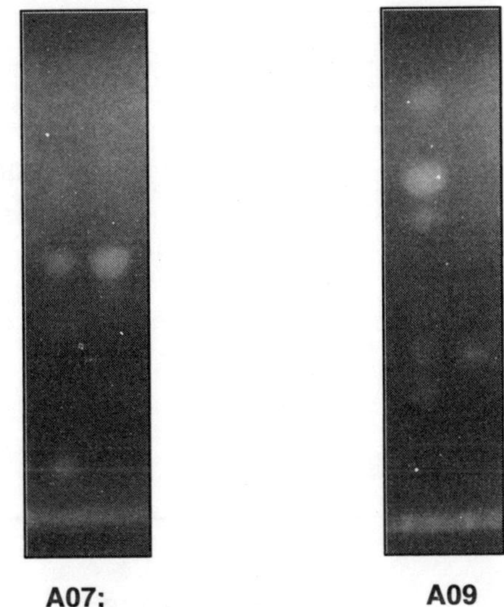

Figure 31.1: RAPD amplifications with primer A07 and A09 to a single aphid.

bands varied among individuals in the two different clones of the different host plants with the same primer as well as the same individual with different primers.

Some RAPD bands were host-plant specific. Four loci from primer A07 and one loci from primer A09 appeared to discriminate between two groups of populations: those collected from taro and those collected from brinjal. Genetic distance when calculated showed that *A. gossypii* clones if taro and brinjal are 75 per cent (as revealed from primer A07)–80 per cent (as revealed from primer A09) different genetically.

Recent morphological, biochemical, and host plant preference studies have shown that a number of aphid species, notably polyphagous species, consist of genetically different forms, *i.e.*, host races, or even appear to represent complexes of several separate species (Inaizumi, 1981). Guldemond *et al.* (1994) recorded significant differences in biological performance of *A. gossypii* on cotton, cucumber and okra. The present study has shown that *A. gossypii* populations on taro and brinjal host plants showed profound differences in most of the biological characters studied. The results of host plant-transfer showed that although the cotton aphid had the wide host plant range but it still appeared preference to some host plants (Blackman and Eastop, 1985; Guldemond *et al.*, 1994; Wool *et al.*, 1995). This study indicated that the populations of *A. gossypii* from taro and brinjal host plants have differentiated genetically.

Host environments of taro and brinjal plants are very different yet they offer choice to essentially morphologically similar *A. gossypii* aphids to select these hosts. The results of this study has shown that aphid-host plant interactions between *A. gossypii* clones produced different fitness on different hosts, fitness being manifested by ability to reproduce in response to preferred host cues and showing different rates

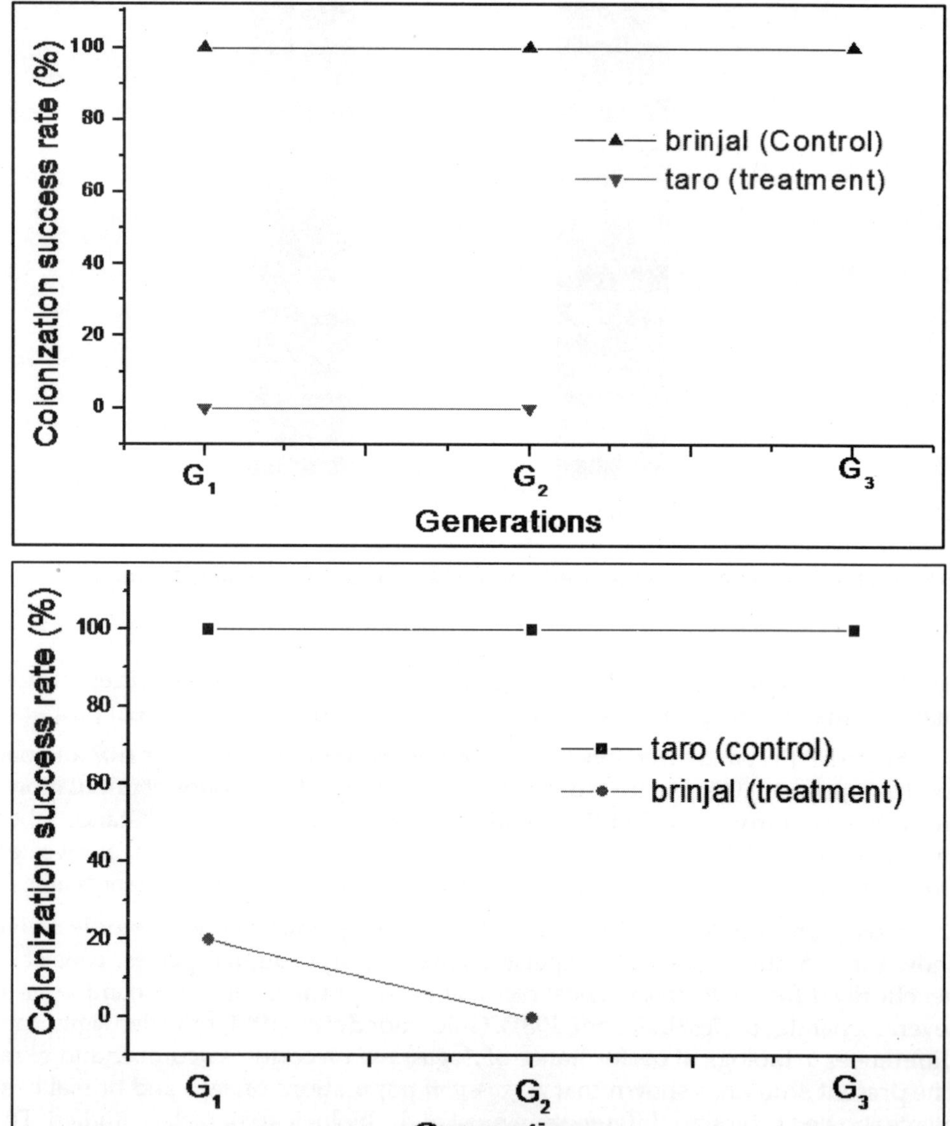

Figure 31,2L Success of colonization of *A. gossypii* aphids through generations on their field hosts (control) and across host plants, treatment I: *A. gossypii* of brinjal transferred to laboratory host, taro (A), treatment II: *A. gossypii* of taro transferred to laboratory host, brinjal (B).

of increase. These different *A. gossypii* forms are evidently specialized genotypes. This also implies that the effects of aphid-host plant interaction produce plasticity in phenotypes *i.e.* polyphenism (Agarwala, 2007). Using random amplified polymorphic DNA markers it was found that populations of *Aphis gossypii* collected from taro and brinjal plants were multiclonal.

Earlier, Agarwala and Das (2007) showed distinguishable variations in morphology, ecological performance and esterase pattern of *A. gossypii* populations from different hosts. Therefore, *A. gossypii* should be regarded as a heterogenous species infesting various host plants at different rates, that is, *A. gossypii* shows host plant specialization. This implies that little or no infestation of *A. gossypii* populations will occur from one plant species to another. Recent studies on *A.gossypii* (Wool and Hales, 1996; Vanlerberghe-Masutti and Chavigny, 1998; Fuller *et al.*, 1999) suggest that its evolutionary potential to adapt to newer host plants might be quite large and it holds the potential of becoming pest on an increasing number of crops.

Acknowledgements

We are thankful to Department of Biotechnology, New Delhi for financial support.

References

Agarwala, B.K and Das, K., 2007. Host-plant based morphological, ecological and esterase variations in *Aphis gossypii* Glover populations (Homoptera Aphididae). *Entomon.*, 32: 89–95.

Agarwala, B.K. and Ghosh, M.R., 1985. Biogeographical considerations of Indian Aphididae (Homoptera). *Ins. Matsum.*, 31: 81–96.

Blackman R.L., 1987. Rearing and handling aphids. In: *Aphids: Their biology, Natural Enemies and Control 2B*, (*ed. Minks, A.K and Harrewijn, P.*) Elsevier, Amsterdum, p. 59–69.

Black, W.C.I.V., Du Teau, N.M. and Puterka, G.J., 1992. Use of random amplified polymorphic DNA polymerase chain reaction (RAPD–PCR) to detect DNA polymorphisms in aphids (Homoptera, Aphididae). *Bull. Entomol. Res.*, 82: 151–159.

Blackman, R.L. and Eastop., 1984. *Aphids on the World's Crop: An identification Guide.* John Wiley, p.446.

Cenis, J.L., Perez, P. and Fereres, A., 1993. Identification of aphid (Homoptera, Aphididae) species and clones by random amplified polymorphic DNA. *Anna. Entomol. Soc. Amer.*, 86: 545–540.

Deb, D.B., 198. *The flora of Tripura State.* Vol. I. Today and Tomorrow's Printers and Publishers, New Delhi.

Diehl, S.R. and Bush, G.L., 1984. An evolutionary and applied perspective in insect biotypes. *Annu. Rev. Entomol.*, 29: 471–501.

Furk, C. and Hines, C.M., 1993. Aspects of insecticide resistance in the melon and cotton aphid, *Aphis gossypii* (Hemiptera: Aphididae). *Anna. Appl. Biol.*, 123: 9–7.

Guldemond, J.A., 1990. Choice of host plant as a factor in reproduction isolation of the aphid genus *Cryptomyzus* (Homoptera, Aphididae). *Ecol. Entomol.*, 15: 43–51.

Guldemond, J.A., Tiggers, W.T. and Vrijer, P.W.F., 1994. Host races of *A. gossypii* Glover on cucumber and chrysanthemum. *Entomol. Soc. Amer.*, 23: 1235–1340.

Han, Z., Moores, G.D., Denholm, I. and Devonshire, A.L., 1998. Association between biochemical markers and insecticide resistance in the cotton aphid, *Aphis gossypii* Glover. *Pesti. Biochem. and Physiol.*, 62: 164–171.

Inaizumi, M., 1981. Life cycle of *Aphis gossypii* (Homoptera, Aphididae), with special reference to biotype differentiation in various host plants. *Kontyu*, 49: 219–240.

Jaenike, J., 1981. Criteria for ascertaining the existence of host races. *Am. Nat.*, 117: 830–834.

Komazaki, S. and Toda, S., 2008. Difference in host preference, life cycle pattern, and insecticide susceptibility among *Aphis gossypii* clones and genetic relationships inferred from internal transcribed spacer 2 sequences of rDNA. *Anna. Entomol. Soc. of Amer.*, 101: 562–572.

Muller, F.P., 1985. Genetic and evolutionary aspects of host choice in phytophagous insects, especially in aphids. *Biologisches Zentralblatt.*, 107: 225–237.

Peng, G., Zhang, X.X., Yang, X.W. and Chen, X.F., 2000. RAPD–PCR detection of polymorphism of cotton aphid. *Entomol. Sin.*, 7: 41–46.

Pureka, G.H., Black, W.C.I.V. and Steiner, W.M., 1993. Genetic variation and phylogenetic relationships among worldwide collections of the Russian wheat aphid, *Diuraphis noxia* (Mordvilko), inferred from allozyme and RAPD–PCR markers. *Heredity*, 70: 604–618.

Saito, T., 1989. Insecticide resistance of the melon aphid, *Aphis gossypii* Glover (Hemiptera: Aphididae). I. Susceptibility to several insecticide and esterase activity of field populations collected in the Shizuoka Prefecture. *Jap. J. Appl. Entomol. Zool.*, 33: 204–210.

Watt, M. and Hales, D.F., 1996. Dwarf phenotype of the cotton aphid, *Aphis gossypii* Glover (Hemiptera: Aphididae). *Aus. J. Entomol.*, 35: 153–159.

Wool, D. and Hales, D.F., 1996. Components of variation of morphological characters in Australian *Aphis gossypii*: host–plant effects predominate. *Entomol Exp. et Appl.*, 80: 166–168.

Yang, X.W., Zhang, X.F. and Chen, X.F., 1999. On the RAPD–PCR analysis of DNA polymorphism of green peach aphid from different host plants. *J. Plant Prot., 26: 145–152.*

Zhang, G. X. and Zhong, V., 1993. Experimental studies on some aphid life cycles and the hybridization of two sibling species. In: *Aphid–plant genotype interactions* (ed. Cambell, R.K. and Eikenbary) Elsevier, p. 37–50.m

Environment, Biodiversity and Traditional System, 2014 *Pages* 323–332
ISBN 978-81-7622-307-2
Editors: D.R. Khanna, G.S. Solanki and S.K. Pathak
Published by: BIOTECH BOOKS, NEW DELHI

Chapter 32

Bioefficacy of *Cinnamomum tamala* as a Potential Source for the Management of Water-Borne Bacterial Pathogens

Amritesh C. Shukla[1], Awadhesh Kumar[1], Rohit K. Misra[2], Anand K. Pandey[2] and Anupam Dikshit[2]*

[1]*Department of Horticulture, Aromatic and Medicinal Plants, Mizoram University, Tanhril, Aizawl – 796 004*

[2]*Biological Product Laboratory, Department of Botany, University of Allahabad, Allahabad – 211 002*

ABSTRACT

In the present investigation an attempt has been made to test the bio-efficacy of *Cinnamomum tamala* (Buch.-Ham.) Nees and Eberm, family Lauraceae; as an antimicrobial agent against water-borne bacterial pathogens. The essential oil was extracted from the leaves of *C. tamala,* using Clevenger's apparatus and analyzed through GC/MS. Out of sixty one compounds; β-caryophyllene (25.5 per cent), linalool (13.6 per cent) and caryophyllene oxide (10.5 per cent) were identified as the major compounds. The oil thus extracted was investigated against some common bacterial pathogens *Escherichia coli* (MTCC723), *Salmonella typhimurium* (MTCC98), *Shigella dysenteriae* (ATCC23513), *Klebsiella pneumoniae* (MTCC109) and *Pseudomonas aeruginosa (MTCC8077)*; using broth microdilution method (NCCLS-2003). The minimum inhibitory concentrations (MIC) were recorded as 1.53mg/ml, 1.15mg/ml,

* *Corresponding Author:* E-mail: amriteshmzu@gmail.com

1.27mg/ml, 2.23mg/ml and 2.63mg/ml, respectively. The oil's toxicity was highly effective against *S. dysenteriae*, which causes dysentery in human beings. The results were also inferred with phylogenetic analysis of tested pathogens.

Keywords: Water-borne bacterial pathogens, Cinnamomum tamala, GC/MS, NCCLS.

Introduction

Management of water using the synthetics is not only expensive but their side effects are well known, hence alternative arrangement for safe drinking water is the global need on high priority basis. Literature reveals that there are number of plants having medicinal values for preventing various water borne bacterial diseases *viz.*, diarrhoea, cholera dysenteries, bacillary dysentery, vomiting, asthma, urinary tract infections (UTI), septicemia, neonatal-meningitis, typhoid fever, osteomyelitis, gasteroenteritis, bone/lung abscesses, food poisoning, sickle cell disease, shigellosis, Reiter's syndrom, dental disorders etc.[1, 2]. The bacteria commonly responsible for such water-borne diseases are *Escherichia coli, Vibrio cholerae, Salmonella typhi, Salmonella typhimurium, Shigella dysenteriae, Staphylococcus aureus, Klebsiella pneumoniae* and *Bacillus subtilis* etc.[3] According to the World Health Organization, medicinal plants could be an affordable source for developing medicines for primary healthcare needs[4]. Therefore, putative medicinal plants should be investigated to understand their properties, safety and efficacy[5]. Botanicals, such as essential oils and plant extracts, have been studied in the past for their antimicrobial activities[6]. Similarly, in the present investigation, an attempt has been made to investigate the bioefficacy of *C. tamala* against the common bacterial pathogens: *E. coli, S. typhimurium, S. dysenteriae, P. aeruginosa* and *K. pneumoniae*.

C. tamala is a medium sized plant, up to 8 m in height and distributed in the north-western Himalayas, Sikkim, Assam, Mizoram and Meghalaya[7]. The plant is commonly known as tejpat, and the oil has hypoglycemic and hypolipidemic properties. *Cinnamon* spp. can be used as spice because of its sweet flavoring and spicy characteristics; however various volatile constituents have an important role in pharmaceuticals especially as anti-inflammation, antimicrobial, antioxidant, antidiabetes type-2 antispasmodic, antiulcer, and cytotoxic properties[8–10]. Besides this, phylogeny of the test pathogens was also analyzed using Clustal-W programme.

Material and Methods

Test Organisms

Bacterial strains *viz., E. coli* (MTCC 723), *S. typhimurium* (MTCC 98), *P. aeruginosa* (MTCC8077) and *K. pneumoniae* (MTCC109) obtained from Microbial Type Culture Collection (MTCC), IMTECH, Chandigarh, India and *S. dysenteriae* (ATCC 23513) obtained from American Type Culture Collection (ATCC), Bangalore, India were used for the present investigation. The pure cultures were maintained in nutrient agar/MHA slants at the regular interval of one-week and stored at 4°C.

Plant Essential Oils

The aerial parts of *Cinnamomum tamala* (Buch.-Ham.) Nees and Eberm, family Lauraceae was collected from Almora, India. The shed dried fresh leaves of medicinal plant C. *tamala* were subjected to hydro-distillation for 8-10 hrs using Clevenger's apparatus[11]. The extracted essential oils dehydrated through anhydrous sodium sulphate. The oils thus obtained were subjected for antibacterial activity.

GC-MS Analysis

Gas Chromatographic analysis of the oil of C. *tamala* was performed on a Perkin-Elmer GC 8500, using a fused capillary column (25m x 0.55 mm i.d., film thickness 0.25um), coated with dimethtyl siloxane (BP-1). The oven temperature was adjusted at 60°C to 220°C at 5°C/min followed by isothermal at 220°C for 15 min; injector temperature 250°C; detector temperature 300°C; carrier gas 'nitrogen' at a linear velocity of 10 psi: split, 1: 80. GC-MS data were obtained on a Shimazdu QP-2000 mass spectrometer at 70Ev and 250°C. GC column: Ulbon HR-1 equivalent to OV-1, fused silica capillary column 0.25 mm x 50mm, film thickness 0.25 μm. The 100°C for 7 min. was initial temperature and heated at 5°C/min to 250°C. Helium was used as carrier gas at a flow rate of 2ml/min. The percentage composition of C. *tamala* leaves oil clearly indicated β-caryophyllene as its major constituent as shown in the table 1.

Antibacterial Investigation

Preparation and Standardization of Inoculum

Standardized inoculums of each bacterium *i.e.* 1.5×10^7 cells CFU/ml (colony forming unit) with 0.5 McFarland standard was taken for antimicrobial assay.

Determination of Minimum Inhibitory Concentration (MIC)

The MIC was defined as lowest concentration of essential oil, showing no visible bacterial growth after the incubation time[12]. This was determined using the broth microdilution method of NCCLS (2003)[13] with slight modification[34]. In order to test the various concentrations, the stock solution of the essential oils (50mg/ml) was prepared in the dimethylsulfoxide (DMSO).

Determination of Minimum Bactericidal Concentration

The bactericidal concentration was lowest concentration at which bacteria in at least five of the six replicates failed to grow in MHB and were not cultured when plating onto MHA/NA[14].

Phylogenetic Relationship between Pathogenic Bacterial Strains

Phylogenetic relationship between the pathogenic bacterial strains *viz., E. coli, S. typhimurium, S. dysenteriae, P. aeruginosa and K. pneumoniae* was determined using the ClustalW computer program[15] and GENETYX-MAC 10.1 software (Software Development Co., Ltd., Tokyo, Japan). Further, the phylogenetic trees were constructed by the DNA maximum-likelihood (ML) method in the PHYLIP program (Phylogeny Inference Package), version 3.5c[16] and the neighbor-joining (NJ)[17] method in the NJPLOT program[18] Bootstrap analysis with the ClustalW program was performed.

Protocols used to determine the MIC:

The experiment was designed in Sterile 96-well Microtiter Plates with Lids (SPL) using ELISA reader (Spectramax plus384, Molecular Devices Corporation, USA).

1) Fill the column 4 to 11 with 180ìl MHB and 20ìlof EO, so that final volume 200ìl can be made.
2) Serial dilutions were carried out from column 4 to 11 and excess broth (100ìl) was discarded from column 11.
3) Now, in each well from column 4 to 11, 100ìl inoculum was added.
4) Column 1 (which contains 200ìl EO only, and no inoculum) served as negative control);
5) Column 2 (contains 200ìl sterile MHB) served as media control;
6) Column 3 (contains 190ìl MHB with 10ìl essential oil) served as drug control;
7) Column 12 (which contains 200ìl inoculum but no EO) served as positive control.
8) Mixtures of the wells were mixed thoroughly
9) Further, these micro plates were incubated at 37ºC for 22–24 hrs.
10) After incubation, microorganism growth inhibition was evaluated by measuring absorbance at 630 nm.

Each experiment was repeated in triplicate.

Based on this information, the relationship between the effectiveness of essential oils on tested pathogens was determined.

Nucleotide Sequence Accession Numbers

The data of nucleotide sequences were obtained from GenBank nucleotide sequence database[19]. The ITS1 sequences of the standard strains were used in the present study. The accession no. for *E. coli* was J01859[19], for *S. typhimurium* X8068119; for *S. dysenteriae* GU27188320;*for P. aeruginosa JQ83914621* and for *K. pneumoniae* the accession no. was JQ389606[22] used for alignment.

Results

The GC-MS analysis of the volatile oil of *Cinnamomum tamala* leaves shows the presence of sixty one compounds; out of which β-caryophyllene (25.5 per cent), linalool (13.6 per cent) and caryophyllene oxide (10.5 per cent) were identified as the major compounds (Table 32.1).

In-vitro antibacterial activity of the essential oils of *C. tamala* was assessed against tested gram-negative water-borne bacterial pathogens. The observation was analyzed quantitatively and qualitatively according to the presence or absence of minimum inhibitory concentrations (MIC) as well as IC_{50} values obtained, as shown in Table 32.2.

Data of the study clearly indicated that the MIC of essential oil of *C. tamala* recorded against *E. coli* was 1.53 mg/ml and IC_{50} value was 0.53; against *S. typhimurium* MIC was 1.15 mg/ml and IC_{50} value 0.57; against *S. dysenteriae* MIC was

1.27 mg/ml and IC_{50} value 0.49; against *P. aeruginosa* MIC value was 2.63 mg/ml with IC_{50} value 0.65, and in case of *K. pneumoniae* the minimum inhibitory concentration was 2.23 mg/ml and IC_{50} value was 1.06 respectively. Besides, their relative average per cent (per cent) growth inhibitions have also been recorded in Figure 32.1.

Table 32.1: Major chemical compounds present in *C. tamala* leaf essential oil.

Sl.No.	*Name of Chemical Compounds*	*Per cent Obtained*
1	Linalool	13.6
2	α-pinene	3.10
3	β-caryophyllene	25.5
4	Camphene	1.8
5	Myrcene	4.6
6	Caryophyllene oxide	10.5
7	α-gurjunene	4.7

Table 32.2: Antibacterial action of *C. tamala* essential oil against water-borne bacterial pathogens.

Sl.No	*Name of Pathogens*	*Antibacterial Activity of C. tamala*		
		IC_{50}	*MIC*	*MBC*
1.	*E. coli (MTCC723)*	0.53	1.53	2.93
2.	*S. typhimurium* (MTCC98)	0.57	1.15	2.50
3.	*S. dysenteriae* (ATCC23513)	0.49	1.27	2.78
4.	*P. aeruginosa* (*MTCC*8077)	0.65	2.63	4.07
5.	*K. pneumoniae* (MTCC109)	1.06	2.23	3.89

IC_{50}: 50 per cent Inhibitory conc.; MIC: Minimum Inhibitory Concentration; MBC: Minimum Bactericidal Conc.

In the present study, the phylogenetic study of ITS1 sequences of 16S rRNA of the above mentioned pathogenic bacteria showed in the Figure 32.2.

All above tested gram negative bacterial pathogens were employed for their phylogenetic study. The phylogenetic tree was constructed using UPGMA clustering method and results were shown that origin of pathogens *viz., E. coli, S. typhimurium, S. dysenteriae* and *K. pneumoniae* were same and they are very close to each other, however the *P. aeruginosa* representing the same origin but it was far from the other pathogens. The distances between each pathogen also correlate the phylogenetic evidences.

Discussion

In vitro investigation in the form of MIC values of *C. tamala* essential oil, the percentage of inhibition was better for *S. dysenteriae*. Comparatively the efficacy of essential oils was determined on the basis of percentage of inhibition which recorded

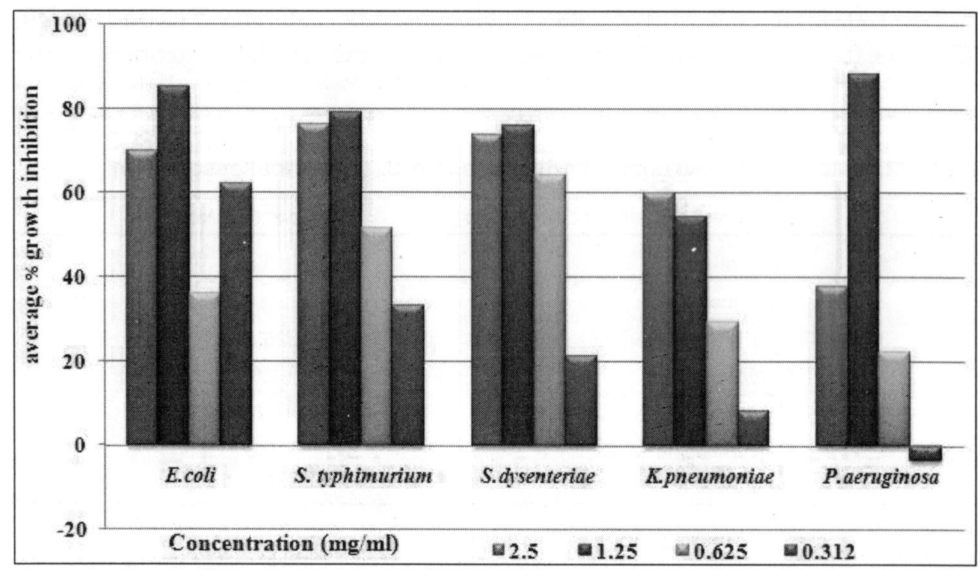

Figure 32.1: Average per cent (per cent) growth inhibition of tested water-borne bacterial pathogens.

Figure 32.2: The rooted phylogenetic tree was constructed with data for standard strains of tested water borne bacterial pathogen.

in form of minimum inhibitory concentration through Elisa Reader (model: Spectramax Plus[384], procured from Molecular Devises, U.S.A.). The oil of *C. tamala* was found to be most effective against the tested bacterial pathogens (Table 32.1).

Based on their phylogenetic study the differences found in between the same class of pathogens represented the hierarchy and their cell wall chemistry. These differences were also occurred *due to the differences in their physiochemical and environmental conditions which were* responsible for the alteration in MIC values of essential oil. According to previous study the correlation between phylogeny of water borne pathogens and essential oil of *T. ammi* L. was very helpful to know the antibacterial activity[23]. Based on shape and mode of nutrition of *E. coli, S. typhimurium,*

S. dysenteriae, K. pneumoniae were facultative anaerobic. This similarity also reflected in the Phylogram as by the one node there two clades originated and showed the closeness in between *E. coli* and *S. typhimurium*. Further one clade which represented the *S. typhimurium* divided into two clades by one node showed the closeness between *S. typhimurium* and *S. dysenteriae*. The same thing happened in case of *S. dysenteriae* and *K. pneumoniae*. However Phylogram (Figure 32.2) clearly showed a much variation in phylogeny between *P. aeruginosa* and rest four pathogens.

According to the results obtained from the present study the variation in antibacterial activities of essential oil may be related to the concentration and nature of constituents, their respective composition, the functional groups, the structural configuration of the components and their possible synergistic interaction[24]. The hydrophobicity, an important characteristic of essential oils (EOs) which enables them to associate in the lipids of the bacterial cell membrane and mitochondria, disturbing the structures and rendering them more permeable leading to leakage of ions and other cell contents[25,26]. Although, a certain amount of leakage from bacterial cells may be tolerated without loss of viability, extensive loss of cell contents or the exit of critical molecules and ions will lead to death[27]. Babu *et al.*,[28] also reported that *Campylobacter jejune* and *Escherichia coli* were found to be more sensitive to cinnamon essential oil. According to another same study it confirmed that phylogenetically closed pathogens have been possessed nearly same result as found in present investigation[29]. The leaves of *C. tamala* are carminative and are used for colic and diarrhoea[30]. Mechanism of action of essential oils which is generally considered to be the disturbance of the cytoplasmic membrane, disrupting the proton motive force (PMF), electron flow, active transport and coagulation of cell contents[31,32]. The previous work reported that Gram-negative bacteria were more resistant to the essential oils present in plants[33] because lipopolysaccharide (LPS) avoids the accumulation of the oils on the cell membrane.

Therefore, on basis of present antibacterial study, the conclusion can be drawn that the sensitivity of the LPS of tested bacterial pathogens were found to be high in *S. dysenteriae* (ATCC23513).

Conclusion

Although, there are number of synthetics are available in the market for the management of water impurities but there is no such type of report, where herbs are used as an antibacterial agent for the water management. Hence, the present findings (*Cinnamomum tamala* possessed strong inhibitory activity against the tested water-borne bacterial pathogens) can be a potential source for natural management of water after successful pharmacological investigations.

Acknowledgements

The authors are thankful to the Head, Department of Horticulture Aromatic and Medicinal Plants, Mizoram University, Aizawl, and the Head, Department of Botany, University of Allahabad for providing the laboratory facilities. Financial assistance of the Department of Science and Technology (DST) is thankfully acknowledged.

References

1. Lai PK, Rov J., 2004. Antimicrobial and chemopreventive properties of herbs and spices. *Curr. Med. Chem*, **47(2):** 234–238.
2. Chaudhry NMA, Tariq P., 2006. Anti–microbial activity of *Cinnamomum cassia* against diverse microbial flora with its Nutritional and medicinal impacts *Pak. J. Bot*, **38(1):** 169–174.
3. Ashbolt, N.J., 2004. Microbial contamination of drinking water and disease outcome in developing regions. *Toxicology*, **198**: 229–238.
4. WHO. The Global Water Supply and sanitation Assessment 2000, Retrieved 28 December 2004, 2003.
5. Ellof, J.N., 1998. Which extractant should be used for the screening and isolation of antimicrobial components from plants. *J Ethnopharmacol* **60:** 1–6.
6. Tepe, B., Donmez, E., Unlu, M., Candan, F., Daferera, D., Vardar–Unlu, G., 2004. Antimicrobial and antioxidative activities of the essential oils and methanol extracts of *Salvia cryptantha* (montbret et aucher ex benth.) and *Salvia multicaulis* (vahl). *Food Chemistry*, **84:** 519–525.
7. Wealth of India:., 1992. *Raw Materials*, vol II. Publication and Information Directorate, CSIR: New Delhi, 580.
8. Husain, A., Virmani, O.P., Sharma, A., Kumar, A., Misra, L.N., 1988. Major Essential Oil–bearing Plants of India. Central Institute of Medicinal and Aromatic Plants: Lucknow, 213.
9. Sharma, S.R., Dwivedi, S.K. and Swarup, D., 1996. *Ind. J. Exp. Biol.*, **34:** 372.
10. Javaprakash, G.K., Rao, L.J. Sakariah. K.K. and Naturforsch, Z., 2002. 57,990.
11. Clevenger, J.F., 1928. Apparatus for the determination of volatile oil. *J Am Pharm. Assoc.* **17:** 346–346.
12. Smith–Palmer, A., Stewart, J. and Fyfe, L., 1998. Antimicrobial properties of plant essential oils and essences against five important food–borne pathogens, *Letters in Food Microbiology*, **26:** 118–122.
13. NCCLS., 2003. Methods for dilution antimicrobial susceptibility tests for bacteria that grow aerobically (5th Ed.)," *Approved standard, M7–A5*. Pennsylvania: Wayne.
14. Delaquis, P.J., Stanich, K., Girard, B., and Mazza, G., 2002. Antimicrobial activity of individual and mixed fractions of dill, cilantro, coriander and eucalyptus essential oils. *Int.J. Food Microbiol* **74:** 101–109.
15. Higgins, D., Clustal W, Version 1.6. Cambridge, United Kingdom," The European Bioinformatics Institute 1996.
16. Felsenstein, J. PHYLIP (Phylogeny Inference Package), version 3.5c. Seattle," University of Washington 1995.
17. Saito, N., and Nei, M., 1995. The neighbor–joining method: a new method for reconstructing phylogenetic trees. *Mol Biol Evol*, **4:** 406–425.

18. Maidak, B.L., Larsen, N., McCaughey, M.J, Overbeek, R., Olsen, G.J, Fogel, K., Blandy, J., and Woese, C.R., 1994. The Ribosomal Database Project. *Nucleic Acids Res,* **22:** 3485–3487.
19. Cilia, V., Lafay, B., and Christen, R., 1996. Sequence heterogeneities among 16S ribosomal RNA sequences, and their effect on phylogenetic analyses at the species level. *Mol. Biol. Evol,* **13(3):** 451–461.
20. Zerzghi, H., Brooks, J.P., Gerba, C.P. and Pepper, I.L., 2010. Influence of long–term land application of Class B biosolids on soil bacterial diversity. *J. Appl. Microbiol,* **109(2):** 698–706.
21. Sen, S. and Sarkar, P., 2012. Polymer Sc. and Technology, University of Calcutta, 92, a.P.C Road, Kolkata, West Bengal 700124, India.
22. Rajput, M., Rajkumar, S., Vyas, Y., Joshi, M.N., Saxena, A.K. and Bagatharia, S.B., 2011. Gujarat Biodiversity Gene Bank, Gujarat State Biotechnology Mission, 9th floor, 11th Block, Udyogbhavan, Gandhinagar, Gujarat 382011, India.
23. Kumar, A., Mishram R.K., Tiwarim A.K., Pandey, A., Shukla, A.C. and Dikshit, A., 2011. Role of Phylogenetic Analysis for Anti–bacterial Activity of Essential Oil of *Trachyspermum ammi* L. Against Water Borne Pathogens. *Advan. in Envt Bio,* **5(6):** 1271–1278.
24. Chang, S.T., Chen, P.F. and Chang, S.C., 2001. Antibacterial activity of leaf essential oils and their constituents from *Cinnamomum osmophloeum. Journal of Ethnopharmacology,* **77:** 123–127.
25. Knobloch, K., Weigand, H., Weis, N., Schwarm, H.M., Vigenschow, H., 1986. Action of terpenoids on energy metabolism," In: Brunke, E.J. (Ed.), Progress in Essential Oil Research: 16th International Symposium on Essential Oils. *De Gruyter, Berlin,* 429– 445.
26. Sikkema, J., De Bont, J.A.M. and Poolman, B., 1994. Interactions of cyclic hydrocarbons with biological membranes. *Journal of Biological Chemistry,* **269(11):** 8022–8028.
27. Denyer, S.P. and Hugo, W.B., 1991b). Mechanisms of antibacterial action–A summary," In: Denyer, S.P., Hugo, W.B. (Eds.), Mechanisms of Action of Chemical Biocides. *Blackwell, Oxford,* 331– 334.
28. Babu, A.J., Sundari, A.R., Indumathi, J., Surjan, R.V.N. and Sravanthi, M., 2011. *Veterinary World,***4:** 311.
29. Kumar, A., Gupta, R., Mishra, R.K. Shukla, A.C. and Dikshit, A., 2012. Pharmaco Phylogenetic Investigation of *Micromeria biflora* Benth. and *Citrus reticulata* Blanco. Natl. Acad. Sci. Lett. (Springer) DOI 10.1007/s40009–012–0029–7.
30. Wealth of India., 1950. *Raw materials,* **2:** 178.
31. Sikkema, J., De Bont, J.A.M. and Poolman, B., 1995. Mechanisms of membrane toxicity of hydrocarbons,"*Microbiological Reviews,* **59(2):** 201–222.
32. Davidson, P.M., 1997. Chemical preservatives and natural antimicrobial compounds," In: Doyle, M.P., Beuchat, L.R., Montville, T.J. (Eds.), *Food Microbiology*: Fundamentals and Frontiers, ASM, Washington, 520–556.

33. Smith–Palmer, A., Stewart, J. and Fyfe, L., 1998. Antimicrobial properties of plant essential oils and essences against five important food–borne pathogens/*Letters in Food Microbiology* **26:** 118–122.

34. Shukla, A.C., 2010. Bioactivities of the major Active Constituent Isolated from the essential oil of *Cymbopogon flexuosus* (Steud) Wats and *Trachyspermum ammi* (L) Sprague as a herbal grain protectants, *D.Sc. Thesis,* Allahabad University, Allahabad.

Environment, Biodiversity and Traditional System, 2014 *Pages* 333–349
ISBN 978-81-7622-307-2
Editors: D.R. Khanna, G.S. Solanki and S.K. Pathak
Published by: BIOTECH BOOKS, NEW DELHI

Chapter 33

Butterfly Diversity in Wildlife Sanctuaries and Natural Forests of Tripura

B.K. Agarwala

Department of Zoology, Ecology and Bio-diversity Laboratories, Tripura University, Suryamaninagar – 799 022, Tripura
E-mail: bagarwala00@gmail.com

Introduction

Recent studies of biodiversity in relation to ecosystem functioning have revealed that species diversity enhances productivity and stability of ecosystem (Naeem *et al.*, 1994; Tilman *et al.*, 1996). Butterflies are known to contribute substantially to the growth, maintenance and expansion of flora. This is particularly true of tropical regions where these insects show high abundance and species diversity (Kunte 1997; Price 1997). Due to richness in vegetation, northeastern region of India is home to rich diversity of butterflies among other insects (Alfred *et al.*, 2002) and it is also part of one of the mega biodiversity hotspots of the world (Myers *et al.*, 2000). A literature review suggests that 76 species of butterflies were previously recorded from the state of Tripura (10,490 sq km), north-east India (Mandal *et al.*, 2002; Agarwala *et al.*, 2010, Majumder *et al.*, 2011). Among other north-eastern states, 104 species from Meghalaya (22429 sq. km), 695 species from Sikkim (7096 sq. km), and 962 species of butterflies from Assam (78438 sq. km) have been recorded (Wynter-Blyth 1957; Haribal 1992). Evidently, the knowledge base of lepidopteran fauna and its distribution in different habitats is uneven and still scanty from this part of India. Given this background, a detailed inventory was carried out to document the butterfly species richness and diversity of different Wildlife Sanctuaries and natural forests of Tripura which comprise of mixed moist deciduous and evergreen forests with patches of long grasses.

Considering the fact that flora of Tripura comprise of more than 530 tree species, 260 species of shrubs, 500 species of herbs, and 250 species of climbers (http://www.incredible-northeastindia.com), the assumption was that new data on spatial scale over a period of time would help to ascertain the true status of species richness and diversity of butterflies in diverse habitats of the state which is home to the Asian population of bison (*Bos gorus* Smith), more than 200 species of resident and migratory birds, reptiles like cobra and king cobra snakes, and mammals like wild boar (*Sus scrofa* L.), spectacle langur (*Trachypithecus phayrie* Blyth), capped langur (*Trachypithecus pileatus* Blyth), slow loris (*Nycticebus coucang* Boddaert) and wild cat (*Felis chaus* Schreber) (http://tripura.nic.in/trishna/intro.html).

Materials and Methods

Study sites

Twenty seven locations were identified for the purpose of survey of butterfly fauna. These were based on the sites identified by the Forest Department, Govt. of Tripura. At each study site, sample location was determined on the basis of richness of floral kinds. GPS readings of each location were taken for latitude, longitude and altitude (Table 33.1).

Butterfly fauna was surveyed using modified Pollard walk methods which is a combination of line transect walk and point counts (Pollard, 1977; Royer *et al.*, 1998). In this method, walks were made along the 1000-m long transect and points were located at 10-m interval. At each point, observers stood and recorded any butterfly detected in circular area of 5-m range (vertical and horizontal) for approximately 10 minutes. Walk census were held on transect lines between the hours of 8 am to 11 am. Walking pace was kept slow but at uniform speed for recording of butterflies in maximum number. Halts were made along transect lines when it was necessary to collect or identify the species. Butterflies found resting on plants and those in flight were counted and most of these could be identified in the fields. Some of the specimens where an exact determination of the species was not possible were captured by sweeping nets or butterfly net for closer examination and their field identification. These were released soon after. A few of the specimens that could not be identified in the field were brought to the laboratory for their identification.

A total of 94 transects were made in 27 location on 61 days during field visits from 01- 08- 2009 to 31- 05- 2010. During the period, a total of 183 man-days were spent in the field (Table 33.2).

Identification of butterflies in field: Butterfly were identified for their species status in the field on the basis of external characters- like wing size, shape, color and pigmentation pattern, type of antennae and leg characters with the help of Pictorial field identification guides (Kehimkar, 2008; Haribal, 1992). Larger specimens that were clearly visible at close distance Were identified with naked eye; smaller or larger specimens that were found flying at distance were identified with the help of binoculars and Tele micro EF 75–300 mm f/4 – 5.6 III USM lens.

Table 33.1: Name of study sites and GPS data of sample locations.

Sl.No.	*Name of Sites*	*GPS Data*		
		Lat.	*Lon.*	*Alt. (ft.)*
1.	Kalapania	23°30.195' N	91°20.900' E	161
2.	Unakoti	24°29.901' N	92°21.687' E	285
3.	Juri RF	24°14.589' N	92°11.426' E	275
4.	Ballianchip	23°59.079' N	92°16.649' E	1853
5.	Rowa WLS	24°17.480' N	92°09.903' E	253
6.	Churaibari	24°26.451' N	92°15.062' E	95
7.	CCRF	23°51.613' N	91°59.862' E	246
8.	Baramura	23°49.540' N	91°16.350' E	353
9.	Chankumabari	23°29.023' N	91°24.721' E	190
10.	Harbang	23°56.232' N	91°33.846' E	138
11.	Subalshing	23°55.302' N	91°27.660' E	131
12.	Garje	23°29.194' N	91°43.103' E	289
13.	Trishna WLS	23°26.137' N	91°28.184' E	269
14.	Kalshi	23°15.768' N	91°39.456' E	112
15.	Malumbari	23°08.291' N	91°32.889' E	256
16.	Harbatali	22°58.709' N	91°38.645' E	258
17.	Baishnabpur	23°01.048' N	91°44.739' E	249
18.	Gumati WLS	23°25.462' N	91°49.655' E	249
19.	Paharpur	23°29.194' N	91°43.103' E	233
20.	Maharani	23°30.184' N	91°34.330' E	158
21.	Killa	23°33.590' N	91°30.10' E	229
22.	Sephahijala	23°39.791' N	91°18.50' E	223
23.	Gatirampara	23°39.361' N	91°18.46' E	218
24.	Tulashikar	24°01.278' N	91°40.72' E	220
25.	Atharamura N.C	24º07.691' N	91º46.643' E	601
26.	Debbari	23º31.576' N	91º33.523' E	321
27.	Gandhari	23°29.265' N	91°36.241' E	144

Table 33.2: Details of exploration for butterfly fauna of Tripura State.

District	*No. of Locations*	*No. of Transects*	*No. of Visits*	*No. of Man Days Spent*
West Tripura	8	24	10	42
South Tripura	12	52	17	87
North Tripura	5	13	5	39
Dhalai Tripura	2	5	4	15
Total	**27**	**94**	**36**	**183**

Results

Family-wise List of Butterflies

World-wide 15000 species of butterflies are recorded. These belong to six families. In the present study 174 species were collected from all of the six families (Table 33.3). About 44.48 per cent of the recorded species belonged to the two families, namely Nymphalidae (49 species) and Lycaenidae (45 species). Another 29.21 per cent of the species belonged to families Hesperiidae (28 species) and Pieridae (22 species). Rest of the 26.31 per cent of species belonged to the families Papilionidae and Satyridae, which represent 14 species and 16 species respectively.

Table 33.3: No. of butterfly species and genera recorded in different families in the state of Tripura.

Sl.No.	*Family*	*No. of Genera*	*No. of Species*
1.	Papilionidae	6	14
2.	Pieridae	12	22
3.	Nymphalidae	31	49
4.	Lycaenidae	34	45
5.	Satyridae	6	16
6.	Hesperiidae	23	28
	Total	**112**	**174**

Family: Papilionidae

These are generally large size butterflies commonly known as *swallowtails* but few species are without swallowtails. All swallowtail caterpillars have a forked organ known as osmeterium (defensive organ) at the base of back of the head. These butterflies are very active fliers and are often seen mud-puddling in large groups. These butterflies generally prefer diverse habitats but few species are restricted to specific habitat with little or no disturbance. They are active pollinators. Worldwide, there are 550 known species of swallowtails, of these 107 species are found in the Indian region, and 14 species have been recorded from Tripura in this study (Table 33.4).

Family: Nymphalidae

Commonly known as brush-footed butterflies, this is the largest family with around 6,000 species occurring worldwide. Family contains many well-known, common and widespread species. A few are migrants capable of extending their range over a large geographical area during summer months. The first pair of legs of nymphalids is much. Unlike other families, nymphalids vary greatly in appearance in larval, pupal, and adult stages. Larvae are often conspicuously coloured and adorned with spectacular spiny protuberances. This family prefers diverse habitats. Forty nine species were recorded from Tripura (Table 33.5).

Table 33.4: List of Papilionidae butterflies recorded from Tripura.

Sl.No.	*Common Name*	*Scientific Name*
1.	Common Rose	*Atrophaneura aristolochiae* (Fabricius)
2.	Common Batwing	*Atrophaneura varuna* (White)
3.	Common Mormon	*Papilio polytes* Linnaeus
4.	Yellow Helen	*Papilio nephelus* Linnaeus
5.	Red Helen	*Papilio helenus* Linnaeus
6.	Great Mormon	*Papilio memnon* Linnaeus
7.	Lime Butterfly	*Papilio demoleus* Linnaeus
8.	Common Bluebottle	*Graphium sarpedon* (Linnaeus)
9.	Tailed Jay	*Graphium agamemnon* (Linnaeus)
10.	Common Jay	*Graphium doson* (C. and R. Felder)
11.	Five Bar Swordtail	*Graphium antiphates* (Cramer)
12.	Common Mime	*Chilasa clytia* (Linnaeus)
13.	Common Birdwing	*Troides helena* (Linnaeus)
14.	White Dragontail	*Lamproptera curius* (Fabricius)

Table 33.5: List of Nymphalidae butterflies recorded from Tripura.

Sl.No.	*Common Name*	*Scientific Name*
1.	Striped Tiger	*Danaus genutia* (Cramer)
2.	Plain Tiger	*Danaus chrysippus* (Linnaeus)
3.	Glassy Tiger	*Parantica aglea* (Stoll)
4.	Common Indian Crow	*Euploea core* (Cramer)
5.	Common Castor	*Ariadne merione* (Cramer)
6.	Angled Castor	*Ariadne ariadne* (Linnaeus)
7.	Blue Spotted Crow	*Euploea midamus* (Linnaeus)
8.	Striped Blue Crow	*Euploea mulciber* (Cramer)
9.	Jewelled Nawab	*Polyura delphis* (Doubleday)
10.	Common Nawab	*Polyura athamas* (Drury)
11.	Great Duffer	*Discophora timora* Westwood
12.	Leopard Lacewing	*Cethosia cyane* (Drury)
13.	Cruiser	*Vindula erota* Fabricius
14.	Common Yeoman	*Cirrochroa tyche* C. and R. Felder
15.	Common Leopard	*Phalanta phalantha* (Drury)
16.	Commander	*Moduza procris* (Cramer)
17.	Common Sergeant	*Athyma perius* (Linnaeus)
18.	Studded Sergeant	*Athyma asura* Moore
19.	Black Vein Sergeant	*Athyma ranga* Moore

Contd...

Table 33.5–*Contd...*

Sl.No.	*Common Name*	*Scientific Name*
20.	Colour Sergeant	*Athyma nefte* (Cramer)
21.	Common Lascar	*Pantoporia hordonia* (Stoll)
22.	Common Sailor	*Neptis hylas* (Linnaeus)
23.	Chestnut-Streaked Sailor	*Neptis jumbah* Moore
24.	Dingiest Sailor	*Neptis harita* Moore
25.	Clear Sailor	*Neptis clinia* Moore
26.	Sullied Sailor	*Neptis soma* Moore
27.	Clipper	*Parthenos sylvia* (Cramer)
28.	Knight	*Lebadea martha* (Fabricius)
29.	Common Baron	*Euthalia aconthea* (Cramer)
30.	Gaudy Baron	*Euthalia lubentina* (Cramer)
31.	Grey Count	*Tanaecia lepidea* (Butler)
32.	Common Earl	*Tanaecia julii* (Lesson)
33.	Marbled Map	*Cyrestis cocles* (Fabricius)
34.	Popinjay	*Stibochiona nicea* (Gray)
35.	Courtesan	*Euripus nyctelius* (Doubleday)
36.	Common Jester	*Symbrenthia hippoclus* (Cramer)
37.	Yellow Pansy	*Junonia hierta* (Fabricius)
38.	Chocolate Pansy	*Junonia iphita* (Cramer)
39.	Peacock Pansy	*Junonia almana* (Linnaeus)
40.	Grey Pansy	*Junonia atlites* (Linnaeus)
41.	Lemon Pansy	*Junonia lemonias* (Linnaeus)
42.	Great Eggfly	*Hypolimnas bolina* (Linnaeus)
43.	Danid Eggfly	*Hypolimnus misipus* (Linnaeus)
44.	Blue Tiger	*Tirumala limniace* (Cramer)
45.	Black Prince	*Rohana parisatis* (Westwood)
46.	Club Beak	*Libythea myrrha* Godart
47.	Red Spot Duke	*Dophla evelina* (Stoll)
48.	Tawny Rajah	*Charaxes bernardus* (Fabricius)
49.	Dark Archduke	*Lexias dirtea khasiana* (Swinhoe)

Family: Satyridae

Satyrids are generally brown coloured and inconspicuous butterflies of medium size. They have marked seasonal forms namely wet season form (WSF) and dry season form (DSF). WSF is characterized by wings with eye-like markings (ocelli) and dark in colour. In DSF, wings are generally paler and eye-like markings are either replaced by spots or striations. Caterpillars are slender but broader in the middle, and bear soft hairs all over the body. Some have two horns and a bifurcated tail. Most of these butterflies are weak fliers and fly very close to the ground during dawn and dusk. They generally fly in the secondary undergrowths and a few of them visit flowers. Most of them feed on rotten fruit, tree sap, animal droppings etc. Their food plants are mainly grasses and bamboo. Fourteen species were recorded from Tripura (Table 33.6).

Table 33.6: List of Satyridae butterflies recorded from Tripura.

Sl.No.	*Common Name*	*Scientific Name*
1.	Bamboo Tree Brown	*Lethe europa* (Fabricius)
2.	Angled Red Forester	*Lethe chandica* (Moore)
3.	Common Palmfly	*Elymnias hypermnestra* (Linnaeus)
4.	Blue-Striped Palmfly	*Elymnias patna* (Westwood)
5.	Common Bush Brown	*Mycalesis perseus* (Fabricius)
6.	Dark-Brand Bush Brown	*Mycalesis mineus* (Linnaeus)
7.	Long-Brand Bush Brown	*Mycalesis visala* (Moore)
8.	Nigger	*Orsotrioena medus* (Fabricius)
9.	Common Four Ring	*Ypthima huebneri* Kirby
10.	Common Five Ring	*Ypthima baldus* (Fabricius)
11.	Common Three Ring	*Ypthima asterope* (Klug)
12.	Common Red Forester	*Lethe mekara* (Moore)
13.	Wood-Mason's Bushbrown	*Mycalesis suaveolens* Wood-Mason and de Nicéville
14.	Common Evening Brown	*Melanitis leda* (Linnaeus)
15.	White Bar Bushbrown	*Mycalesis anaxias* Fruhstofer
16.	Spotted Palmfly	*Elymnias malelas* (Hewitson)

Family: Pieridae

About a thousand species of pierids are known to occur throughout the world, out of which 109 species occur in India. These are commonly known as 'whites and yellows' because their wing colour is generally white, with black, yellow and orange spots. Caterpillars are green in colour. They are sun loving insects especially males, sometimes found in large congregate on wet patches near rivers or stream beds for mud puddling. They visit various species of flowers. Caterpillars feed on plants belonging to families Capparaceae, Cruciferae, Salvadoraceae, Fabaceae and Caesalpiniaceae. This family is represented by 22 species in Tripura (Table 33.7).

Family: Lycaenidae

Another family with more than 6,000 species recorded in the world, known by some of the prettiest and smallest butterflies. In India, there are 521 species. Majority of these butterflies have blue upper sides, though there are species with other colours too. A good number of species have threadlike tails at hind wings. Some of these butterflies prefer undergrowth, others keep to high canopy. Few species are carnivorous in larval stage. Caterpillars are typical flat woodlouse-like and are usually well-camouflaged to larval host-plant. Many lycaenids in the larval and pupal stages are attended by ants. The family is represented by 43 species in Tripura (Table 33.8).

Table 33.7: List of Pieridae butterflies recorded from Tripura.

Sl.No.	*Common Name*	*Scientific Name*
1.	Common Grass Yellow	*Eurema hecabe* (Linnaeus)
2.	One Spot Grass Yellow	*Eurema andersoni* (Moore)
3.	Three Spot Grass Yellow	*Eurema blanda* (Boisduval)
4.	Spotless Grass Yellow	*Eurema laeta* (Boisduval)
5.	Common Emigrant	*Catopsilia pomona* (Fabricius)
6.	Mottled Emigrant	*Catopsilia pyranthe* (Linnaeus)
7.	Great Orange Tip	*Hebomoia glaucippe* (Linnaeus)
8.	Crimson Tip	*Colotis danae* (Fabricius)
9.	Common Wanderer	*Pareronia valeria* (Cramer)
10.	Striped Albatross	*Appias libythea* (Fabricius)
11.	Chocolate Albatross	*Appias lyncida* (Cramer)
12.	Green-Veined White	*Pieris napi* (Linnaeus)
13.	Indian Cabbage White	*Pieris canidia* (Sparrman)
14.	Psyche	*Leptosia nina* (Fabricius)
15.	Common Jezebel	*Delias eucharis* (Drury)
16.	Red-Spot Jezebel	*Delias descombesi* (Boisduval)
17.	Red-Base Jezebel	*Delias pasithoe* (Linnaeus)
18.	Painted Jezebel	*Delias hyparete* (Linnaeus)
19.	Yellow Jezebel	*Delias agostina* (Hewitson)
20.	Common Gull	*Cepora nerissa* (*Fabricius*)
21.	Lesser Gull	*Cepora nadina* (Lucas)
22.	Tree Yellow	*Gandaca harina* (Horsfield)

Table 33.8: List of Lycaenidae butterflies recorded from Tripura.

Sl.No.	*Common Name*	*Scientific Name*
1.	Common Gem	*Poritia hewitsoni* Moore
2.	Ape Fly	*Spalgis epius* (Westwood)
3.	Indian Sunbeam	*Curetis thetis* (Drury)
4.	Indian Oakblue	*Arhopala atrax* (Hewitson)
5.	Spotless Oakblue	*Arhopala fulla* (Hewitson)
6.	Large Oakblue	*Arhopala amantes* (Hewitson)
7.	Western Centaur Oakblue	*Arhopala pseudocentaurus* (Doubleday)

Contd...

Table 33.8–*Contd...*

Sl.No.	*Common Name*	*Scientific Name*
8.	Common Acacia Blue	*Surendra quercetorum* (Moore)
9.	Yamfly	*Loxura atymnus* (Stoll)
10.	Chocolate Royal	*Remelana jangala* (Horsfield)
11.	Fluffy Tit	*Zeltus amasa* (Hewitson)
12.	Common Tit	*Hypolycaena erylus* (Godart)
13.	Copper Flash	*Rapala pheretima* (Hewitson)
14.	Slate Flash	*Rapala manea* (Hewitson)
15.	Common Tinsel	*Catapaecilma elegans* (Druce)
16.	Club Silverline	*Spindasis syama* (Horsfield)
17.	Long-Banded Silverline	*Spindasis lohita* (Horsfield)
18.	Purple Sapphire	*Heliophorus epicles* Godart
19.	Common Pierrot	*Castalius rosimon* (Fabricius)
20.	Banded Blue Pierrot	*Discolampa ethion* (Westwood)
21.	Angled Pierrot	*Caleta caleta* Hewitson
22.	Zebra Blue	*Leptotes plinius* Fabricius
23.	Common Line Blue	*Prosotas nora* (C. Felder)
24.	Dark Cerulean	*Jamides bochus* (Stoll)
25.	Common Cerulean	*Jamides celeno* (Cramer)
26.	Metalic Cerulean	*Jamides alecto* (C. and R. Felder)
27.	Pea Blue	*Lampides boeticus* (Linnaeus)
28.	Dark Grass Blue	*Zizeeria karsandra* (Moore)
29.	Tiny Grass Blue	*Zizula hylax* (Fabricius)
30.	Lesser Grass Blue	*Zizina otis* (Fabricius)
31.	Quaker	*Neopithecops zalmora* (Butler)
32.	Plains Cupid	*Chilades pandava* (Horsfield)
33.	Lime Blue	*Chilades lajus* (Stoll)
34.	Malayan	*Megisba malaya* (Horsfield)
35.	Transparent 6-Lineblue	*Nacaduba kurava* (Moore)
36.	Common Imperial	*Cheritra freja* (Fabricius)
37.	Punchinello	*Zemeros flegyas* (Cramer)
38.	Gram Blue	*Euchrysops cnejus* (Fabricius)
39.	Rosy Oakblue	*Narathura selta* (de Nicéville)
40.	Silverstreak Blue	*Iraota timoleon* (Stoll)
41.	Broadtail Royal	*Creon cleobis* (Godart)
42.	Bigg's Brownie	*Miletus nymphis* (Fruhstorfer)
43.	Pale Grass Blue	*Pseudozizeeria maha* (Kollar)

Family: Hesperiidae

Hesperiids, commonly called 'skippers', are very dull looking butterflies with thick body and resemble moths. Their antennae are hook-like. They are generally brown in colour with various spots and streaks present on wings. Few species are brightly coloured. They are usually very swift fliers. Most of them visit flowers. Altogether 3,500 species of skippers occur throughout the world, and India has 321 species. The family is represented by 28 species in Tripura (Table 33.9). Larvae are of green colour and cylindrical with large head. These generally live inside the rolled leaves. Their larval food plants consist of grass, palm, and bamboo. But some also feed on dicotyledonous species.

Table 33.9: List of Hesperiidae butterflies recorded from Tripura.

Sl.No.	*Common Name*	*Scientific Name*
1.	Brown Awl	*Badamia exclamationis* (Fabricius)
2.	Common Small Flat	*Sarangesa dasahara* Moore
3.	Spotted Small Flat	*Sarangesa purendra* Moore
4.	Common Snow Flat	*Tagiades japetus* (Stoll)
5.	Fulvous Pied Flat	*Pseudocoladenia dan* (Fabricius)
6.	Suffused Snow Flat	*Tagiades gana* (Moore)
7.	Pale Palm Dart	*Telicota colon* (Fabricius)
8.	Straight Swift	*Parnara guttatus* (Bremer and Grey)
9.	Rice Swift	*Borbo cinnara* (Wallace)
10.	Small Branded Swift	*Pelopidas mathias* (Fabricius)
11.	Paintbrush Swift	*Baoris farri* (Moore)
12.	Common Redeye	*Matapa aria* (Moore)
13.	Chestnut Bob	*Iambrix salsala* (Moore)
14.	Dark Velvet Bob	*Koruthaialos butleri* de Nicéville
15.	Bright Red Velvet Bob	*Koruthaialos sindu* (C. and R. Felder)
16.	Chocolate Demon	*Ancistroides nigrita* (Latreille)
17.	Restricted Demon	*Notocrypta curvifascia* (C. and R. Felder)
18.	Grass Demon	*Udaspes folus* (Cramer)
19.	Bush Hopper	*Ampittia dioscorides* (Fabricius)
20.	Indian Skipper	*Spialia galba* (Fabricius)
21.	Pygmy Scrub Hopper	*Aeromachus pygmaeus* (Fabricius)
22.	Indian Ace	*Halpe homolea* (Hewitson)
23.	Moore's Ace	*Halpe porus* (Mabille)
24.	Branded Orange Awlet	*Bibasis oedipodea* (Swainson)
25.	Chestnut Angle	*Odontoptilum angulata* (C. and R. Felder)
26.	Indian Awlking	*Choaspes benjaminii* (Guerin–Meneville)
27.	Indian Palm Bob	*Suastus gremius* (Fabricius)
28.	Blank Swift	*Caltoris kumara* (Moore)

Abundance and Ranking of Species

Field records of number of specimens encountered during sampling at each hotspot were considered to represent abundance of respective species at that location. 174 species of butterflies showed their distribution into 91 ranks in order of their abundance. Common emigrants, common grass yellow and mottled emigrants occupied ranks 1, 2 and 3, respectively, and each of these species showed abundance of > 4.5 per cent and between these shared 15.56 per cent or 1952 individuals of the total abundance of 12546 butterflies encountered in this study. Next 23 butterfly species are ranked between 4 and 23 and showed specific abundance of > 1 per cent to about 3.2 per cent and between these shared about 53 per cent or 6663 individuals of the total abundance. Another 70 butterfly species are ranked between 24 and 60 and these showed species abundance of <1 per cent to (about 3627 individuals). Rest of 77 species are ranked between 61 and 91 (30 no. Of ranks) and these showed specific abundance of <0.1 per cent to 0.01 per cent and comprised of 305 individuals. Of these, 5 species occupied 88th rank, 6 species occupied 89th rank, 14 species occupied 90th rank and another 27 species occupied rank 91th. These species showed distribution of 0.03 per cent (= 4 individuals) in 88th rank, 0.02 per cent (= 3 individuals) in 89th rank, 0.016 per cent (=2 individual only) in 90th rank and 0.01 per cent (=14 individuals only) respectively. Ranking of 174 species according to their abundance clearly showed that distribution of these insect in the state of Tripura is very unequal.

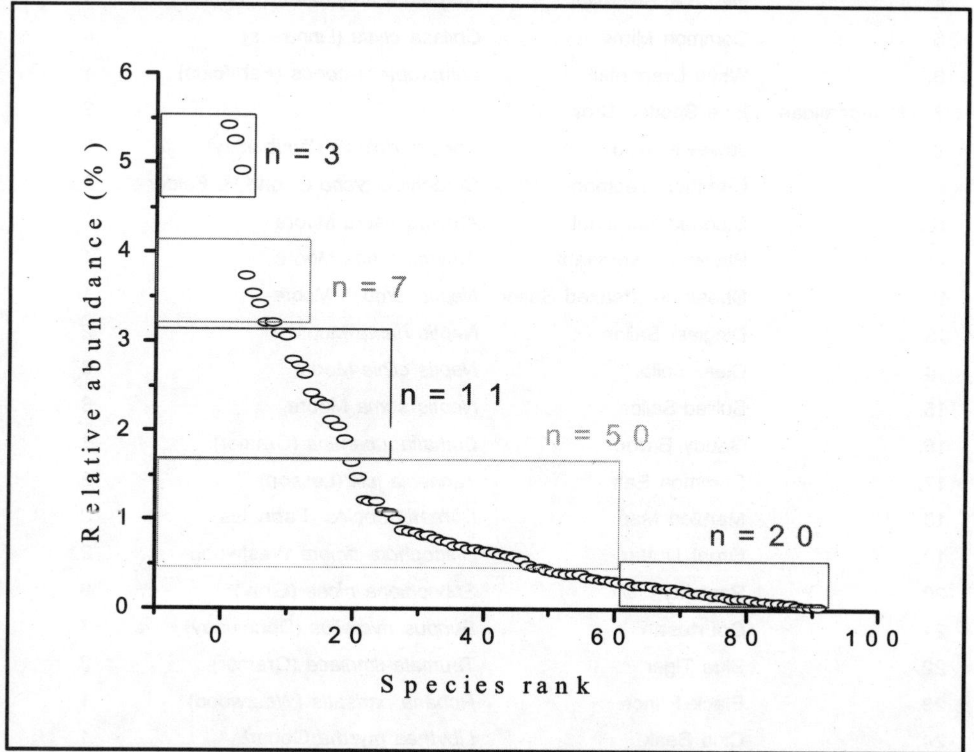

Figure 33.1: Rank-abundance of butterfly species recorded in the state of Tripura.

The result implies that some species are widely distributed and occurred commonly in most of the locations explored in the study and a majority of the other species showed poor distribution and low abundance.

List of Rare Species of Butterflies from Tripura

Rare species are those for which distribution of individuals in natural populations are scarce in a given area or in its distribution range. Rare species are not necessarily endangered species as many of such species are considered to show tolerance to narrow range of environmental variations. However, several of the rare species are considered to be endangered because these species face ecological disturbance that inhibit their natural rate of growth and reproduction. In this study 70 butterfly species were considered to be rare species due to their low (<10 individuals) encounter rate (Table 33.10).

Table 33.10: List of rare species of butterfly recorded in Tripura.

Sl.No.	*Family*	*Common Name*	*Scientific Name*	*No. of Specimens Encountered*
1.	Papilionidae	Common Batwing	*Atrophaneura varuna* (White)	1
2.		Red Helen	*Papilio helenus* Linnaeus	4
3.		Common Blue Bottle	*Graphium sarpedon* (Linnaeus)	9
4.		Five Bar Swordtail	*Graphium antiphates* (Cramer)	4
5.		Common Mime	*Chilasa clytia* (Linnaeus)	4
6.		White Dragontail	*Lamproptera curius* (Fabricius)	1
7.	Nymphalidae	Blue Spotted Crow		2
8.		Jewell Nawab	*Polyura delphis* (Doubleday)	2
9.		Common Yeoman	*Cirrochroa tyche* C. and R. Felder	3
10.		Studded Sergeant	*Athyma asura* Moore	2
11.		Blackvein Sergeant	*Athyma ranga* Moore	1
12.		Chestnut-Streaked Sailor	*Neptis jumbah* Moore	1
13.		Dingiest Sailor	*Neptis harita* Moore	2
14.		Clear Sailor	*Neptis clinia* Moore	8
15.		Sullied Sailor	*Neptis soma* Moore	8
16.		Gaudy Baron	*Euthalia lubentina* (Cramer)	1
17.		Common Earl	*Tanaecia julii* (Lesson)	8
18.		Marbled Map	*Cyrestis cocles* (Fabricius)	2
19.		Great Duffer	*Discophora timora* Westwood	2
20.		Popinjay	*Stibochiona nicea* (Gray)	9
21.		Courtesan	*Euripus nyctelius* (Doubleday)	7
22.		Blue Tiger	*Tirumala limniace* (Cramer)	2
23.		Black Prince	*Rohana parisatis* (Westwood)	1
24.		Club Beak	*Libythea myrrha* Godart	1

Contd..

Table 33.10–*Contd...*

Sl.No.	Family	Common Name	Scientific Name	No. of Specimens Encountered
25.		Red Spot Duke	*Dophla evelina* (Stoll)	2
26.		Tawny Rajah	*Charaxes bernardus* (Fabricius)	1
27.		Dark Archduke	*Lexias dirtea khasiana* (Swinhoe)	2
28.	Satyridae	Blue-Striped Palmfly	*Elymnias patna* (Westwood)	4
29.		Wood-Mason's Bushbrown	*Mycalesis suaveolens* Wood-Mason and de Nicéville	1
30.		Long Brand Bush Brown	*Mycalesis visala* Moore	1
31.		Angled Red Forester	*Lethe chandica* (Moore)	6
32.		Common Red Forester	*Lethe mekara* (Moore)	1
33.		Crimson Tip	*Colotis danae* (Fabricius)	1
34.		Green-Veined White	*Pieris napi* (Linnaeus)	2
35.		Common Jezebel		3
36.		Common Gull	*Cepora nerissa* (*Fabricius*)	3
37.		Lesser Gull	*Cepora nadina* (Lucas)	1
38.	Lycinidae	Chocolate Royal	*Remelana jangala* (Horsfield)	9
39.		Angled Pierrot	*Caleta caleta* Hewitson	9
40.		Zebra Blue	*Leptotes plinius* Fabricius	2
41.		Indian Sunbeam	*Curetis thetis* (Drury)	1
42.		Plains Cupid	*Chilades pandava* (Horsfield)	1
43.		Transparent 6-Lineblue	*Nacaduba kurava* (Moore)	6
44.		Common Imperial	*Cheritra freja* (Fabricius)	2
45.		Fluffy Tit	*Zeltus amasa* (Hewitson)	1
46.		Silverstreak Blue	*Iraota timoleon* (Stoll)	1
47.		Tufted Whit Royal	*Pratapa deva* (Moore)	4
48.		Broadtail Royal	*Creon cleobis* (Godart)	1
49.		Quaker	*Neopithecops zalmora* (Butler)	3
50.		Indian Oakblue	*Arhopala atrax* (Hewitson)	1
51.		Common Acacia Blue	*Surendra quercetorum* (Moore)	6
52.		Pale Grass Blue	*Pseudozizeeria maha* (Kollar)	5
53.		Common Tinsel	*Catapaecilma elegans* (Druce)	9
54.		Large Oakblue	*Arhopala amantes* (Hewitson)	1
55.	Hesperiidae	Suffused Snow Flat	*Tagiades gana* (Moore)	2
56.		Fulvous Pied Flat	*Pseudocoladenia dan* (Fab.)	6
57.		Spotted Small Flat	*Sarangesa purendra* Moore	7
58.		Bright Red Velvet Bob	*Koruthaialos sindu* (C. and R. Felder)	2
59.		Pygmy Scrub Hopper	*Aeromachus pygmaeus* (Fab.)	1

Contd...

Table 33.10–*Contd...*

Sl.No.	*Family*	*Common Name*	*Scientific Name*	*No. of Specimens Encountered*
60.		Moore's Ace	*Halpe porus* (Mabille)	1
61.		Branded Orange Awlet	*Bibasis oedipodea* (Swainson)	1
62.		Chestnut Angle	*Odontoptilum angulata* (C. and R. Felder)	4
63.		Indian Ace	*Halpe homolea* (hewitson)	3
64.		Brown Awl	*Badamia exclamationis* (fab.)	5
65.		Grass demon	*Udaspes folus* (Cramer)	1
66.		Straight Swift	*Parnara guttatus* (Bremer and Grey)	1
67.		Indian Awlking	*Choaspes benjaminii* (Guerin–Meneville)	1
68.		Indian Palm Bob	*Suastus gremius* (Fabricius)	3
69.		Dark Valvet Bob	*Koruthaialos butleri* de Niceville	1

Discussion

This is the first study on the distribution and abundance of butterflies in the four wildlife sanctuaries and 23 quashi-evergreen and moist deciduous forests of Tripura. Records of 59 species in Trishna sanctuary, the biggest of the four sanctuaries in Tripura, compares favourably with 71 species found in Aralam sanctuary in Kerala, south India which share similarity in habitats (Sreekumar and Balakrishnan 2001). All the species recorded in Aralam sanctuary (55 sq. km) are 'generalist' and none of these is considered to be threatened or of rare occurrence. In contrast, eight species *viz. Lethe europa* (F.), *Cepora nerissa* (F.), *Castalius rosimon* (F.), *Narathura selta* (de Niceville), *Pantoporia hordonia* (stoll), *Megisba malaya* (Horsfield), *Arhopala fulla* (Hewitson) and *Baoris farri* (Moore) recorded in Trishna sanctuary are enlisted in the Indian Forest Act 1972 under Schedule I and II, and another species, *Troides helena* L., is listed in Appendix -II of CITES and these are protected by law.

Overall species abundance richness revealed that Nymphalidae was the most speciose and individualized family, and Papilionidae was the least speciose family in the study area. In the context of tropical environment, similar patterns of species abundance and richness were reported from western ghat and western coast of India (Kunte 1997; Eswaran and Pramod 2005; Padhye *et al.*, 2006; Krishna Kumar *et al.*, 2007; Raut and Pendharkar 2010), from the Silent Valley National Park, Kerala, south India (Mathew and Rahamathulla 1993), and in Parambikulam sanctuary in south India (Sudheendrakumar *et al.*, 2000). The dominance of Nymphalidae species may be attributed to their being polyphagous which help these butterflies to live in a variety of habitats and also because many species of this family are active fliers that help them to forage larger areas. However, in the present study, members of Papilionidae and Pieridae were poorly represented in all the four sanctuaries as well as in natural forests in comparison to studies made in other sanctuaries of India which reported members of Hesperiidae to be least represented. One of the possible reasons for this difference could be the difficulties in the observation of Hesperiidae butterflies due to their dull colour, and ability to fly rapidly following any disturbance. Several studies have reported the influence of habitat disturbance on abundance of butterflies (Hill *et al.*, 1995). Habitat selection in butterflies is directly related to availability of preferred food plants for larval and adults (Grossmueller and Lederhouse 1987; Thomas 1995). In the present study, maximum number of species and individuals were observed in mature mixed moist deciduous forest of Debbari. Presence of plenty of woods and dense understorey delimited by profuse growth of diverse herbs and shrubs in this habitat provide rich nectar source to adult butterflies and food to larvae. Results showed that nearly 30 per cent of the butterfly species recorded in all the sampling sites were the same despite differences in habitat features. High similarity of butterfly fauna between different habitats an indicator of low beta diversity over short distances between the four transects in this forest and bear resemblance to the study conducted in tropical forests where the herbivorous insects showed low beta diversity even over large areas (Novotny *et al.*, 2007). It is generally accepted that rich floral diversity in tropical forests promotes herbivores many of which are generalists (Price 1997).

Many studies have reported the relationship between habitat heterogeneity and species diversity (Atauri and Lucio 2001; Tews *et al.*, 2004). In most habitats, plant communities determine the physical structure of the environment, and therefore, have considerable influence on the distributions and interactions of animal species (Lawton 1983; McCoy and Bell 1991). Rosenzweig (1981) stated that diversity is enhanced by the presence of specialists that exhibit distinct habitat preferences. High butterfly diversity in Debabari, Mahalaxmi, and Ballianchip forests and Trishna sanctuary could be attributed to the edge effects of divergent landscapes, and additional feature of occurrence of large number of pits caused by uprooting of plants by wild boars and other wild animals. Pits provide sites for deposition of mineral rich feces and urines perfect for mudpuddler butterflies [*Papilio polytes* L., *Papilio demoleus* L., *Castalius rosimon* Fruhstofer; *Eurema hecabe* (Moore)]. This finding is in agreement with the findings of Ramos (1996).

Acknowledgements

Authors are thankful to the Japan International Co-operation Agency (JICA) and Department of Forest, Government of Tripura for partial financial support. Thank is also due to Koushik Majumder, Department of Botany, Tripura University for helps in the identification of some of the host plants recorded in the study area. Special thank are due to Joydeb Majumder, Rahul Lodh and Dipankar Kishore Sinha for manifold helps in the study.

References

Agarwala, B.K., Roy Choudhury, S. and Roy Chaudhury, P., 2010. Species richness and diversity of butterflies in urban and natural locations of north-east India. *Entomon.*, 35: 87–91.

Alfred,J.R.B., Das, A.K. and Sanyal, A.K., 2002. Ecosystem of India. *ENVIS–Zoological Survey of India*, Kolkata, pp. 410.

Atauri, J.A. and Lucio, J.V. de., 2001. The role of landscape structure in species richness distribution of birds, amphibians, reptiles and lepidopterans in Mediterranean landscapes. *Landscape Ecology*, 16: 147–159.

Eswaran, R. and Pramod, P., 2005. Structure of butterfly community of Anaikatty hills, Western Ghats. *Zoo's Print Journal*, 20(8): 1939–1942.

Grossmueller, D.W. and Lederhouse, R.C., 1987. The role of nectar source distribution in habitat use and oviposition by the tiger swallowtail butterfly. *Journal of Lepidopteran Society*, 41(3): 159–165.

Haribal, M., 1992. *The Butterflies of Sikkim Himalaya and their Natural History*. Sikkim Nature Conservation Foundation.

Hill, J.K., Hamer, K.C., Lace, L.A. and Banham, W.M.T., 1995. Effects of selective logging on tropical forest butterflies on Buru, Indonesia. *Journal of Applied Ecology*, 32: 754–760.

Incredible North East India, 2011. Available online, http://www.incrediblenorth-eastindia.com/tripura/trishna–wildlife–sanctuary.html.

Kehimkar, I., 2008. *The Book of Indian Butterflies*. Oxford University Press, USA.

Krishnakumar, N., Kumaraguru, A., Thiyagesan, K. and Asokan, S., 2008. Diversity of papilionid butterflies in the Indira Gandhi Wildlife Sanctuary, Western Ghats, Southern India. *Tiger Paper*, 35: 1–8

Kunte, K., 1997. Seasonal pattern in butterfly abundance and species diversity in four tropical habitats in northern Western Ghats. *Journal of Bioscience*, 22(5): 593–603.

Lawton, J.H., 1983. Plant architecture and the diversity of phytophagous insects. *Annual Review of Entomology*, 28: 23–39

Majumder, J., Lodh, R. and Agarwala, B.K., 2011. Butterfly fauna of Rowa wildlife sanctuary, Tripura, North-East India. *Proceedings of National Conference on Water, Energy and Biodiversity with Special Reference to North-east Region*. Excel India Publishers, New Delhi, 1: 266–271.

Mandal, D.K., Ghosh, S.K. and Majumder, M., 2002. State Fauna Series 7: Fauna of Tripura, (Insecta: Lepidoptera). Part–3. *Zoological Survey of India*, Kolkatta, pp. 283–334.

McCoy, E.D. and Bell, S.S., 1991. Habitat structure: The evolution and diversification of a complex topic. In: *Habitat Structure: The Physical Arrangement of Objects in Space*, (Eds.) S.S. Bell, E.D. McCoy and H.R. Mushinsky. Chapman and Hall, London, pp. 3–27.

Myers, N., Mittermeier, R.A., Mittermeier, C.G., da Fonseca Gustavo, A.B. and Kent, J., 2000. Biodiversity hotspots for conservation priorities. *Nature*, 403: 853–858.

Naeem, S., Thompson, L.J., Lawler, S.P., Lawron, J.H. and Woodfin, R.M., 1994. Decaling biodiversity can alter the performance of ecosystem. *Nature*, 368: 734–736.

Novotny, V., Miller, S.E., Hulcr, J., Drew, V., Basset, Y., Janda, M., Setliff, G.P., Darrow, K., Stewart, A.J.A., Auga, J., Isua, B., Molem, K., Manumbor, M., Tamtiai, E., Mogia,

M. and Weiblen, G.D., 2007. Low beta diversity of herbivorous insects in tropical forests. *Nature*, 448(9): 692–695.

Padhye, A.D., Dahanukar, N., Paingankar, M., Deshpande, M. and Deshpande, D., 2006. Season and Landscape wise distribution of butterflies in Tamhini, Northern, Western Ghats, India. *Zoo's Print Journal*, 21(3): 2175–2181.

Pollard, E. 1977. A method for assessing changes in the abundance of butterflies. *Biological Conservation*, 12: 115–134.

Price, P.W., 1997. *Insect Ecology*, 3rd edn. Wiley, New York.

Ramos, F.A., 1996. Nymphalid butterfly communities in an Amazonian forest fragment. *Journal of Research on the Lepidoptera*, 35: 29–41.

Raut, N.B. and Pendharkar, A., 2010. Butterfly (Rhopalocera) fauna of Maharashtra Nature Park, Mumbai, Maharashtra, India. *Journal of Species Lists and Distribution*, 6(1): 22–25.

Rosenzweig, M.L., 1981. A theory of habitat selection. *Ecology*, 62: 327–35.

Spitzer, K., Novotny, V., Tonner, M. and Leps, J., 1993. Habitat preferences, distribution and seasonality of the butterflies (Lepidoptera, Papilionidae) in a montane tropical rain forest, Vietnam. *Journal of Biogeography*, 20: 109–121.

Sudheendrakumar, V.V., Binoy, C.F., Suresh, P.V. and Mathew, G., 2000. Habitat association of butterflies in the Parambikulam Wildlife Sanctuary, Kerala, India. *Journal of Bombay Natural History Society*, 97: 193–201.

Tews, T., Brose, U., Grimm, V., Tielborger, K., Wichmann, M.C., Schwager, M. and Jeltsch, F., 2004. Animal species diversity driven by habitat heterogeneity/ diversity: the importance of keystone structures. *Journal of Biogeography*, 31: 79–92.

Tilman, D., Wedin, D. and Knops, J., 1996. Productivity and sustainability influenced by biodiversity in grassland ecosystem. *Nature*, 379: 718–720.

Thomas, J.A., 1995. The ecology and conservation of Maculinea arion and other European species-of large blue butterfly. In: *Ecology and Conservation of Butterflies*, (Ed.) A.S. Pullin. Chapman and Hall, London, pp. 180–210.

Wynter-Blyth, M.A. 1957. *Butterflies of the Indian Region*. Bombay Natural History Society, Bombay.

Environment, Biodiversity and Traditional System, 2014 *Pages* 351–363
ISBN 978-81-7622-307-2
Editors: D.R. Khanna, G.S. Solanki and S.K. Pathak
Published by: BIOTECH BOOKS, NEW DELHI

Chapter 34

Spawning Density and Intensity of Horseshoe Crab *Tachypleus gigas* (Müller) along Balasore Coast of Odisha

A.C. Sahu[1], L. Dey[2] and U.K. Sahoo[3]*

[1]Department of Zoology, Belavoomi Mahavidyalaya, Avana, Balasore, Odisha
[2]Department of Zoology, K.K. Women's College, Balasore, Odisha
[3]Department of Forestry, Mizoram University, Aizawl – 796 004, Mizoram

ABSTRACT

The spawning density and intensity of Asian horseshoe crab *Tachypleus gigas* (Müller) along Balasore coast of Odisha was studied during 2008-2011. The population of *T. gigas* was found throughout the year, although the breeding season was restricted from March to September. The spawning density of *T. gigas* varied significantly ($P<0.05$) between high tide of full moon and low tide of new moon. The year-wise variation of the males and females arriving different shores of the Balasore coast of Bay of Bengal reveal somewhat of a decreasing trend when compared to the base year 2008. The mean spawning density of each of the 4 transects at all sites was 1.44 crabs m^{-2} (ranges from 1.22 at Dhamara beach during new moon to 1.66 at Dhamara beach during full moon). The spawning indices of *T. gigas* in the

* *Corresponding Author:* E-mail: anatasahu_2007@rediffmail.com

Balasore coast of Bay of Bengal ranged from 0.12 to 0.64 and showed a decreasing trend over the years when moved from 2008 to 2012 in all the beaches except for the year 2011 at Dhamara. No crab below 12 cm carapace width was noticed during the survey period.

Keywords: ***Tachypleus gigas, Balasore coast, Spawning density, Spawning intensity, Sex ratio.***

Introduction

The Asian horseshoe crab, *Tachypleus gigas* (Müller), also known as Asian king crabs, belongs to the family Limulidae is the only marine primitive invertebrates which are widely distributed on earth. There are only four species of horseshoe crabs in existence in the world today. These are *Limulus polyphemus, Tachypleus gigas, Tachypleus tridentatus, and Carcinoscorpius rotundicauda*. These creatures are sometimes called "living fossils" because they have changed little from their fossilized relatives; the earliest species identified is approximately 450 million years old. They evolved from trilobites about 550 million years ago, and keep similar shapes for more than 200 million years. They have been classified in Phylum Arthropoda; subclass Xiphosura and phylogenetically more related to Arachnida than to Crustaceans (Osaki and Kawabata 2003, Veseley and Veseley, 1999).

The Bay of Bengal, Orissa coast has a commercial fishery which uses king crabs for fertilizer and animal feed (Fowler 1908, Shuster 1960) but it ceased in the recent years. However, at present there are large fishery activities for various marine diets, at the Orissa coast. Human activities nevertheless threaten the crabs' survival in various parts of the world. Beach development and shoreline modifications are some of the important reasons which prevent the crabs from reaching spawning areas. Oil spills from barges or tankers have been reported to pose threats to breeding and young ones. Although many efforts are seen in various countries to increase the population of the species by implementing better management practices, the population is showing immediate increase. The crabs do not breed until nine to ten years and this is another reason for the reduced population size of the species. There is little documentation on king crab population size in other estuaries of the world based on tagging. The study conducted at Cape Cod Bay estimated the total population of this species to be from 50 to 100 thousand adults (Shuster 1950). Size and distribution have been studied for Plum Island Sound area of Massachusetts. Besides, some studies are devoted on longevity (Ropes 1961). In other studies, patterns of movement have also been determined in the vicinity of breeding beaches in the Gulf of Mexico (Rudloe 1980). Combined with tagging programs, such shoreline surveys could contribute to a better understanding of the ranges, population size, and population dynamics for any Limulidae population. Further, life table statistics are not available, especially for the several age classes of young *Tachypleus.* Should the need for management of any of the populations of Limulidae arise, statistics on the fate of the young will be required in order to predict the effect of fishing and other pressures on populations. While there have been numerous studies concerning the general biology and life history of *T.gigas*, information concerning its status and population biology is not sufficiently known. As the demand of the species intensifies from different user groups, information about its status becomes increasing necessary. There are concerns that the population is declining in abundance and for any future management, an accurate data on the population biology of the species is very essential. The present paper deals with the spawning migration of *Tachypleus gigas* and spawning intensity at Balasore coast of Odisha.

Materials and Methods

The state of Odisha has a coastline of about 480 sq km and is one of the most dynamic coastal environments in India due to its location, and physical factors especially its network of large and powerful rivers with their delta and estuarine systems, each with a variety of ecological niches and habitats. The coastline traverses six coastal districts of Odisha *viz.*, Balasore (80 km), Bhadrak (50 km), Kendrapada (68 km), Jagatsinghpur (67 km), Puri (155 km) and Ganjam (60 km). The present study includes first two coastal districts of Odisha covering a total of 130 sq km (Figure 34.1). Balasore district is located in the northeast of the state of Odisha and lies between 21°3' to 21°59' north latitude and 86°20' to 87°29' east longitude. The average altitude of the district is 19.08 m and has a total area of 3634 km^2. The coastal belt is about 81 km wide and shaped like a strip. In this region, sand dunes are noticed along the coast with some ridges. Two important rivers namely Budhabalanga and Subarnarekha pass through this district from west to east before surging into the Bay of Bengal. Chandipur (Balaramgarhi), about 10 km from Balasore was selected for the survey under Balasore coast of Bay of Bengal. Avana Estuary, located 3km from Avana village and/or Belavoomi Mahavidyalaya is relatively less exploited for fishery activities. The Panchubisa river drains into the Bay of Bengal forming an estuary at low tide. The coast has intertidal zones of about 6 km into the sea that provides wide opportunity for intertidal organisms like king crabs. Dhamra sea beach is located at 20°50' N and 86°58' E in Bhadrak district near the mouth of Dhamra river and surrounding areas. The area is a known business centre with a fishing jetty and trawler base and is about 10 km away from Dhamra town. The Dhamra river itself is formed by the confluence of the Brahmani and Baitarani rivers. Dhamara discharges to sea through two distinct river channels, a north and southern channel, which are further separated by an island in the sea named Udabali or Kanika Sands. The site is a unique habitat in that it has long stretches of inter-tidal mudflats the site up till river mouth. The inter-tidal zone in some places is as wide as two kilometers and serves as an important breeding ground for king crabs. The area is also strategic as it is about 13 km from the nestling beaches of the Gahirmatha Marine Sanctuary and about 5 km from the northern boundary of the Bhitarkanika Sanctuary.

To estimate the density of spawning population of *T.gigas*, surveys were made in the above mentioned three estuaries during 2008-2011 using quadrate/transect sampling. To standardize the minimum area to be surveyed, we examined several protocols for the survey adopted by various workers. Initially the quadrat size was restricted to 1mx1m and the number of quadrate in each site was 25 (twenty five), randomly laid to assess the spawning population. However, the data collected from these quadrates showed a lot of variations and we missed many a times to record the availability of *T.gigas* from the small size of the quadrate. Since the spawning density

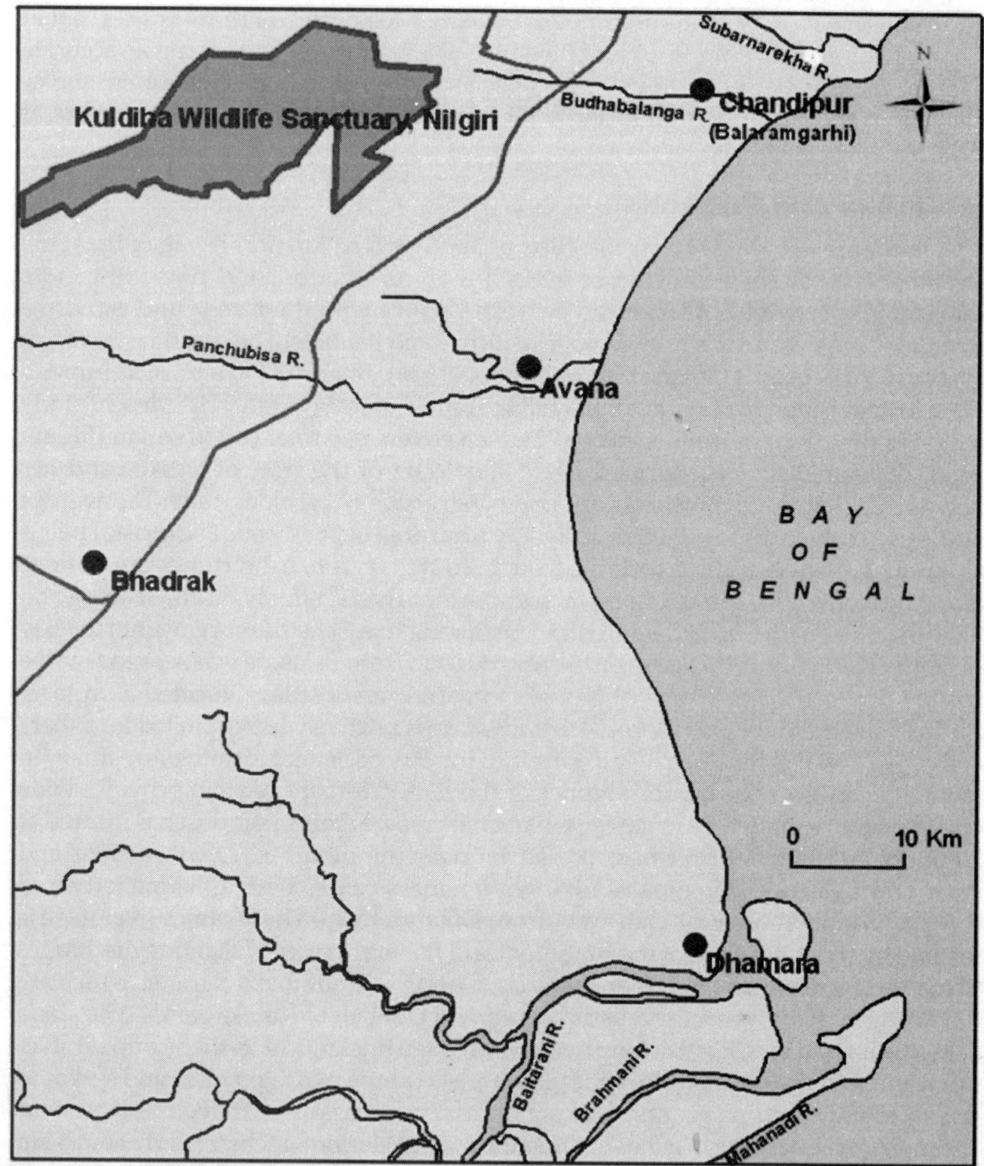

Figure 34.1: Map showing the study sites.

was observed to be relatively lower compared to many other studies reported, instead of using quadrates, we laid out four transects each measuring 25 m2 (1m width x 25m long) in each site to increase the area covered by the survey. Transects were laid adjacent to water edge and extended to 1 m into the water and through 100 m length along the beach. A gap of 10 m was maintained between two transects. Spawning survey was carried out regularly twice a month, coinciding with full and new moon phases from January 2008 to December 2011. The samples were collected one hour prior to and one hour following the predicted high tides except under extreme weather conditions. Since the breeding activity was more pronounced in the daylight high tide than the night tides for *T. gigas* (Chatterjee *et al.*, 1992), in our present study, we restricted all our observations to during day light hours at the highest high tides of full or new moon. All animals within transects were counted and recorded the number of single males, single females, pairs (female and amplexed male), females with amplexed males and satellite males.

The spawning index defined as the average number of spawning female crabs per m2 was calculated. The solitary females were not included for calculation of spawning index. A yearly average of spawning indices was calculated by averaging the spawning female density for each site as suggested by Mattei *et al.* (2010). The levels of spawning activity were categorized into four groups like no spawning activity (0 crabs), low spawning activity (less than 5 crabs per m), moderate (5-10 crabs per m) and high spawning activity (more than 10 crabs per m).

The population data of *T.gigas* between the sites were subjected to analysis of variance (ANOVA, fixed effects method) to know the variations between the sites and seasons. An attempt was also made to draw frequency distribution curve of the spawning individuals arriving at different sites so as to depict the size of animals that occurs along the Balasore coast of Bay of Bengal.

Results and Discussion

Spawning Density

The population of *T. gigas* has been found throughout the year, though the breeding season of this species is found to be restricted from March to September. The spawning density of *T. gigas* varied between high tide of full moon and low tide of new moon and between different months and years (Figures 34.2–34.4). At Avana beach, during 2008, a total of 28 males/transect (=25 m2) and 34 females were observed in new moon while a total of 22 males and 32 females area were observed during full moon. In 2009, 14 males and 21 females during new moon and 14 males and 27 females were reported and in 2010, the corresponding figures were 14 males and 23 females during new moon and 15 males and 23 females during full moon. In 2011, a total of 15 males and 18 females were observed during new moon and 12 males and 15 females during full moon (Figure 34.2). June was found to the peak spawning month at Avana beach showing the arrival of 9 crabs (4 males and 5 females) per 25m2 during new moon and 10 crabs (4 males and 6 females) per 25m^2 during full moon in the year 2008. At Chandipur (Balaramgarhi), the population density of the spawning crabs showed variations during the survey period (Figure 34.3). In 2008, a total of 15 males and 26 females were observed in new moon while a total of 12 males and 24 females were observed during full moon. In 2009, 13 males and 17 females during new moon and 14 males and 16 females were reported and in 2010, the corresponding figures were 13 males and 17 females during new moon and 11 males and 21 females during full moon. In 2011, a total of 11 males and 15 females were observed during new moon and 11 males and 17 females during full moon. April to July was found to the peak spawning period at Balaramgarhi showing the arrival of

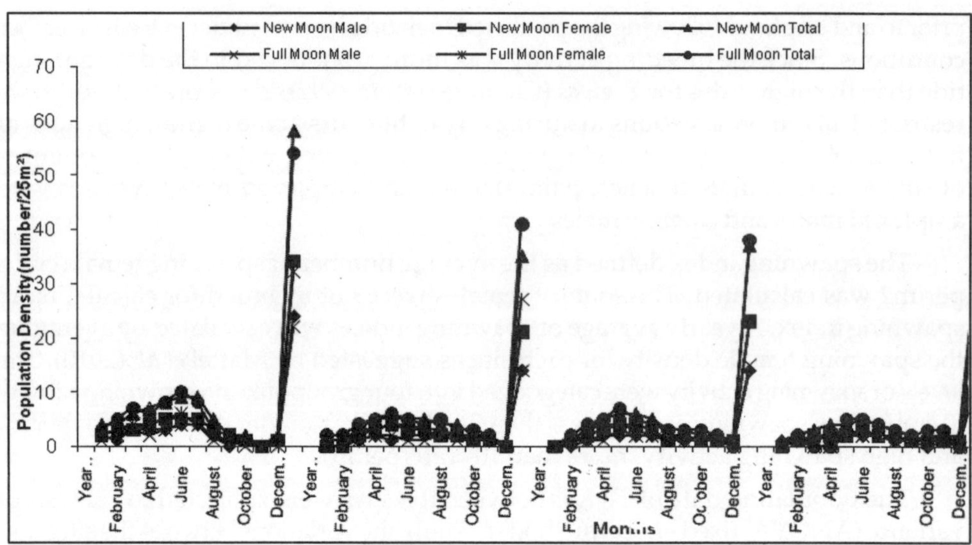

Figure 34.2: Occurrence of *T. gigas* at the Avana estuary during different times of the study period.

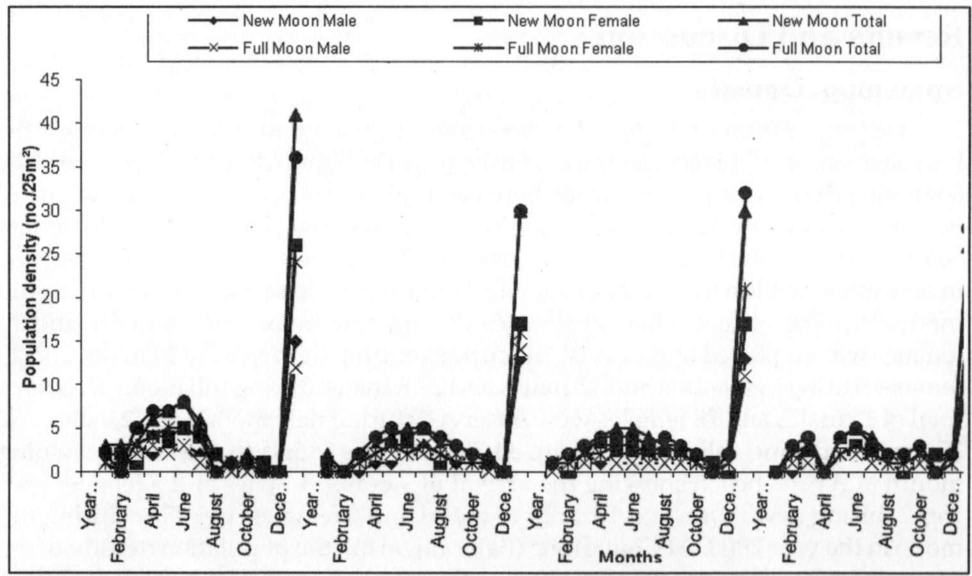

Figure 34.3: Occurrence of *T. gigas* at the Chandipur seashore during different times of the study period.

more number of crabs during this period than the other months. At Dhamara beach, during 2008, a total of 14 males and 20 females were observed in new moon while a total of 14 males and 21 females were observed during full moon. In 2009, 11 males and 16 females during new moon and 13 males and 15 females were reported and in

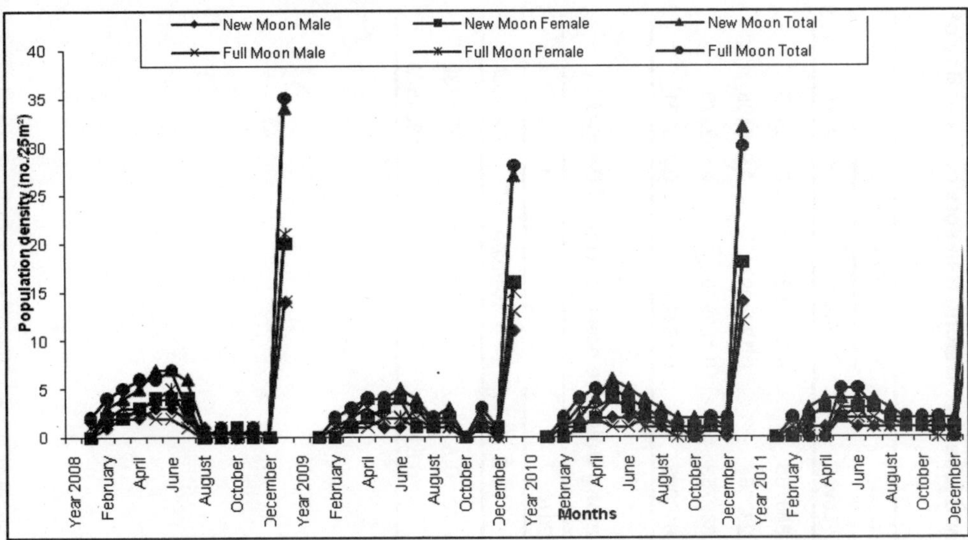

Figure 34.4: Occurrence of *T. gigas* at the Dhamara estuary during different times of the study period.

2010, the corresponding figures were 14 males and 18 females during new moon and 12 males and 18 females during full moon. In 2011, a total of 11 males and 19 females were observed during new moon and 12 males and 11 females during full moon. Some variations were noticed with respect to the peak spawning activities of the crab when compared with Avana and Chandipur, however, the lowest number of female to male (1: 3) was reported in the month of November 2011. Except for the year 2008, in other three years there were significant variations in the spawning crab populations that were observed between new moon and full moon in the study sites (Table 34.1). The year-wise variation of the males and females arriving different shores of the Balasore coast of Bay of Bengal (Table 34.2) reveal somewhat of a decreasing trend when compared to the base year 2008 which is applicable to both new moon and full moon period. Nevertheless high spawning activities in all the three beaches were found to be in April spreading over to September with peak during June.

In the present study, although the densities of the spawning crabs varied from transect to transect and over time, the mean spawning density of each of the 4 transects at all sites was 1.44 crabs m^{-2} (ranges from 1.22 at Dhamara beach during new moon to 1.66 at Dhamara beach during full moon). The transition in the monsoon seems to have no marked effect on population density. These are probably minimum figures, and more individuals could have been found had the search included thorough probing of the mud for buried crabs. We had to rely on the visual data as probing through mud was practically very difficult and a most laborious one. From a comparison of these figures using analysis of variance, the overall density in the 4 transects probably did not change much between January 2008 and December 2009 (Table 34.1), however, in subsequent two years some significant variations ($P<0.05$) was noticed between the spawning populations. The time from May to July was

Table 34.1: Analysis of variance on the spawning population of *T. gigas* due to sites and lunar cycle along Balasore coast of Bay of Bengal.

Lunar cycle	*Sex*	*Avana*		*Chandipur*		*Dhamara*	
		F-Value	*P-Level*	*F-Value*	*P-Level*	*F-Value*	*P-Level*
New Moon	Male	2.079492	0.116630	0.264264	0.850766	0.331658	0.802475
	Female	2.292648	0.91189	1.312423	0.282366	0.138439	0.936495
Few Moon	Male	1.591460	0.204982	0.218905	0.882758	0.109900	0.953870
	Female	2.010334	0.126337	0.612356	0.610605	0.899216	0.449221

Table 34.2: Analysis of variance on the spawning population of *T. gigas* due to years along Balasore coast of Bay of Bengal.

Lunar cycle	*2008*		*2009*		*2010*		*2011*	
	F-Value	*P-Level*	*F-Value*	*P-Level*	*F-Value*	*P-Level*	*F-Value*	*P-Level*
New Moon	1.55010	0.22728	0.400446	0.673233	0.312000	0.734117	0.492140	0.615732
Full Moon	1.03454	0.36662	1.395168	0.262026	0.489726	0.617177	0.181890	0.834523

found to the peak spawning months in the Balasore coast of Bay of Bengal. A total of 163 crabs were noticed during new moon and 160 crabs during the full moon at Avana beach during the study period, the corresponding figures for Chandipur was 127 and 126 respectively and for Dhamara, it was 122 and 116 respectively (Figures 34.2–34.4). No discernable variation was noticed with respect to the peak spawning activities of the crab between all the sites except that the lowest number of female to male (1: 3) was reported in the month of November 2011 at Dharmara beach (Figure 34.4). A decreasing trend on the number of males and females arriving different shores of the Balasore coast of Bay of Bengal could be related to various factors. The little variations that were observed on the spawning population between new moon and full moon could be related to height and energy of tides. During new moon, the shores experiences low energy tides which probably facilitated more spawning crabs to come ashore for breeding purpose than the full moon. This was true for all the beaches. The spawning population were in the order of Avana>Chandipur>Dhamara. This could be related to a combination of both sand size and fishing activities. Avana seashores were found comparatively less exploited for fishing purposes while Dhamara was the most affected due to some construction activities of the future Dhamara port. The variations in spawning populations of *T. gigas* reported during months were obviously related to the breeding season (April to June) when most animal prefer to breed during this time (Chatterji *et al.*, 1992, Chaterji and Sahj 1993). Lunar phase is reported to influence the marine organism significantly (Barnwell 1968, Saigusa and Hidaka 1978, Greenspan 1982). The abundance of horseshoe crab population in relation to full moon tides has also been reported in Apalche Bay, Florida (Rudloe 1980) and at Sea-horse key (Cohen and Brockmann 1983). Similarly, a definite correlation has been reported between abundance and mating activity (Howard *et al.*, 1984) and lunar phase. The reproductive behaviour of adult *L. Polyphemus* has been found to be greatly influenced by lunar, tidal and circadian rhythms with maximum activity during the spring tides, particularly at full moon (Rudloe, 1980).

Spawning Intensity

The spawning indices of *T. gigas* in the Balasore coast of Bay of Bengal ranged from 0.12 to 0.64 (Table 34.3). The spawning intensity showed a decreasing trend over the years when moved from 2008 to 2012 in all the beaches except for the year 2011 at Dhamara where maximum spawning intensity was observed. At Avana shore out of 648 *T. gigas*, 33 amplexed ones were available *i.e.*at the rate of 19.6 crabs a couple. At Chandipur (Balaramgarhi) breeding beach out of 442 crabs 18 coupled or amplexed were observed with 24.5 individual crabs a couple and in Dhamara shoreline, among the 414 horseshoe crabs 28 coupled are recorded i.e an amplexed is formed within 14.8 crabs. The study showed that the Chandipur beach which had exhibited lesser number of couples but showed maximum spawning index. Avana beach, on the other hand, had large number of crages and also a fairly higher number of amplexus resulting in moderate spwaing index. Dhamara experienced lowest number of crabs but amplexus were high resulting in least spawning index.

Table 34.3: Spawning index of *T. gigas* in Balasore coast of Bay of Bengal during different years.

Year	Site		
	Avana Estuary	*Chandipur*	*Dhamara Estuary*
2008	0.44±0.001	0.20±0.003	0.24±0.00
2009	0.28±0.13	0.16±0.012	0.12±0.00
2010	0.28±0.001	0.16±0.001	0.12±0.01
2011	0.24±0.031	0.12±0.00	0.64±0.022

±S.E.m, n=4

Table 34.4: Sex ratio of the visiting crabs for spawning at different beaches of Balasore coast of Bay of Bengal during NM (new moon) and FM (full moon) in different years.

Period	Avana				Chandipur				Dhamara			
	2008	2009	2010	2011	2008	2009	2010	2011	2008	2009	2010	2011
NM	0.70	0.66	0.60	0.83	0.57	0.76	0.76	0.73	0.70	0.68	0.77	0.57
FM	0.68	0.51	0.65	0.80	0.50	0.87	0.52	0.64	0.66	0.86	0.66	1.09
Yearly Average	0.696	0.583	0.63	0.818	0.54	0.818	0.631	0.687	0.682	0.774	0.722	0.766
4-year Average	0.621				0.653				0.731			

Sex Ratio and Frequency Distribution

The number of crabs observed during the lunar cycle and the sex ratio (Male: Females) during different years at different sites reveal maximum (0.731) at Dhamara and minimum (0.621) at Avana over a 4-years period (Table 34.4). There was no clear trend as to whether the more number of males/females were attracted with the high tides. However, in some months like December 2008 and October 2010 at Avana, August 2009 and September 2011 at Chandipur and July/September 2010 and November 2011 at Dhamara beach. However, these extremes may not have been affected by the low numbers collected during the visits and were not repeated for the same months in other years. The sex ratio of different months in all the three study sites showed a similar pattern except in July, September/November when the number of males outnumbered females which might suggest a period of high breeding activity as has been reported for *Limulus polyphemus* in temperate climate during breeding season (Smith *et al.*, 2002). *Tachypleus gigas* in the present study showed a sex ratio of 1: 1 in most of the year, similar findings were also observed by (Chaterji *et al.*, 1992) with an occasional increase in the number of males. The findings of the present study also concur with the population structure of reported elsewhere (Cartwright-Taylor and Hsu, 2012).

The frequency distribution of the spawning horseshoe crab based on the carapace width (Dorsal Prosomal Breadth, DPB) at three different sites during the survey period is given in Figures 34.5–34.7. No crab below 12 cm carapace width was noticed during the survey period. A very small proportion of the crab (0.77-1.45 per cent) was noticed at 12 cm DPB in all the sites. Similarly, no crab was noticed having DPB of 13-14 cm. A majority of the crabs (17.69-20.0 per cent) of the crab population was found in 22-23 cm DPB class. This was true in all the sites (Figures 34.5–34.7). A very proportion of the crabs were also reported beyond 24 cm DPB. There was no significant

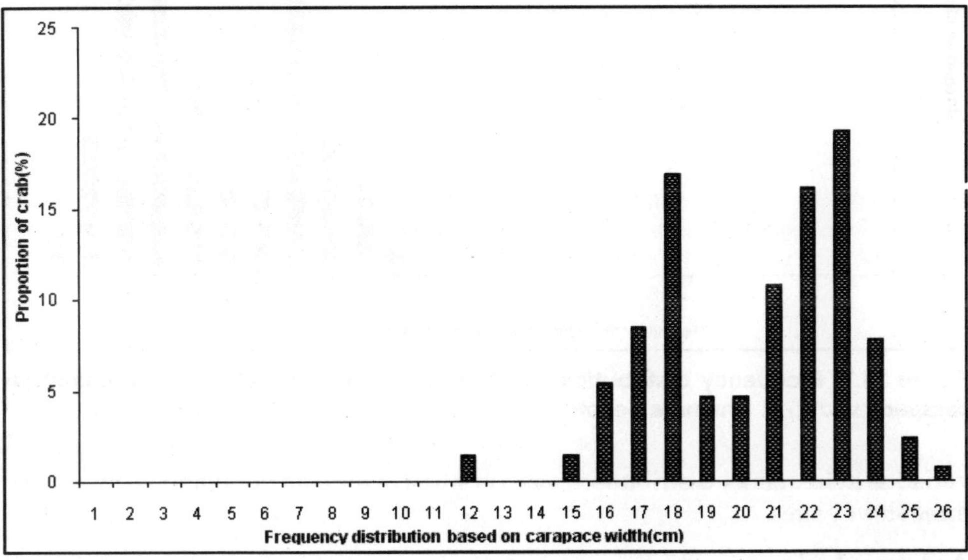

Figure 34.5: Frequency distribution of spawning population of *T. gigas* (based on carapace width) at Avana beach.

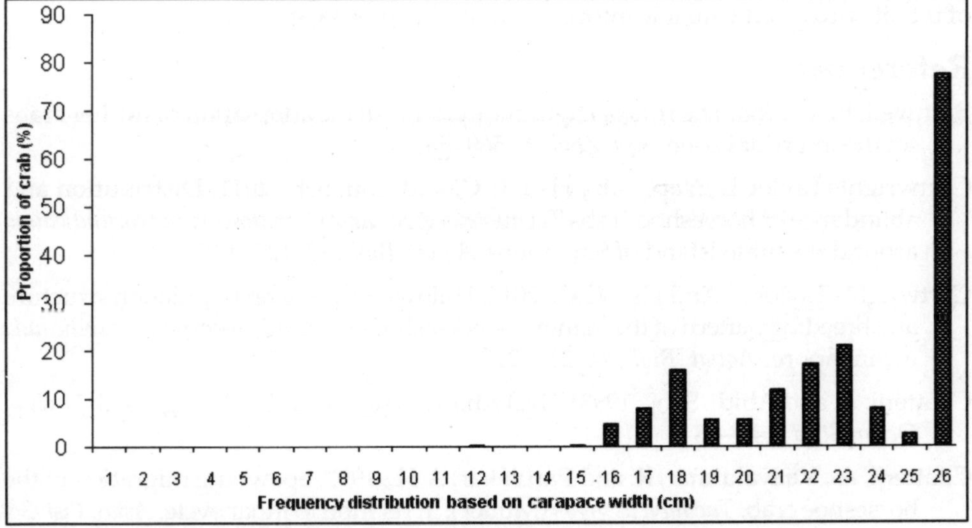

Figure 34.6: Frequency distribution of spawning population of *T. gigas* (based on carapace width) at Chandipur beach.

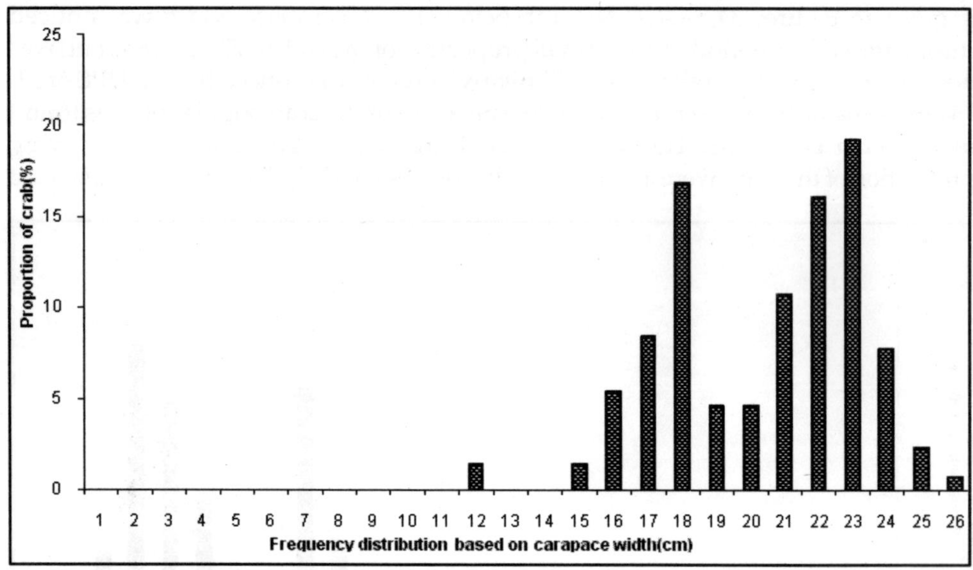

Figure 34.7: Frequency distribution of spawning population of *T. gigas* (based on carapace width) at Dhamara beach.

variation in frequency distribution/size variation of the crab population between the three survey sites.

Acknowledgements

The first author (ACS) wishes to thank the Principal, Belavoomi Mahavidyalaya, Avana, Balasore for providing necessary facilities to carry out this work. Fishermen of the site provided valuable information used in this paper.

References

Barnwell, F.H., 1968. The role of rhythmic systems in the adaptation of fiddler crabs tot the intertidal zone. *Am. Zool.*, 8: 569–583.

Cartwright–Taylor L., Yap, V.B., Hsu, C.C. and Lou, S.T., 2011. Distribution and abundance of horseshoe crabs *Tachypleus gigas* and *Carcinoscorpius rotundicauda* around the main island of Singapore. *Aquat. Biol.*, 13: 127–136.

Cartwright–Taylor L. And Hsu, C.C., 2012. Follow–up study on population structure and breeding pattern of the mangrove horseshoe crab *Carcinoscorpius rotundicauda* in Singapore. *Aquat. Biol.*, 14: 217–222.

Chatterji, A. and Abidi, Sahj., 1993).The Indian horseshoe crab: a living fossil. *J Indian Ocean Stud* 1: 43–48.

Chatterji, A., Vijayakumar, R. and Parulekar, A.H., 1992. Spawning migration of the horseshoe crab, *Tachypleus gigas* (Müller), in relation to lunar cycle. *Asian Fish Sci* 5: 123–128

Cohen, J. A. and Brockmann, H.J., 1983. Breeding activity and mate selection in the horseshoe crab *Limulus polyphemus. Bull. Mar. Sci.* 33: 274–28 1.

Fowler, H. W., 1908. *The king crab fisheries in Delaware Bay*, p. 11 l–l 19 + 4 pl. In Ann. Rep. N. J. State Museum, *1907.*

Greenspan, B.N., 1982. Semi–montly reproductive cycles in male and female fiddler crabs, Uca pugnax. *Anim. Behav.*, 80: 1084–1092.

Howard, H.A., Flordalice, R.W., Camara, M.D., Kass, L., Powera, M.K. and Barlow, R.B. Jr., 1984. Mating behaviour of Limulus: relation to lunar phase, tide height and sunlight. *Biol. Bull.* 167: 527.

Mattei, J.H., Beekey, M.A., Rudman, A. and Woronil ((2010).Reproductive behaviour in horseshoe crabs: Does density matter?. *Curr. Zool.*, 56(5): 634–642.

Osaki, T. and Knwabata, S., 2003. Structure and function of cagulin, a clottable protein in horseshe crabs. *Cellular and Molecular Life Science*, 61: 1257–1265.

Ropes, J. W., 1961. Longevity of the horseshoe crab, *Limulus polyphemus* L. *Trans. Am. Fish. Sot.*, 90: 79–80.

Rudloe, A., 1980. The breeding behavior and patterns of movement of horseshoe crabs, *Limulus polyphemus*, in the vicinity of breeding beaches in Apalachee Bay, Florida. *Estuaries*, 3(3), 177–183.

Saigusa, M. and Hidaka, T., 1978. Semilunar rhythms in the zoea–release activity of the land crab, Limulus polyphemus. *Bull. Mar. Sci.*, 36: 388–395.

Shuster, C.N. JR., 1960. Horseshoe crabs in former years, during the month of May these animals dominated Delaware Bay shores. *Estuarine Bulletin*, 5: 1–9.

Shuster, C.N., Jr., 1950. Observations on the natural history of the American hlorseshoe crab, *Limuluspolyphemus. Woods Hole Oceanogr. Inst. Contr*,.564: 18–23.

Smith, D.R., Pooler, P.S., Swan, B.L., Michels, S.F., Hall, W.R., Himchak, P.J. and Millard, M.J., 2002. Spatial and temporal distribution of horseshoe crab (*Limulus polyphemus*) spawning in Delaware Bay: implications for monitoring. *Estuaries*, 25: 115–125

Veseley, M.D. and Veseley, D.L., 1999. Environmental up–regulation of the ortial ontriutetic peptide gene in the living fossil, *Limulus polyphemus. Biochemical and Biophysical Research Communication*, 254: 751–756.

Environment, Biodiversity and Traditional System, 2014 Pages 365–378
ISBN 978-81-7622-307-2
Editors: D.R. Khanna, G.S. Solanki and S.K. Pathak
Published by: BIOTECH BOOKS, NEW DELHI

Chapter 35

Lichens in Mizoram and their Medicinal Uses in Traditional Tribal System

M. Chinlampianga[1], Amritesh C. Shukla[1] and D.K. Upreti[2]*

[1]Department of Horticulture, Aromatic and Medicinal Plants, Mizoram University, Aizawl – 796 004

[2]Lichenology Laboratory, National Botanical Research Institute, Lucknow, U.P.

ABSTRACT

Since ancient times lichens have a household item in India. Lichen produce a diverse range of chemical products and several species of lichens enjoyed a good position in ancient and traditional system of medicine like Ayurveda and Unani. Still a number of species of lichens are used by the different ethnic groups all over the world. The lichens are utilized for different purposes depending on their nutritive, medicinal, decorative, brewing, distilling, dyeing, cosmetics and perfumery properties. These different uses are substantiated by the complex lichen secondary metabolism, producing about 750 secondary compounds referred as "Lichen substances".

Globally, lichens are in use for various purposes. In Ireland, *Peltigera aphthosa* is used as vermifuge. In Arabian medicine, *Alectoria usneoides* is used in the treatment of splenomedaly (enlarged spleen). In China, *Lethariella* sp. of lichens is used as medicated teas. In Sweden, *Parmelia saxatilis* is used to treat warts. In India, *Lecanora sp.* and *Cetraria islandica* are important source of nutritive food in arid region; *Lobaria pulmonaria, Parmotrema sp.* and *Evernia prunastri* are used as dye stuff; *Evernia prunastri* and *Evernia furfuracea* (tree moss) are used for perfumery industry.

* *Corresponding Author:* E-mail: amriteshmzu@gmail.com

Further, in Mizoram, *Lecanora tropica, Usnea* sp. and *Porina* sp. are used as diuretic, indigestion, dysentery, however, some local herbal practitioners reported that many lichen species in Mizoram can be used as important ingredients for formulating drugs against immunomodulator, sexual debility for men and used for treating chronic diseases and they are more effective than other medicinal plant in preventing and curing the diseases in human beings.

Besides this, various secondary metabolites of lichens have also been reported; *Lobaria pulmonaria* has a gastroprotective effect; *Parmelia physodes* as an antibiotic; *Xanthoparmelia pokornyi* and its constituents (gyrophoric and stenosporic acid) as potential antimicrobials; Usnic acid isolated from *Telschisters chrysophthalmus* and prietin isolated from *Ramalina celastri* demonstracted antiviral activity against the arena viruses Junin and Tacaribe; Betaquinone, glycosides, colleflaccinosides isolated from *Collema flaccidum* as an antitumor activity; however, Heteroglycans and β – glucan isolated from *Thamnolia vermicularis* var. *subuliformis* tested for *in vitro* immune-modulating activity and reported to have various infiuences on the immune system.

Although, number of biological and pharmacological investigations have already been reported globally as well as in India also but there is no any such type of report in Mizoram, where many lichens species are growing luxuriantly. Hence, there is an urgent need for biological investigations as well as conservation of such traditional information on lichens before they lost forever.

Keywords: Lichens, Medicinal uses, Traditional system, Mizoram.

Introduction

India is a rich centre of biodiversity contributing nearly 15 per cent of the 13500 species of lichens (Negi, 2000). Many lichen species of the Himalayan region are said to effectively cure dyspepsia, bleeding piles, bronchitis, scabies, stomach disorders and many disorders of blood and heart (Saklani and Upreti, 1992; Negi and Kareem, 1996; Sochting, 1999). Even though manifold activities of lichen metabolites have now been recognized their therapeutic potential has yet not been fully explored and thus remains pharmaceutically unexploited (Turk *et al.*, 2003). It is known that microorganisms have developed resistance to many antibiotics. This creates enormous problems in the treatment of infectious disease, and investigators therefore seek new antimicrobial substances from different sources such as higher plants and lichens (Crittenden and Porter, 1991)

Mizoram is situated between: 92°.15′E to 93°.29′ E Longitude and 21°.58′ N to 24°.35′ N Latitude in the northeastern Indo-Burma region of India. The altitudinal range of this hilly region varies from 40 to 2157 m above msl. This wide array of climatic and altitudinal zones favor the luxuriant growth of diversified and rich vegetation. This region is also the abode of many endemic elements and a number of species which have become rare, threatened and endangered, Karaman *et al.*, 2003). Some local herbal practitioner's in Mizoram used some lichen species (commonly called Bawnghar) as important ingredients for their formulation. It was reported that decoction of some lichen species as effective remedy against diuretic and kidney problem (Rozika, 2005). It is the intention of this paper to present some of the traditional medicinal uses of lichens and their importance for the modern pharmaceutical industry through literature review.

Lichens are the symbiotic organisms of fungi and algae, synthesize numerous metabolites, the "lichen sub-stances," which comprise amino acid derivatives, sugar alcohols, aliphatic acids, macrocyclic lactones, mono-cyclic aromatic compounds, quinones, chromones, xanthones, dibenzofuranes, depsides, depsidones, depsones, terpenoids, steroids, carotenoids and diphenyl ethers (Clix *et al.*, 1984; Fiedler *et al.*, 1986). The secondary metabolites have shown to have manifold biological and pharmaceutical activity such as antimicrobial, antimycobacterial, immuno-modulatory, antiviral, cytotoxic, antitumor, allergic, plant growth inhibitory, antiherbivore, ecological roles and enzyme inhibitory (Dülger *et al.*, 1997, 1998; Huneck 1999; Ozturk 1999; Aslan *et al.*, 2001; Perry *et al.*, 1999; Manojlovic *et al.*, 2002, Chand *et al.*, 2009). Antibiotic and antifungal activity screenings of Indian lichens have been initiated recently (Shahi *et al.*, 2001; Balaji and Hariharan 2007; Sati and Joshi, 2011; Shukla *et al.*, 2011). Their efficacy is due to the synthesis of unique secondary compounds, a number of which have important biological roles (Perry *et al.*, 1999).

The traditional uses of lichen as medicine open and widen the research field in various directions for the present and future generation. Lichens have been known as plant resources for various purposes in the past, present especially as an important source for dyes, medicines, food, fodder and perfumes as well as an important pharmaceutical agent. There is a vast treasure of medicines hidden in a small group of plant and in future, it can be a potential and alternative source for commercial industries. According to Divakar and Upreti 2008, lichen *Parmelia perlata*, an ingredient listed in some of the ayurvedic and siddha medicines is extinct now in India, because of over collection. The same case may be happen in the state of Mizoram if the local people and foreign traveler realized the economic importance of lichens.

Materials and Methods

Collection and Identification of Lichen Samples

Following lichen specimens *Ramalina sinensis* Jatta, *Ramalina hossei* Vain, *Ramalina conduplicans* Vain, *Parmotrema tinctorum* (Nyl.), *Cladonia fruticulosa* Kremp., *Parmotrema stuppeum* (Taylor) Hale, *Cryptothecia lunulata* (Zahlbr.), *Pyrenula interdusta* (Nyl.) Zahlbr. *Lepraria* sp., *Garphis scripta* (L.) Ach.G. Prosperns Vain., *Pertusaria quassiae* (Fée) Nyl, *Lecanora tropica* Zahlbr., *Usnea* sp. *Porina* spp. were collected during the preliminary survey from the undisturbed forest area of Rei-ek village under Mamit, Mizoram in year 2009-11. Identification was done using relevant key and monographs (Yazýcý and Aslan, 2003; Yazici *et al.*, 2004; Divakar and Upreti 2005; Awasthi 2007). Samples were dried at room temperature for 48 h. Herbarium was prepared and kept at the department and the same were deposited at the lichen herbarium, National Botanical Research Institute (NBRI), Lucknow, India for taxonomical confirmation.

Method of Data Collection

To compile the present investigation, data were gathered through personal interview with knowledgeable elder persons and local herbal practitioners and repeated literature survey on the related topics from major databases including Science Direct, Chemical Abstracts, Pub-Med and Directory of Open Access Journals, Conference proceeding, published edited books available in the study area was done time to time and cross-checked them properly. All the information extracted from the identified knowledgeable persons and local herbal practitioner from selected village the through open ended interview and verbal discussion were documented and correlate with the available literature for scientific confirmation.

Results and Discussion

The local herbal practitioner mostly use *Parmotrema* sp, *Parmelia* sp and *Usnea* sp collected on different strata like tree bark and rock surface. They are collected as and when required from nearby to distance undisturbed forest area. The available information on traditional use of lichen as medicine including the mode of preparation, formulation and application.

Mode of Preparation, Formulation and Application

The collected lichens used for medicines were collected in fresh form and then shade dried for 10-15 minute after washing. The raw drugs are chopped into pieces before soaking (usually for 24-48hrs or directly boil without soaking) and then boil with equal volume of water for 2- 4hrs till the final volume reaching 1/3 of the initial volume (for better concentration and higher effectiveness in action, as Plate-1(a-d). Now, the product is ready for use in concentrated form. However, before using as house hold remedy, it has to make the further dilution with water (@1: 3 dilutions, (see the protocol as Scheme-1.) Depending up on containers and storage condition, the drugs can be used for several days to months.

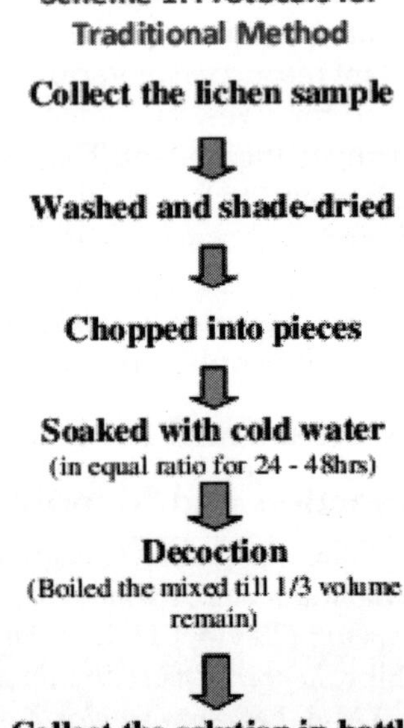

In order to formulate the drugs using lichen as single or as combination with other ingredients, the practitioner mixed together at a time or prepared individual raw drug and then mixed together to form a single drugs. The proportion for each components/ingredients depends on the type of disease to be treated. The mode of application and dose is depending on the type of disease and its severity.

In addition to the above listed fourteen identified lichens species from Mizoram during the survey, 30 genus with 45 nos. of lichens species used in traditional medicine as well as their pharmaceutical importance were compiled in this paper (Tables 351–35.3). In the area of Ruby Valley, Nevada "One cure for smallpox is prepared by boiling together the leaves of Cowania, powdered (rock lichens) and 'kah seep' [a black, pitch-like substance.the dried urine of mountain rats]. The solution is taken internally at morning and night in doses of a half-cup (Llano, 1951; Train, *et al.*, 1974; Turner, 1983). In Tonopah, Nevada "The black, orange and green lichens (cructose lichens) are scraped from rocks and soaked overnight in cold water. The solution then is taken internally to stop diarrhea. Similarly, lichens collected from barks of pine tree were collected and the decoction is given for household remedy against stomach problems like indigestion and gastric problems.

Plate 1(a): *Usnea barbata* **(in natural condition)**

Plate 1(b): *Decoction of Usnea (using traditional method)*

Plate 1(c): Usnea Extract

Plate 1(d): (+) Usnic acid (using GC-MS)

Lichen bioactivities including, antimicrobial, antifungal, antiviral, anticancer, antioxidant actions, antipyretic and immune-regulator were reviewed and summarized (Tables 35.2 and 35.3). Recent progress in lichens and lichen-forming fungi was discussed with emphasis on their potential to accelerate commercialization of lichen-based products.

Earliest mention of *Lecanora esculenta* in Arabic works was 9th to 13th centuries as an ingredient of wine made from honey and of medicinal compounds. There is a tradition in Sistann (in eastern Persia) that Alexander's army escaped starvation in 330-327 B.C. by eating *Lecanora esculenta*, a famine food (Turner, *et al.*, 1983; Crum, 1993). A decoction of *Cetraria islandica* is recommended as a tonic for convalescents (Schneider, 1904). Significant amounts of *Cetraria islandica* species are still sold annually in European pharmacies for concoction of herbal tonics and laxatives (Richardson, 1991). It was used as "re-constituent after tuberculosis, anticatarrhal" ("present traditional use") in the area of Venezia Giulia region, northeast Italy (Lokar, and Poldini, 1988).

In Sweden, *Cetraria islandica* is used to treat asthma, treat diabetes, nephritis, lung diseases, and wasting diseases, and thought to be effective treatment for whooping cough and colds (Ahmadjian, Nilsn, 1963). Infusion of *Evernia vulpine* and marrow was taken for stomach disorders like ulcers (Ahmadjian,.; Nilsn, 1963) Boiled and taken in a weak solution for internal problems in Okanagan-Colville, British Columbia (Turner, *et al.*, 1980). Hill men from Darjeeling and Sikkim use *Lobaria pulmonaria* for curing eczema on the head, cleaning and strengthening hair, remedy for lung troubles, haemorrhages and asthma (Biswas, 1956). *Parmelia karatschadalis* used as a poultice, applied to renal and lumbar regions, which causes a copious flow of urine. *Parmotrema* sp. was used to treat kidney problems in Mizoram (Rozika, 2005).*Pertusaria communis* was used in Europe to cure intermittent fever; more useful on men than women (Llano, 1951).

Table 35.1: Traditional use of some lichen species.

Lichens	*Family*	*Medicinal use*	*References*
Everniastrum cirrhatum (Fr.) Hale ex Sipman	Parmeliaceae	Wound healing, antiseptic, bronchitis	Saklani and Upreti (1992)
Flavoparmelia caperata (L) Hale	Parmeliaceae	Burns, fever, pain, and in Siddha system of medicine	Saraswathy *et al.* (1990)
Heterodermia leucomela (L) Poelt	Physciaceae	Wound healing	Saklani and Upreti (1992)
Lecanora flavidorufa Hue	Lecanoraceae heart	Disorders of blood and	Sochting (1999)
Lobaria isidiosa (Bory) Trevisan	Stictaceae	Skin diseases, chest complaints	Sochting (1999)
Ramelia reticulata (Taylor) Hale and Fletcher	Parmeliaceae	Kidney disorder, veneral disease	Sochting (1999)
Stereocaulon foliolosum Nyl.	Stereocaulaceae	Urinary trouble, blister of the tongue	Saklani and Upreti (1992)

Table 35.2: List of lichen species used against microbial species.

Lichen Species	*Microbial Species*	*Reference*
Antibacterial Activity		
Ramalina dendriscoides	(*Bacillus subtilis*)	Krystle *et al.*, 2010
Ramalina farinacea	*Bacillus, E. coli, Enterococcus, Mycobacterium*	Esimone and Adikwn, 1999, Perry *et al.*, 1999
Cladonia foliacea *Hypogymnia physodes*	Gram +ve bacteria	Mitrovic *et al.*, 2011
Antifungal Activity		
Parmelia caperata, Umbilicaria cylindrical	*Candida albicans*	Krystle *et al.*, 2010
Evernia prunastri	Strongest antifungal-effect on yeast	Mitrovic *et al.*, 2011
Hypogymnia physodes	Strong antifungal effect- on filamentous fungi	Mitrovic *et al.*, 2011
Ramalina sp. and *Stereocaulon* sp.	Strong efficacy against *Alternaria alternata* *Aspergillus flavus* *Penecillium italicum*	Shukla *et al.*, 2011

As a liniment it is applied to head in cases of headaches and powder is applied to promote healing of wounds." (Nadkarni, 1976. Saklani; Upreti, 1992) reported that crude drug 'chharila', extracted from 3 spp. of *Parmelia* is sold in Indian bazaars and used in Ayurvedic and Unani systems of medicine. Decoction (one cup daily) of *Ramalina bourgeana* used in the areas of Viso and Nijar (Spain) as a diuretic for the treatment of kidney stones (Gonzalez-Tejero *et al.*, 1995) *Stereocaulon himalayense.* Lepchas from

Sakyong Valley, North Sikkim pounded *Stereocaulon himalayense* lichen species and boiled in water. About 100 ml of decoction twice a day after meals used to treat urinary troubles. Small amounts prescribed for blisters of the tongue (Saklani and Upreti. 1992). *Usnea barbata* used as contemporary alternative medicine in U.S for the treatment of fungal infections of the mouth, stomach, intestines, anus, vagina, nose, ear, and skin, as well as systemic fungal infections." Administered internally, topically, and as a douche (Herbal Resources, 1998). The same species is also used by the some tribal community in Mizoram to treat skin disease. Tribal people from East Africa, chewed fresh and juice of *Usnea africana* Hewas (Iraqw) to treat stomach ache and decoction of *Xanthoria parietina* subsp. *ectanea* used to treat toothache, kidney disorders, menstrual complaint and various pains in Funete del Escribano, Puebloblanco and Campohermoso (Gonzalez-Tejero *et al.*, 1995.Rowe *et al., 1989*) reported that the Turkey lichens, *Evernia prunastri, Pseudoevernia furfuracea* and *Alectoria capillaris* were active against Gram-positive bacteria and the *Candida albicans*. Recently, lichen has attracted attention because they contain compounds that can be used as medicinal agents or as neutraceuticals in functional foods. *U. esculenta* has been used in traditional preventatives and remedies for bloody vomit and diarrhoea, skin disease, epilepsy, and yellow jaundice (Jang *et al.*, 2003). They are ingredients of many medicines of plant origin since they possess a broad spectrum of biological activities, including anti-bacterial, anti-inflammatory, anti-tumorous, purgative, astringent, anti-viral (Cyong *et al.*, 1987) antioxidant and antifungal

Scheme 2- Lichens: Tradition to Technology

Traditional information

- Collection
- Categorization
- Identification
- Documentation
- Conservation

Scientific validation

- Bioprospects/ Biological screening
- Physico-chemical analysis & their synthesis
- Detail *in vitro* investigations
- Detail *in vivo* investigations
- Mechanism of action
- Pre-clinical investigations
- Clinical investigations
- Drug formulation

Pharmaceutical companies

- Multicentral Trials
- Commercialization

(Agarwal *et al.*, 2000; Manojlovic *et al.*, 2002; Manojlovic *et al.*, 2008). Some of the lichen substances can be contact allergens in people who are susceptible and allergic reaction to human depending on their chemical constituents (Table 35.4).

Table 35.3: Bioactive constituents of 11 lichen species and their therapeutic activities in human beings.

Lichen Species	*Bioactive Constituents*	*Therapeutic Activity*
Cetraria islandica	Protolicheterinic acid	Antitumor: Inhibition of growth-of breast cancer cell lines and mitogen stimulated lymphocytes
Collema flaccidum colleflaccinosides	Betaquinone, glycosides,	Antitumor activity
Hypopymnia physodes	Phenolic	Antioxidant
Ramalina celastri	Prietin	Antiviral activity (arena)
Telschisters chrysophthalmus	Usnic acid	Antiviral activity (arena virus)
Thamnolia vermicularis-var. *subuliformis*	Heteroglycans and -a β-glucan	Immunomodulating activity
Usnea diffracta	Usnic acid (Scheme-1)	Antipyretic and analgesic
Usnea sp.	Usnic acid, phenolic compound, etc.	Wide-spectrum antibiotic
Xanthoparmelia pokornyi	Gyrophoric and stenosporic acid	Potential antimicrobials

Recent study report that antibacterial activity of methanol extracts of five lichen species (*Flavoparmelia caperata, Evernia prunastri, Hypogymnia physodes and Cladonia foliacea)*. Two lichen species were tested for the first time (*Evernia prunastri* and *Cladonia foliacea*). The analysis of their antibacterial potential were performed on 15 strains of bacteria and revealed the strongest inhibitory effect, especially on Gram (+) bacteria, of *Hypogymnia physodes* and *Cladonia foliacea* (Mitroviæ *et al.*, 2011*)* (Table 35.1). Also, Mitroviæ *et al.* (2011) determined strong antifungal effect of *Evernia prunastri* and *Hypogymnia physodes.* While *Evernia prunastri* exerted the best effect on yeasts, *Hypogymnia physodes* were better on filamentous fungi. Antiviral properties have been attributed to specific lichen secondary metabolites. For instance, Perry *et al.*, 1999) shows antiviral activity of usnic acid against Herpes simplex type 1 and Polio type 1 virses. Parietin extracted from *Teloschistes chrysophthalmus* proved as virucidal for Junín and Tacaribe arena viruses (Fazio *et al.*, 2007). Lichenan, widely distributed in lichens, demonstrated inhibition of tobacco mosaic virus (Stubler and Buchenauer, 1996). Antitumor activities of lichens are of major importance as source of potential drugs for lethal malignant diseases. Usnic acid (Scheme 1) the most extensively studied lichen metabolite since its first isolation in 1844, exhibited an antiproliferative effect on human leukemia cells (K562) and endometrial carcinoma (HEC-50) cells (Carderelli *et al.*, 1997, Ingolfsdottir, 2002, Krismundsdottir *et al.*, 2002). Furthermore, antipyretic and analgesic effects of lichen components were demonstrated on animal studies. For instance, usnic acid (see chemical structure as Plates 351d from *Usnea* *diffracta* inhibited acetic-acid-induced writhing in mice and raised the pain threshold in dose dependent manner (Okuyama *et al.*, 1995).

Table 35.4: Lichens and their chemical substances responsible for allergic reactions.

Lichen Substance	*Reference*
Atranorin	Dahlquist and Fregert, 1980; Thune and Solberg, 1980; Goncalo *et al.*, 1988; Hausen *et al.*, 1993; Stinchi *et al.*, 1997; Aalto-Korte *et al.*, 2005; Cabanillas *et al.*, 2006
Diffractaic acid	Thune and Solberg, 1980
Evernic acid	Dahlquist and Fregert, 1980; Thune and Solberg, 1980; Goncalo *et al.*, 1988; Hausen *et al.*, 1993; Aalto-Korte *et al.*, 2005; Cabanillas *et al.*, 2006
Fumarprotocetraric acid	Dahlquist and Fregert, 1980; Thune and Solberg, 1980; Goncalo *et al.*, 1988; Hausen *et al.*, 1993.
Lobaric acid	Thune and Solberg, 1980
Perlatolic acid	Hausen *et al.*, 1993
Physodalic acid	Thune, 1977; Thune and Solberg, 1980
Physodic acid	Thune, 1977; Thune and Solberg, 1980
Salazinic acid	Thune and Solberg, 1980
Stictic acid	Thune and Solberg, 1980; Hausen *et al.*, 1993
Usnic acid	Mitchell and Champion, 1965; Thune and Solberg, 1980; Goncalo *et al.*, 1988; (see chemical structure Hausen *et al.*, 1993; Stinchi *et al.*, 1997; Aalto-Korte *et al.*, 2005; Cabanillas *et al.*, 2006)

Finally, antioxidant properties, already mentioned previously concerned with phenolic content of lichens. Jayaprakasha and Rao, 2000) examined antioxidant capacities of methyl orsellinate, atranorin, osellinic acid and lecanoric acid. Bhattarai *et al.* (2008) noticed stronger antioxidant activities in lichens from Antarctic that the one in lichens from native to temperate or tropical regions. Mitroviæ *et al.*, 2011) compared the chemical content of lichen extracts (*Flavoparmelia caperata, Evernia prunastri, Hypogymnia physodes* and *Cladonia foliacea*) and their free radical scavenging ability.

Conclusion

Traditional medicine is the synthesis of therapeutic experience of generations of practicing physicians of indigenous systems of medicine. Until the discovery of modern medicines, any system of medicine that relieved the patient and their ailment(s) was considered to be a therapeutic system without further investigation. Whether the system was properly analyzed, researched or organized did not matter. But with the rapid development of conventional medicines, this ethno-medicinal knowledge required to be proved scientifically with proper evidence to prove the therapeutic efficacy of the ethnic system of medicine to be used as complementary therapies. An increasing amount of the scientific research on the flora and fauna of Mizoram is being done with a view to document and identify commercially useful genetic and biochemical resources.

The present paper revealed that some of the commonly found lichens in Mizoram like *Parmotrema tinctorum* (Nyl.), *Cladonia fruticulosa* Kremp. and *Usnea* sp, shows significant potential against various human and phytopathogens. Further, in Indian context out of the present 2000 lichen species so far known, hardly less than 1 per cent of the species are screened out for their biological activity. More than 50 per cent lichen species known from the world exhibit presence of peculiar antibiotic substances. The study shows that there is negligible report and no proper documentation of traditional use of lichens and study on bioprospects of lichens from Mizoram so far. Thus there is a lot of scope for carrying out the researches on systematic investigation of lichens and their bioprospects, as per the procedure given in Scheme 2.

Acknowledgements

The authors are thankful to the peoples of Mizoram especially the Mizo-tribal communities, who have shared their information's and the authorities of the Mizoram University, Aizawl as well as Lichenology Laboratory, NBRI, Lucknow for providing various kinds of supports during the compilation of the work.

References

Aalto–Korte K., Lauerma A. and Alanko K., 2005), Occupational allergic contact dermatitis from lichens in present–day Finland. *Contact Derm.* 52: 36 – 38.

Agarwal, K. S, Singh S.S, Verma S. and Kumar, S., 2000. Antifungal activity of anthraquinone derivatives from *Rheum Emodi. J Ethnopharmacol* 72: 43–46.

Ahmadjian, V.; Nilsn, S., 1963. Swedish lichens. Year book (American Swedish Historical Foundation) Hellson, J.C. and Gadd, M., 1974. Ethnobotany of the Blackfoot Indians. National Museum of Man Mercury Series, *Canadian Ethnology Service* 19: 76.

Aslan, A., Güllüce, M. and Atalan E., 2001. A Study of Antimicrobial Activity of Some Lichens. *Bull. Pure Appl. Sci.*, 20: 23–26.

Awasthi, D.D., 2007. A Compendium of the Macrolichens from India, Nepal and Sri Lanka.–*Bishen Singh Mahendra Pal Singh*, Dehra–Dun, India. pp 580.

Balaji P. and Hariharan, G.N., 2007. In vitro antimicrobial activity of *Parmotrema praesorediosum* thallus extracts. *Research Journal of Botany* 2(1): 54–59.

Bhattarai, H.D., Paudel, B., Hong, S.G., Lee, H.K. and Yim, J.H., 2008): Thin layer chromatography analysis of antioxidant constituents of lichens from Antarctica. *J. Nat. Med.*, 62: 481–484.

Biswas, K., 1956. Common Medicinal Plants of Darjeeling and the Sikkim Himalayas.edt. pp. 90.

Cabanillas, M., Fernandez–Redondo V. and Toribio J., 2006), Allergic contact dermatitis to plants in a Spanish dermatology department: a 7–year review. *Contact Derm.* 55: 84 – 91.

Cardarelli, M.A., Serino, G., Campanella, L., Ercole, P., De Cicco–Nardone, F., Alesiani, O. and Rossiello, F., 1997. Antimitotic effects of usnic acid on different biological systems. *Cell. Mol. Life. Sci.*, 53: 667–672.

Chand, P., Singh, M, and Rai, Mayank., 2009. Antibacterial activity of some Indian Lichens. *Journal of Ecophysiology and Occupational Health* 923–29.

Clix, J.A., Whitton, A.A., Sargent, M.V., 1984. Recent progress in the chemistry of lichen substances. *Prog. Chem. Org. Nat. Prod.* 45: 207–211.

Crittenden, P.D. and Porter, N., 1991. Lichen–forming fungi: potential sources of novel metabolites. *Trends in Biotechnolology* 409–14.

Crum, H., 1993. A lichenologist's view of lichen manna. Contributions to the University of Michigan Herbarium 19: 293–306.

Cyong J.C, Matsumoto, T., Arakawa, K., Kiyohara, H., Yamoda, H. and Otsuka, Y., 1987. Antibacteroides Fragilis substances from Rhubarb. *J Ethnopharmacol* 19: 279–283.

Dahlquist, I. and Fregert, S., 1980), Contact allergy to atranorin in lichens and perfumes. *Contact Derm.* 6: 111 – 119.

Dülger, B., Gücin, F., Kara, A. and Aslan, A., 1997. *Usnea florida* (L) wigg. Likeninin Antimikrobiyal Aktivitesi. *Tr J. Biol.* 21: 103–108.

Dülger, B., Gücin, F. and Aslan, A., 1998. *Cetraria islandica* (L.) Ach. Likenin Antimikrobiyal Aktivitesi. *Tr J. Biol.* 22: 111–118.

Divakar, P.K. and Upreti, D.K., 2005. Parmelioid Lichens in India: A Revisionary Study. *Bishen Singh Mahendra Pal Singh*. Dehradun. pp 488.

Esimone, C.O. and Adikwn, M.U., 1999. Antimicrobial activity of the cytotoxicity of *Ramalina farinacea*. *Fitoterapia*, 7: 428–431.

Fazio, A.T., Adler, M.T., Bertoni, M.D., Sepulveda, C.S., Damonte E.B., Maier and M.S., 2007. Lichen secondary metabolites from the cultured lichen mycobionts of *Teloschistes chrysophthalmus* and *Ramalina celastri* and their antiviral activities. Z. *Naturforsch*, 62: 543–549.

Fiedler, P., Gambaro, V., Garbarino, J.A. and Quihot, W., 1986. Epiphorellic acids 1 and 2, two diaryl ethers from the lichen *Cornicularia epiphorella*. *Phytochemistry* 25: 461–465.

Goncalo, S., Cabral F. and Goncalo M., 1988), Contact sensitivity to oak moss. *Contact Derm.* 19: 355 – 357.

Gonzalez–Tejero, M.R., Martinez–Lirola, M.J., Casares–Porcel, M. and Molero–Mesa, J., 1995. Three lichens used in popular medicine in eastern Andalucia (Spain). *Economic Botany* 49 (1): 96–68. pp. 97

Hausen, B. M., Emde L. and Marks V., 1993), an investigation of the allergic constituents of *Cladonia stellaris* (Opiz) Pous and Vezda ('silver moss,' 'reindeer moss' or 'reindeer lichen'). *Contact Derm.* 28: 70 – 76.

Herbal Resources, Inc., 1998. Spilanthes–Usnea Compound: Anti–Fungal Remedy.

Huneck, S., 1999. The Significance of lichens and their metabolites. *Naturwissenschaften* 86: 559–576.

Ingondolfsdottir, K., 2002. Usnic acid. *Phytochemistry*, 61: 729–736.

Jang, C. H., Eun, J. S., Park, H. W., Seo, S. M., Yang, J. H., Leem, K. H., Oh, S. H., Oh, C. H., Baek, N. I. and Kim, D. K., 2003. An acetylcholinesterase inhibitor from the leaves of *Securinega suffruticosa*. *Kor. J. Pharmacong*. 34: 14–17.

Jayaprakasha, G.K. and Rao, L.J., 2000): Phenolic constituents from the lichen *Parmotrema stuppeum* (Nyl.) Hale and their antioxidant activities. *Z. Naturforsch*. 55: 1018–1022.

Karaman, I., Sanin, F., Gulluce, M., Ogtc, H., Sengul, M. and Adiguzel A., 2003. Antimicrobial activity of aqueous and methanol extracts of *Funiperus oxicedrus*. *Journal of Ethnopharmacology* 37: 1–5.

Kokwaro, J.O., 1976. Medicinal Plants of East Africa. East African Literature Bureau, Lampala, Nairobi, Dar es Salaam. pp. 10.

Kristmundsdottir, T., aradottir, H.A.E., Ingolfsdottir, K. and Ogsmundsdottir, H. M., 2002. Solubilization of the lichen metabolite (+) –usnic acid for testing in tissue culture. *J. Pharm. Pharmacol*. 54: 1447–1452.

Llano, G.A., 1951. Economic uses of lichens. *Ann. Rep. Smiths. Inst.*: 385–422.

Lokar, L.C., Poldini, L., 1988. Herbal remedies in the traditional medicine of the Venezia Giulia Region (north east Italy). *Journal of Ethnopharmacology* 22: 231–278.

Manojlovic, T. N., Solujic, S. and Sukdolak, S., 2002. Antimicrobial activity of an extract and anthraquinones from *Caloplaca schaereri*. *Lichenologist* 34: 83–85.

Manojlovic, T. N., Vasiljevic, P., Bogdanovic–Dusanovic, G. and Manojlovic, I., 2008. Antioxidant activity of some lichen growing in Serbia. *Planta Med* 74: 996–997.

Mitchell, J. C. and Champion, R. H., 1965), Human allergy to lichens. *Bryologist* 68: 116 – 118.

Mitroviæ, T., Stamenkoviæ, S., Cvetkoviæ, V., Tošiæ, S., Stankoviæ, M., Radojeviæ, I., Stefanoviæ, O., Èomiæ, Lj. Ðaèiæ, D., Æurèiæ, M. and Markoviæ, M., 2011: Antioxidant, antimicrobial and antiproliferative activities of five lichen species. *Int. J. Mol. Sci.*, 12: 5428–5448.

Nadkarni, K.M., 1976. Indian Materia Medica. In A.K. Nadkarni, edt.pp. 922

Negi, H.R., 2000): On the patterns of abundance and diversity of macrolichens of Chopta–Tunganath in the Garhwal Himalaya. *J Biosci* 25: 367–378.

Negi, H.R. and Kareem, K., 1996. Lichens: The unsung heroes. *Amrut* 1 3–6. of antiinfectiveagents. *Journal of Natural Products* 50: 1025–40.

Okuyama, E., Umeyama, K., Yamazaki, M., Kinoshita, Y. and Yamamoto, Y., 1995. usnic acid and diffractaic acid as analgesic and antipyretic components of *Usnea diffracta*. *Planta Med*, 61: 113–115.

Ozturk, S., Guvenc, S, Arikan, N. and Yilmaz O.(1999. Effect of usnic acid on mitotic index in root tips of *Allium cepa* L. *Lagascalia* 21: 47–52.

Perry, N.B., Benn, M.H., Brennan, N.J., Burgess, E.J., Ellis, G., Galloway, D. J., Lorimer, S.D. and Tangney S., 1999. Antimicrobial, antiviral and cytotoxic activity of New Zealand lichens. Lichenologist 31: 627–636.

Richardson, D.H.S., 1991. Lichens and man. In Hawksworth, D.L., ed. Frontiers in Mycology. pp. 192.

Rowe, J.G., M.T. Saenz and Garcia, 1989.Contribution a a'l e' tudedel' activite antibactérienne de queques lichens du sudde l'Espagne. *Pharmaceutique francaise*, 47: 89–94.

Rozika, R., 2005).Ramhmul damdawite (Medicinal plants) edt. pp. 43.

Saklani, A. and Upreti, D.K., 1992. Folk uses of some lichens in Sikkim. *Journal of Ethnopharmacology* 37: 229–333.

Saraswathy, A, Rajendiran, A., Sarada, A. and Purushothamam., 1990. Lichen substances of *Parmelia caperata*. Indian Drugs 27: 460–462.

Sati, S.C. and Joshi, S., 2011. Antibacterial Activity of the Himalayan Lichen *Parmotrema nilgherrense*. BMRJ. 1(2): 26–32.

Schneider, A., 1904. A Guide to the Study of Lichens. Knight and Miller, Boston. pp. 24.

Shahi, S.K., Shukla, A.C., Dikshit, A. and Uperti, D.K., 2001. Broad–spectrum antifungal properties of the lichen, *Heterodermia leucomela*. *The Lichenologist* 33(2): 177–179.

Shukla, A.C., Pianga, M., Verma, A., Dikshit, A. and Upreti, D.K., 2011. Efficacy and potency of lichens of Mizoram as antimycotic agents. *Indian Phytopathology* [ISSN: ISSN, 0367– 973X], 16(4): 367–370.

Shukla, A.C., Pandey, K.P., Mishra, R.K., Dikshit, A., and Shukla, N., 2011. Broad spectrum antimycotic plant as a potential source of therapeutic agent. J of Natural Products [ISSN 0974–5211]; 4: 42–50.

Sochting, U., 1999. *Lichens of Bhutan. Biodiversity and Use.* Copenhagen Botanical Institute. Department of Mycology, University of Copenhagen.

Stinchi, C., Guerrini, V., Ghetti, E. and Tosti A., 1997), Contact dermatitis from lichens. *Contact Derm.* 36: 309 – 310.

Thune, P., 1977. Contact allergy due to lichens in patients with history of photosensitivity. *Contact Derm.* 3: 267 – 272.

Thune, P. and Solberg, Y. J., 1980. Photosensitivity and allergy to aromatic lichen acids, Compositae oleoresins and other plant substances. *Contact Derm.* 6: 81 – 87.

Train, P., Heinrichs, J.R., Archer, W.A., 1974. Medicinal uses of plants by Indian tribes of Nevada. Pages 53– In Horr, D.A., ed. American Indian Ethnohistory: Paiute Indians IV. pp.118.

Turk, A.O., Yilmaz, M., Kivank, M. and Turk, H., 2003. Antimicrobial activity of extracts of the lichen *Cetraria aculeata* and its protolichesterinic acid constituent. *Verlag der Zeitschrift fur Naturforschung* (C) 58 (11–12): 850–854.

Turner, N.J., Bouchard, R. and Kennedy, D.I.D.(1980. Ethnobotany of the Okanagan–Colville Indians of British Columbia and Washington. Occasional Papers of the British Provincial Museum No. 21, Province of British Columbia. pp. 15.

Turner, N.J., Thomas, J. and Carlson, R.T., 1983. Ethnobotany of the Nitinaht Indians of Vancouver Island. *British Columbia Prov. Mus. Occas. Pap.* 24: 55–165.

Yazýcý, K. and Aslan, A., 2003. Lichens from the regions of Gümü_hane, Erzincan and Bayburt (Turkey). *Mycology* 24: 287–300.

Yazýcý, K., Aslan, A. and Aptroot A., 2004. Four new lichen species from Turkey. *Mycotaxon* 90: 177–180.

Environment, Biodiversity and Traditional System, 2014 *Pages* 379–389
ISBN 978-81-7622-307-2
Editors: D.R. Khanna, G.S. Solanki and S.K. Pathak
Published by: BIOTECH BOOKS, NEW DELHI

Chapter 36

Ayurveda: The Ancient Science for Human Health

Sukanta Sarkar

Lecturer (Economics), Faculty of Management Studies, ICFAI University, Agartala, Tripura
E-mail: sukantaeco@gmail.com

ABSTRACT

Ayurveda is a system of traditional medicine native to India and a form of alternative medicine. Current practices derived from Ayurvedic medicines are regarded as part of complementary and alternative medicine. It is the most ancient and traditional system of medicine in India. The Ayurvedic system of medication is based on many centuries of experience in medical practice handed down through generations. Ayurvedic medicine originated in the early civilizations of India some 3,000-5,000 years ago making Ayurvedic medicine the oldest surviving healing system in the world. It advocates that the primary and essential factors of the human body that govern our entire physical structure and function, is a combination of any two of the five bhutas with the predominance of one. This is the most fundamental and characteristic principle of Ayurveda and is called "Tridosha" or the Three Humours. They are categorized into Vata, Pitta and Kapha. Vata governs movement, Pitta is concerned with functions of heat, metabolism, and energy production and Kapha, governs physical structure and fluid balance. Ayurveda is regarded as "The Science of Life. The main objectives of this paper are (i) to examine the growth of ayurveda medicine in India and (ii) To completely eradicate the disease and dysfunction of the body.

Introduction

Ayurveda is a system of Indian medicine now practiced all over the world as a complementary system. Ayurvedic practitioners have identified a number of medicinal preparations and surgical procedures for curing various ailments and diseases, not completely curable in other systems (The ancient science of holistic healing, 2012).

Disease free condition is best source of virtue, wealth, gratification and emancipation while the diseases are destroyers of this welfare and life itself (Singh, 2012). Over centuries ayurveda has had a nurturing influence on ancient Chinese systems of medicine, Unani medicine, and the humoral medicine practiced by Hippocrates in Greece. The entire science of ayurveda is based on the *'Five Great Elements'* (Panchabhuta) theory. These five elements are earth (prithvi), water (jal), fire (Agni or tej), air (vayu) and ether or space (akash) (Introduction, 2012). The ultimate goal of ayurveda is to create a state of holistic health for the individual, to create, consequently, a healthy society and environment with its herbal health remedies (Holistic Health Benefits of Ayureva, 2012). Ayurveda, recognized as alternative medicine now, represents the science of life and longevity originating in the Vedic traditions of India (Ayurveda–Basic Concepts of Ayurveda, 2012). Charaka's treatise, the primary and the oldest known ayurvedic text, says that the equilibrium of the three *doshas* in the body means health and their imbalance manifests as disease. To Ayurveda's detriment, the three *doshas* are translated crudely as wind, bile and phlegm (Chopra, 2012).

Concept

The word Ayurveda is made of two words, *'Ayu'* meaning life and *'Veda'* means knowledge, wisdom or essential truth. Ayurveda means knowledge of life. It is the principle of life that if applied results in to Holistic health, Harmony and Happiness. It is the ancient wisdom of India, though a system of medicine but aims at self realization or knowing one's essential nature. Ayurveda is that which deals with –

1. **Hitayu** (Positive Life) – a person is known to be evolved in ancient wisdoms when he never harms any person in thought, speech and action; lives, moves and acts in kindness, humility, goodness and works with single-minded dedication in life; establishes harmony in a family and society and is self disciplined, controls his/her negative traits like anger, agitation, excitement etc.
2. **Ahitayu** (Negative Life) – a person who is always guided by his/her preferences, prejudices, manifests negative traits of anger, agitation in personal, professional and social lives, distracted, and never mind harming others for own benefits in life.
3. **Sukhayu** (Life of Pleasure) – a lifestyle where person enjoys worldly pleasures as he is physically, mentally fit and is young. He is having strength and vitality is having name and fame. He is having intellect and is knowledgeable and his sense is under his control and is able to do their work.
4. **Dukhayu** (Life of Sorrow) a person living in unconsciousness, ignorance and falsehood at every moment of his life which contradicts the ways of living as mentioned above falls into this category of living. He always suffers and lives a painful life as he lives, moves and acts in Prajna Aparadh (actions lacks awareness, clarity, progress and harmony) (What is Ayurveda?, 2012).

The main objectives of this chapter are:

1. To examine the growth of ayurveda medicine in India and
2. To completely eradicate the disease and dysfunction of the body.

Historical Background

Around 1500 BC ayurveda was delineated into two distinct ayurveda schools: (I) Atreya: The School of Physicians, and (II) Dhanvantari: The School of Surgeons (School of Thought, 2012). One view of the early history of ayurveda asserts that around 1500 BC, ayurveda's fundamental and applied principles got organized and enunciated. In this historical construction, Ayurveda traces its origins to the Vedas, *Atharvaveda* in particular, and is connected to Hindu religion. Atharvaveda (one of the four most ancient books of Indian knowledge, wisdom and culture) contains 114 hymns or formulations for the treatment of diseases. Ayurveda originated in and developed from these hymns (Arurveda, 2012). The true history of Ayurveda starts from the time of the Holy books, the Vedas. Ancient mythology contends that the concept and essence of Ayurveda was revealed by the creator of the world himself – Lord Brahma. Aryavarta, the native land of Aryans, covers a large area surrounding the Himalayas and is believed to be the place where the Rishis and Munis lived. At present it covers areas in countries like Nepal, India, Pakistan, Bhutan and Tibet. The civilizations of these countries are deeply influenced by the unique intellectual contributions of these Rishis and Munis (History of Ayurveda, 2012).

The Ancient Science for Human Health

Ayurveda has holistic approach towards life which takes into account person's social, psychological and spiritual layers of existence. It is a science of life and not just a way of drug therapy. Ayurveda has an integral approach unlike modern science which has lopsided, organ oriented approach and concentrates on symptomatic treatment. The most significant part of the Ayurvedic system is consideration of society along with the management and prevention of diseases and promotion of health (Modern perspective of Ayurveda, 2012).

Modern lifestyle is negative and destructive, owing to the philosophy of reductionism, age of anxiety and consumerism. Ayurvedic lifestyle, principles and practices take the person back towards holism, balance and harmony, thereby helping modern man in preventing many diseases, promoting health and inspire to live a life guided by eternal laws.

Ayurveda is the indigenous system of medicine in India. It evolved in India thousands of years ago through the efforts of people endowed with extraordinary powers of observation of nature and its complex processes. For millenniums this comprehensive system tended the healthcare needs of Indian people, rich and poor alike. The quintessence of Ayurveda lies in the unique philosophy that there is a symbiotic relationship between man and nature. This symbiosis, therefore, demands a holistic approach in treating human ailments (Ayurveda: *The Science of Life, 2012)*

Table 36.1: Ayurvedic body types.

	Vata	*Pitta*	*Kapha*
Physical characteristics	Thin. Prominent features. Cool, dry skin. Constipation. Cramps.	builu. Fair, thin hair. Warm, moist skin. Ulcers, heartburn, and hemornrhoids. Acne Average	Large build. Wavy, thickhair. Pale, cool, oily skin.Obesity, allergies, and sinus problems. High cholesterol.
Emotional characteristics	Moody. Vivacious. Imaginative. Enthusiastic. Intuitive.	Intense. Quick tempered. Intelligent. Loving. Articulate.	Relaxed. Not easily angered. Affectionate. Tolerant. Compassionate.
Behavioral characteristics	Unscheduled sleep and meal times. Nervous disorders. Anxiety.	Orderly. Structured sleep and meal times. Perfectionist.	Slow, graceful. Long sleeper and slow eater. Procrastination.

The vata dosha is associated with air and ether, and in the body promotes movement and lightness. Vata people are generally thin and light physically, dry-skinned, and very energetic and mentally restless. Pitta is associated with fire and water. In the body, it is responsible for metabolism and digestion. Pitta characteristics are medium-built bodies, fair skin, strong digestion, and good mental concentration. Kapha dosha is associated with water and earth. People characterized as kapha are generally large or heavy with more oily complexions (Ayurvedic Medicine, 2012).

Growth of Ayurveda Medicine in India

Ayurveda is the oldest existing health care system. Ayurveda's growth graph in the past decade has been impressive. Middle and upper class people who had shifted their allegiance completely to allopathy, are again trying out ayurveda and other therapies, after bad experiences with allopathy. The Indian government, with an eye on nationalistic legitimacy, is also promoting ayurveda. In Europe and America, new ayurveda hospitals are now focusing almost exclusively on *panchkarma* and allied treatments, perhaps because getting medicines ratified by drug authorities is a time-consuming and expensive process (Chopra, 2012). Ayurveda advocates that the primary and essential factors of the human body that govern our entire physical structure and function, is a combination of any two of the five bhutas with the predominance of one. This is the most fundamental and characteristic principle of Ayurveda and is called *"Tridosha"* or the Three Humours. They are categorized into *Vata, Pitta* and *Kapha*. Vata governs movement, Pitta is concerned with functions of heat, metabolism, and energy production and Kapha, governs physical structure and fluid balance (Ayurveda–The Indian Science of Life, 2012).

In the recent years, the interest in medicinal plants has increased in a great deal. Apart form this; people from the west have also taken this matter seriously by conducting various researches on plant based medicines (Ayurvedic Medicinal plants, 2012).

Indian medicinal plants are the essence of Ayurveda and Ayurvedic treatments. When used judicially and clocking with the basic principles they produce miraculous effects. Their role cannot be confined to mere curation of disease but they also used being of human body. Hence, Ayurvedic drugs are rightly called the elixirs of life. Ayurvedic Herbs played important role in Ayurvedic treatment, from ancient time to this most modern time. Indian Medicinal Plants/Herbs shows good result on disease cure. Ayurveda is the medical/Health care system which uses this as treatment base with theoretical principles (Ayurvedic Medicinal Plants, 2012).

Ayurvedic treatments differ from the majority of conventional cures in its unique approach towards healing. Ayurvedic treatments lay emphasis on examining the doshas/prakriti or the natural states of individuals before proceeding. The prakriti or the physical constitution, susceptibility to diseases, mental make-up and lifestyle of an individual is ascertained in accordance to the elemental constitution of the larger prakriti or the universe. Ayurveda takes into consideration the body, mind and soul of an individual as the unit for diagnosis. Hence, it recognizes negative emotions like anger, fear, insecurity, jealousy and greed as incorrect thinking on the part of an individual (Ayureva Treatments, 2012).

Various advanced ayurvedic treatments were born out of a necessity to keep the mind and body in perfect shapes to pursue the path of self-realization. Each individual is believed to possess undefined measures of creative capability, which, ideally, need to be realized. To achieve this, ayurveda emphasizes that the individual has to experience its oneness with the universe. The treatment methods, diet and lifestyle regimen in ayurveda are meticulously planned to heal the body as well as enrich the mind and the soul of each different individual (Holistic Health Benefits of Ayurveda, 2012).

Costs of Ayurvedic treatments can vary, with initial consultations running anywhere from $40 to over $100, with follow-up visits costing less. Herbal treatments may cost from $10 to $50 per month, and are often available from health food or bulk herb stores. Some clinics offer panchakarma, the intensive Ayurvedic detoxification treatment, which can include overnight stays for up to several weeks. The prices for these programs can vary significantly, depending on the services and length of stay (Ayurvedic Medicine, 2012). Rapid reduction in natural resources as a consequence to the expanded urbanization, global warming and reduced natural habitat posed a considerable threat to the sustainability of traditional medicine (Rastogi and Kphle, 2008).

Eradicate of Diseases and Dysfunction

Disease is defined as the state of mind wherein a person experiences discomfort, pain and injury (Diseases, 2011). Ayurveda teaches that disease arises when a person is out of harmony with the universe. Disruptions can be physical, emotional, spiritual or a combination of these. Therefore, physical, spiritual and mental health is intertwined, and all must be in balance to achieve healing and overall well-being.

In Ayurvedic medicine, each human has a prakriti. This makes up the physical and psychological characteristics of a person and the way the body functions to maintain health. Each prakriti is made up of three doshas, or life forces. The balance of the doshas affects the health of a person. Ayurvedic medical treatment focuses on eliminating impurities in the person or diet, reducing illness symptoms, increasing resistance to disease and reducing worry and increasing harmony. According to the Divyajyot Ayurveda Research Foundation, specific herbs for treating brain tumors are effective. The herbs administered for brain tumors or cancer includes Mandukaparni (Bacopa monerea) or Kasturi Bhairav Rasa and divya herbs (Olsen, 2011). Ayurveda is effecting for controlling of various diseases.

1. **Piles:** Piles which is known as haemorrhoids are caused due to swelling and inflammation of the veins present in the rectum and anus. Piles treatment through Ayurveda has been quite effective. The ayurveda treatment is a more holistic method with no or very few side-effects. Ayurveda medicines or herbs useful for reducing constipation are asafoetida, fennel and cumin. Ghee, buttermilk and green leafy vegetables in diet can also reduce constipation. Several ayurvedic remedies for treatment of piles are available. Some of the commonly used and effective ayurvedic herbs are: (a) Carrot (Daucus carota): It is effective in reducing the bleeding from piles (can be taken raw or cooked), (b) Garlic (Allium sativum): It is effective in reducing the bleeding from piles and is helpful to stop the growth of bleeding piles. Insert the garlic high up in the rectum so that it remains inside the rectum overnight, (c) Other herbs prescribed for treatment of piles are trikatu, cassia fistula, mimosa pudica, ailanthus excels, blumea balsamifera, and semicarpus anacardium (Sachdev, 2011).
2. **Heart:** As per Ayurveda, heart problems are essentially vata disorders and it treats the cause, not the symptom. Diet is extremely important in ayurveda along with panchakarma and marma. Panchakarma is needed to flush out toxins accumulated in the body. The heart dhara treatment improves blood circulation, removes blocks, and strengthens cardiac muscles. A warm liquid containing crushed Ayurvedic herbs, water, and oil is poured into dough (made of urad dal flour and water) and placed on the heart area for about 20-30 minutes. Marma chikitsa is a therapy wherein the doctor uses his fingers to stimulate various vital energy points in the body for free flow of energy which helps to remove heart blocks (Ayurveda for Heart Health, 2011).
3. **Cancer:** According to ayurvedic thought, having the three doshas out of sync might be a cause of cancer. One possible explanation for the rise in cancer diagnoses in modern times can be traced to the increased use of chemicals and other synthetic materials in our diets. These chemicals prevent us from obtaining the natural minerals our bodies need in order to function correctly. Also, anxiety and stress lead to increased toxins in the body, which then puts us at higher risk for cancer. Because there are so many different types of cancer, this makes determining the causes and treatments that much more difficult for ayurvedic practitioners. It can be solved by medicines, diet and panchakarma. *Panchakarma* is a set of five ayurvedic remedies that may or may not all be used as part of a therapeutic

cleansing program (Cancer Natural Ayurvedic Cure | Cancer Remedies, 2011).

4. **Skin:** Abhyangam is a popular body massage in ayurveda. The word abhyanga itself means to massage. The Massage is done with a combination of specially prepared Ayurvedic herbal oils and applied all over the body stimulating the vital points. This is very good for the general health of the skin and prevents early aging and relieves muscular aches and pain. After the massage herbal steam bath is also given. Abhyangam can increase the production of white blood corpuscles and antibodies, which provide more resistance against viruses and diseases.
5. **Blood Pressure:** Hypertension is one of the major causes of disability and death all over the world. Hypertension causes heart attacks, strokes, kidney failures and other disorder if it is neglected and remains untreated. According to Ayurveda, high blood pressure involves all the doshas, the heart, and the blood vessels. People with Pitta and Vata predominante constitution and Pitta and Vata imbalance, are more prone to hypertension than any other. Unprocessed anger, frustration, irritability, anxiety and fear leads to mal-adaptation of the endocrine system, which then leads to conditions like hypertension (Ayurvedic Treatment for Hypertension, 2011).
6. **Gastric:** In Ayurveda gas related problems are known as '*Adhmaan.*' Adhmaan is caused in body by the aggravated or disturbed Vata dosha. Yurveda attributes the gastric problems to the vitiation of Vata dosha. Due to the improperly digested food there is retention of wind or gas along with feces in the bowel. Hinguvastaka Powder works well with other Ayurvedic medicine like Aarogyavardhini vati, Avipattikara Churna, Shankha vati, Shankha Bhasma, Kumaryasava and Rasonadi vati (Ayurveda and Gastric Problems, 2012).
7. **Thyroid:** Thyroid disease is mentions in detail by Charaka, one of the writers of the Ayurvedic *materia medica*. There is a mention that milk, barley, green grams, Bengal grams, sugarcane juice, cucumber and milk products are good for thyroid health. Sour products, on the other hand, aggravate thyroid conditions like hyperthyroidism and hypothyroidism. The Ayurvedic cure for specific thyroid problems involves the use of a potent herb, Kaanchanara. Another herb of the genus of East Indian and African trees that yields balsamic products known as Guggulu is widely used for addressing symptoms of thyroid problems like weight gain (Thompson, 2012).
8. **Menstrual cycle:** Menstrual cycle is one thing that decides majority of women's health. Most of the women experience painful monthly menses. A painless and problem free menstrual cycle depends on the proper functioning of the endocrine glands to stimulate hormone secretion. With a healthy lifestyle and food habits, one can experience a healthy menstrual cycle. However, if menstrual problems persist, Ayurveda has a cure in the form of many Indian herbs and home remedies that can give relief to the suffering women. The herbs for treating menstrual problems are asafetida,

henna, ginger, aloe vera, valerian, castor oil, ashwagandha, liquorice etc (Menstrual Problems, 2012).

9. **Breast cancer:** Breast cancer starts with the tissues of the breast. There are two types of breast cancer namely, ductal carcinoma (starts from ducts or tubes that that moves milk to nipples) and lobular carcinoma (starts from lobules). Some of the conventional treatment methods for this cancer are radiation therapy, chemotherapy, hormonal therapy and surgery. All of them have their share of side-effects but when these conventional lines of treatment are used with ayurvedic remedies, one gets faster results with minimum side effects. According to ayurveda, breast cancer is caused due to imbalance in three doshas namely, vata, pitta and kapha. This imbalance leads to the formulation of cysts leading to breast cancer. Ayurvedic medicines that act on the *'Mansa'* dhatus (tissues),*'Rasa'*, and *'Rakta'* are effective for breast cancer treatment (Kafee, 2011).
10. **Tuberculosis:** According to modern medical studies, Tuberculosis is caused by entry of tubercle bacillus though the nasal passage, all the way to the lungs. It multiplies millions of times, and leave tiny spots called tubercles all over the body. However, according to the ayurvedic perspective, tuberculosis is caused by the lowering of the body's resistance to diseases entering the body. The following is a list of Ayurvedic Home Remedies for Tuberculosis. These have been proven to be quite helpful in treating tuberculosis: Milk Diet, Custard Apple, Goose Berry, Pineapple, Orange etc (Tuberculosis Treatment | Ayurvedic Herbal Natural Cure for Tuberculosis, 2011).
11. **Paralysis:** Paralysis is one of those diseases which need quick treatment and care. The body is said to be in paralyzed stage when there is complete loss of muscle function for one or more group. This is mainly caused because of the damage caused to the nervous system. Minor paralysis can be cured easily, but the major ones need more time to be cured. Paralysis can consume a human body completely or partially. Paralysis mainly occurs due to strokes or injuries. Minor paralysis can be cured easily, but the major ones need more time to be cured. Ashwagandha commonly known as Winter Cherry is used to cure paralysis. Bala known as Country Mallow is also considered as an effective herb for the treatment of paralysis. Apart from this, special therapies and other types of treatments are used to cure paralysis in Ayurveda. Apart from this, some therapies involve application of medicated oils in the affected regions (Ayurveda can cure Paralysis, 1999).
12. **Hepatitis:** Kamalahar is an indigenous ayurvedic medicine that is very effective in the treatment of hepatitis, jaundice, fatty liver, cirrhosis, ascites and other liver disorders. Kamalahar prevents hepatic disorders and stimulates appetite and growth. It also helps in the regeneration of hepatic cells. Kamalahar is enriched with Tecoma undulata, Phyllanthus urinaria, Embelia ribes, Taraxacum officinale, Mineral salts, Nyctanthes arbortritis and Terminalia Arjuna (Kamalahar, 2011).

13. **Teeth:** Ayurveda, an ancient Indian medical science has documented various methods by which any one can restore the whiteness of teeth. For clean and white teeth, pungent, bitter and astringent foods must form a regular part of diet. These tastes have a sharp effect on the teeth, thus keeping them clean. They also prevent excess buildup of plaque on the teeth as plaque can make teeth lose their whiteness. Ayurveda offers various types of toothpastes and toothbrushes for getting white healthy teeth. One of them is toothpaste produced by mixing a paste of holy basil leaves with mustard oil. The neem and babul twigs can be used as toothbrushes. They are rubbed strongly on the teeth (at times with charcoal) to clear all accumulated plaque on the teeth and increase their whiteness (Ayurvedic Remedy for Teeth Whitening, 2011).
14. **Blood Sugar:** According to ayurveda, diabetes is a metabolic kapha type of disorder in which diminished functioning of agni leads to a tendency toward high blood sugar. Ayurvedic practitioners also use several herbal preparations for diabetics. Exercise is another cornerstone of ayurvedic treatment of diabetes. Yoga and breathing exercises are traditionally used. The most important herbs for all doshas are shilajit, gudmar turmeric, neem, amalaki, guggul, and arjuna. Turmeric with aloe vera gel (1 to 3 gms./.035 to.1 oz) is best used during the early stages of diabetes for regulating pancreas and liver functions (Ayurvedic treatment for diabetic, 2007).

Concluding Observation

Ayurvedic treatments are primarily dietary and herbal. Ayurvedic medicine is a form of alternative medicine that is based upon traditional remedies that originate in India. Ayurveda says that there is nothing in this world that is not medicine. The entire science of ayurveda is based on the *'Five Great Elements'* (Panchabhuta) theory. These five elements are earth (prithvi), water (jal), fire (Agni or tej), air (vayu) and ether or space (akash).

Indian civilization and ayurveda are interconnected. Growth of ayurvedic treatment facilities are increasing very rapidly in India. Ayurvedic medicines are useful for treatment of various diseases like piles, heart, cancer, skin problem, blood pressure, gastric, thyroid, menstrual cycle, breast cancer, tuberculosis, paralysis, hepatitis, blood sugar etc.

References

Aruveda, 2012, Retrieved from http://en.wikipedia.org/wiki/Ayurveda

Ayurveda can cure Paralysis, 1999. Retrieved from http://www.damthemovie.com/ayurveda–cure–paralysis.php

Ayurveda and Gastric Problems, 2012. Retrieved from http://www.skinsheen.com/skin–ayurveda–and–gastric–problems–974.aspx

Ayurveda–Basic Concepts of Ayurveda, 2012. Retrieved from http://www.lifepositive.com/Body/Ayurveda/Basic_Concepts_of_Ayurveda1 2003.asp

Ayurveda for Heart Health, 2011. Retrieved from http://www.lifepositive.com/body/body–holistic/heart/heart_ayurveda.asp

Ayurvedic Medicinal Plants, 2012. Retrieved from http://ayurvedicmedicinalplants.com/

Ayurvedic Medicinal Plants, 2012. Retrieved from http://www.indianmedicinalplants.info/

Ayurvedic Medicine Plants, 2012. Retrieved from http://medical–dictionary.thefreedictionary.com/ayurvedic+medicine

Ayurvedic treatment for diabetic, 2007. Treatment http://diabetcare.blogspot.in/2007/10/ayurvedic–treatment–for–diabetic.html

Ayurvedic Remedy For Teeth Whitening, 2011. retrieved from http://ayurveda–foryou.com/treat/teeth_whitening.html

Ayurveda: The Science of Life, 2012. Retrieved from http://www.astrologycom.com/ayurveda.html

Ayureva Treatments, 2012. Retrieved from http://www.lifepositive.com/Body/ayurveda/ayurvedic–treatments.asp

Ayurvedic Treatment for Hypertension, 2011. Retrieved from http://ayurveda–foryou.com/treat/hypertension.html

Ayurveda–The Indian Science of Life, 2012. Retrieved from http://www.prokerala.com/health/ayurveda/

Cancer Natural Ayurvedic Cure | Cancer Remedies, 2011. Retrieved from http://www.ayurvedic–medicines.com/cancer.html

Chopra, P, 2012. Ayurveda–Medicine and treatment in India, Detailed available at http://www.lifepositive.com/Body/ayurveda/ayurveda–India.asp

Chopra, P, 2012. Ayurveda–Kerala: A healthy revival of Ayurveda, Detailed available at http://www.lifepositive.com/Body/ayurveda/ayurveda–India–article.asp

Diseases, 2011. Retrieved from http://ayurveda.iloveindia.com/ayurveda–fundamentals/ayurveda–diognosis.html

History of Aruveda, 2012. Retrieved from http://www.medindia.net/ayurveda/index.asp

Holistic Health Benefits of Ayurveda, 2012. Retrieved from http://www.lifepositive.com/Body/ayurveda/holistic–health.asp

Introduction, 2012. Ayurvedic Treatment for Breast Cancer, Retrieved from http://www.lifepositive.com/Body/ayurveda/ayurveda–theory.asp

Kafeel, B, 2011. Retrieved from http://www.onlymyhealth.com/ayurvedic–treatment–breast–cancer–1309929424

Kamalahar, 2011. Retrieved from http://www.khatorepharma.com/products/kamalahar.html

Menstrual Problems, 2012. Retrieved from http://health.indiamart.com/ayurveda/ayurvedic–tips/ayurvedic–cure–menstrual–problems.html

Modern perspective of Ayurveda, 2012. Retrieved from http://www.gdpau.com/ayurveda.html

Rastogi and Kphle., 2008. Sustainable Traditional Medicine: Taking the Inspirations from Ancient Veterinary Science, Evidence–Based Complementary and Alternative Medicine. Volume 2011, 2011. Article ID 151435, p.1, Retrieved from http://www.hindawi.com/journals/ecam/2011/151435/

Sachdev, P, 2011. Ayurvedic Treatment for Piles, Retrieved from http://www.onlymyhealth.com/ayurvedic–treatment–piles–1303103483

School of Thought, 2012. Retrieved from http://www.lifepositive.com/Body/ayurveda/ayurveda–schools.asp

Singh, 2012. Retrieved from http://www.gdpau.com/ayurveda.html

The ancient science of holistic healing(2012. Retrieved from http://www.lifepositive.com/body/ayurveda/ayurveda.asp

Thompson, T, 2012. Retrieved from http://www.nativeremedies.com/articles/ayurvedic–cure–for–thyroid–disease.html

Tuberculosis Treatment | Ayurvedic Herbal Natural Cure for Tuberculosis, 2011. Retrieved from http://www.ayurvedic–medicines.com/tuberculosis.html

What is Ayurveda?, 2012. Retrieved from http://www.gdpau.com/ayurveda.html

Environment, Biodiversity and Traditional System, 2014 *Pages* 391–402
ISBN 978-81-7622-307-2
Editors: D.R. Khanna, G.S. Solanki and S.K. Pathak
Published by: BIOTECH BOOKS, NEW DELHI

Chapter 37

Diversity and Distribution of Predacious Coccinellids (Coleoptera: Coccinellidae) in Agro- and Forest Ecosystems of Tripura, Northeast India

*Joydeb Majumder and B.K. Agarwala**

Ecology and Biodiversity Laboratories,
Department of Zoology, Tripura University,
Suryamaninagar – 799 022, Tripura

ABSTRACT

Present inventory was carried out in two types of habitats namely, agro- and forest ecosystems for predacious coccinellids in Tripura, northeast India. Random collection was made by hand picking and netting during January 2008 to December 2010. A total of 1627 individuals of coccinellids representing 24 species under 17 genera were collected. Of which 870 individuals, 18 species was recorded from agro-ecosystem and 757 individuals, 22 species was recorded from forest ecosystems, respectively. Rank distribution of overall species based on relative abundance showed that 24 species distributed in 21 ranks. Diversity indices and individual based rarefaction curve revealed that species richness and diversity of coccinellids was more in forest ecosystem than the agro-ecosystems. Principle component

* *Corresponding Author:* E-mail: bagarwala00@gmail.com

analysis showed that temperature and relative humidity has negative impact on coccinellids diversity of both ecosystems. Besides, habitat-specific species recorded in this study can be used as bio-indicators.

Keywords: Coccinellidae, Diversity, Agro-ecosystem, Tripura, Northeast India.

Introduction

Worldwide continuous loss of tropical forests shifted the focus and interests of scientists, conservationists and ecologists towards managed agricultural areas along with natural forest ecosystems for biodiversity conservation of flora and fauna (Putz *et al.*, 2001; Donald, 2004; McNeely, 2004).

Among the predatory insects, ladybird beetles are one of the most economically important groups, and are widespread in agricultural and forest ecosystems. Both larvae and adults of coccinellidae (excluding most of the members of the subfamily Epilachninae) feed on a number of distantly related phytophagous insects (Agarwala and Ghosh, 1988; Hodek and Honek, 1996; Iperti and Paoletti, 1999) like aphids, coccids, jassids, scale insects, thrips, leafhoppers, mites and other soft bodied insect pests of the agricultural and horticultural crops (Omkar and Pervez, 2000), and some are very prey specific (Agarwala and Ghosh, 1984). They are frequently observed to forage in unstable habitats with variable prey density (Dixon, 2000; Agarwala *et al.*, 2001) of agro- and forest ecosystems, because in natural environments, prey abundance is controlled by individual or combined actions of natural enemies, deterioration in host quality or abiotic factors (Helden *et al.*, 1994; Dixon, 1998). However, most of the ladybirds synchronizes their population with that of their specific prey population (Dixon, 1998), and have developed various defensive mechanisms to deter their natural enemies (Agarwala and Bhowmik, 2011). Many studies reported that most of the species showed high feeding potential, foraging activity and reproductive efficiency, which are the characteristics of effective biocontrol agent used in IPM (Fernandez-Arhex and Corley, 2003; Kohno, 2003). With regard to biodiversity, the significance of predacious coccinellids lies in the fact that they keep the prey population under control, thereby contribute to the maintenance of the ecosystem balance because control of natural pest reduces dependence on chemical pesticides, which are considered harmful to environment and promote resistance in pests (Murdoch, 1975). World wide 4200 species of coccinellids have been reported, of which about 90 per cent are known to be predaceous on different insect and acarine pests (Iperti and Paoletti, 1999). Out of these, 119 predacious species was known from India (Poorani, 2002). These include 46 species reported from Manipur and Nagaland (Shantibala and Singh, 1991), and 38 species reported from Manipur alone (Chakraborty, 2004). Till date, 35 species each were reported from Arunachal Pradesh (Hemcahandra *et al.*, 2010) and Tripura (Chaktaborty and Biswas, 2002), and, 49 species were from Meghalaya (Chakraborty and Biswas, 2000). Aim of the present study was to record the diversity and distribution pattern of coccinellids in agro- and forest ecosystems with respect to abiotic factors in the state of Tripura, northeast India.

Method and Materials

Study Area

Tripura is a small hilly state of northeast India surrounded by Bangladesh on three sides, and by states of Mizoram and Assam one side. The state geographically lies between 22° 56' to 24°32' N and 91°10' to 92° 21' E (Figure 37.1), and is an integral part of the Indo-Burma biodiversity hotspot and, thus, occupies a critical position in the global biodiversity scene (Myers *et al.*, 2000). Topography of Tripura consists of hilly terrain with undulating slopes, plains and valleys varying from semi-evergreen to moist deciduous and secondary bamboo brakes which, in turn, support diverse faunal communities. It constitutes about 0.3 per cent of the total geographical area of the country (Kshirsagar and Upadhyay, 2009). Climatic condition of Tripura is tropical with annual rainfall of about 2100 mm due to southwest monsoon and temperature ranges between 10 °C in winter to 37 °C in summer, with altitudes varying from 15 to 850 m above sea level. The important vegetation types of the state are the tropical semi-evergreen rain forests, tropical moist deciduous forests, tropical dry deciduous forests, riparian forests, forest plantations, grasslands as well as agro ecosystems comprising of paddy, banana, pineapple, vegetables and plantation crops such as arecanut, coconut, rubber etc. Orange orchards and coffee plantations are present in hills of the state.

Because of increasing population pressure, many patches of natural vegetation in Tripura are under threat. The agro-ecosystem is also subject to disturbances due to filling of paddy fields, adoption of modern agricultural practices leading to large-scale application of chemical fertilizers and pesticides. As a result, the microhabitats of many groups of organisms are affected which has tremendous implications on their survival.

Field Survey and Insect Collection

A study was conducted in eight different locations of the state during January 2008 to December 2010 for collection of predacious coccinellids. Of the locations, four locations namely, Debbari, Baramura, Saikabari, and Jampui hills are forested habitats and remaining four locations namely Bagabasa, Shanmura, Halhali, and Kailasahar are agriculture dominated areas. Prior to field study a reconnaissance survey was carried out in December 2007 to identify the sampling sites, one each in eight locations. Physical, geographical and habitat features of each sampling locations are provided in Table 37.1. Insects were collected by hand picking and netting, twice in a month, at each sampling sites at an interval of 15 days, and total population per month was counted. All the collected specimens from each sampling sites were treated separately and are preserved in the Ecology and Biodiversity laboratories of Tripura University. Collected specimens were identified to species level and few of the specimens were send to subject expert for proper identification. During the study, abiotic factors like temperature and relative humidity was recorded for each sampling day and were averaged for each month.

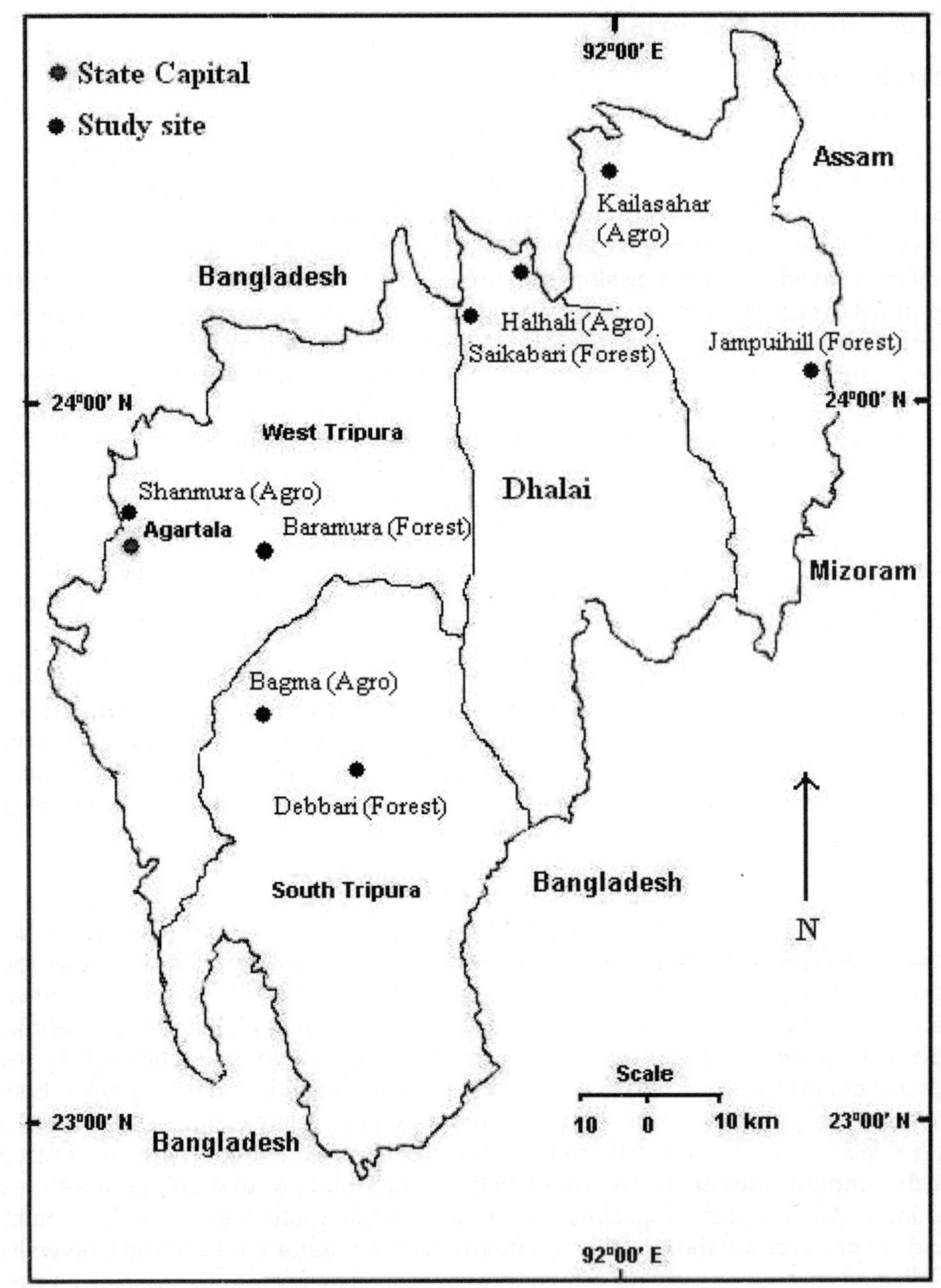

Figure 37.1: Map of Tripura showing study sites.

Data Analysis

Raw data of this study were subjected to statistical analysis to calculate overall Shannon diversity, Simpson's dominance and Pielou evenness index for agro- and

Table 37.1: Geo-coordinate position and vegetation characteristics of the study sites [agro- (A) and forest (F)] of Tripura.

Study Aite	*Geo-coordinates*	*Altitude (m) asl.*	*Physical and Vegetation Structure*
Debbari (F)	N 23°32.0532 E 91°36.5982	35	Most of the area is covered with mature secondary moist deciduous riparian woody forests dominated by *Shorea robusta*, *Dipterocarpus turbinatus*, and *Terminalia belliraca*, *Ficus racemosa*, *saraca asoca* plants. Besides these, a number of bamboo species, epiphytes, climbers and flowering herbs and shrubs are also present. Presence of moderate sign of anthropogenic pressures.
Baramura (F)	N 24°4.052 E 91°44.392	462	Topography is uneven and most of the area is inaccessible due to presence of undulating terrain and rich forest. Vegetations of the area dominated by vascular woody plants like *Shorea robusta*, *Artocarpus chaplasha* and bamboos, representing secondary moist deciduous forests. Plantation of *Tectona grandis* trees along road sides and citrus plant in the valley were also found. Anthropogenic pressures are moderate.
Saikabari (F)	N 24°6.452 E 91°53.152	368	Landscape of the area consists of undulating soft rocky hills. Several small streams flowing across the area. Vegetation profile of the area represents mature moist secondary forests dominated by *Melocanna baccifera*, *Mangifera indica*, *Tectona grandis*, *Duabanga grandiflora*, *Garcinia cowa* etc. Logging, shifting cultivation, bamboo extraction, grazing by domestic cattle is found to be frequent in this location.
Jampuihills (F)	N 23°48.6452 E 92°15.6822	837	Hilly forested area bordered with Mizoram state in the east side. Vegetation is of semi-evergreen moist deciduous type and dominated by woody plants like *Michelia champaca*, *Melocanna baccifera*, *Mangifera indica*, *Saraca asoca*, *Duabanga grandiflora*, *Mitragyna rotundifolia* along with under storey herb, shrub, epiphytes, orchids and bamboo of different kinds. Most of the area is undisturbed with little fragmentation of forest due to shifting cultivation.
Bagabasa (A)	N 23°45.6522 E 91°15.9582	766	Open agricultural field surrounded by patches of fragmented secondary forests. Paddy is the dominant crop and other seasonal crops and vegetables are also cultivated in this area. Excessive use of chemical fertilizer and pesticides are the major threats.
Shanmura (A)	N 23º50.542 E 91°15.082	12	Landscapes of the area are plain, comprising of intensively managed agricultural lands dominated by paddy and other seasonal crops and vegetables with patches of herbs and shrubs.
Halhali (A)	N 24°6.182 E 91°49.162	78	Topographically it is a valley area surrounded by secondary moist deciduous forests. Landscape is heterogeneous, comprising of agricultural plains mostly dominated by paddy, vegetables and patches of *Tectona grandis*, *Hevea brasiliensis* plantations. Adjacent hilly forests understory was found to be poor due to soil erosion in low catchments areas.
Kailasahar (A)	N 24°15.572 E 91°57.492	34	Landscapes of the area are mostly plains consisting of managed agricultural lands, pasture lands and patches of herbs, shrubs at the periphery of agro-ecosystem. Agricultural field is dominated by different seasonal crops. Few patches of afforested *Tectona grandis* and *Areca catechu* plants are other features.

forest ecosystems (Magurran, 1988). Species recorded in this study were ranked based on their relative abundance. Data of species richness counts of two years from agro- and forest sites were pooled to get rarefaction curves for comparison of estimated species richness between the habitats. To show field population trends of coccinellids in agro and forest ecosystem with relation to abiotic factors, double axis line graph was drawn. PAST version 2 (Hamer *et al.*, 2001) was used to determine diversity indices and Biodiversity Pro version 2 (Lambshead *et al.*, 1997) was used to determine rarefaction curves. Rank abundance diagram, population line graph and t-test analysis were performed by Origin version 5 (Micrococal Software, Inc.).

Results

Species Richness and Diversity

A total of 1627 individuals of predacious beetles belonging to the Coccinellidae family, representing 24 different species under 17 genera were collected during the study period. A list of the species recorded from agro- and forest ecosystems are provided in Table 37.2 and showed data on the specie abundance for each habitat, which lead to a separate diversity analysis (Table 37.3). Of the 24 species recorded, 18 species (75.00 per cent) was recorded in agro- and 22 species (91.67 per cent) were recorded at the forest sites. Species overlap between agro- and forest ecosystems was 14 (58.33 per cent) of all recorded species and the similarity value is 0.62. Fourteen species (58.33 per cent) are recorded here as new records for the state (Table 37.2). Six species namely, *Anisolemnia dilatata* (F.), *Harmonia sedecimnotata* (F.), *Jauravia quadrinotata* Kapur, *Oenopia kyrbyi* Mulsant, *Pseudaspidimerus* sp. and *Rodolia* sp.were found only in forested habitats, 4 species *viz. Brumoides suturalis* (F.), *Cryptogonus quadriguttatus* (Weise), *Harmonia dimidiata* (F.) and *Propylea luteopustulata* (Mulsant) were restricted to only agro-ecosystems, and another 14 species were presented in both the ecosystems. High species richness was recorded in moderate to least disturbed forested ecosystems which is evident from the expected number of species generated by individual based rarefaction analysis, than in agro-ecosystem (Figure 37.2)

Diversity analysis showed that coccinellidae species were more diverse over forest ecosystems (*Hs*= 2.33) than in agro-ecosystems (*Hs*= 2.30). The *J* value showed that in forest ecosystem relative abundance or evenness within 24 species was more than agro-ecosystems (*J'*= 0.78) with dominance of 13 per cent (0.13) species (Table 37.3).

Rank Abundance

Rank distribution of recorded coccinellids species showed that 24 species are distributed into 21 ranks. *Micraspis discolor* (F.), *Cheilomenes sexmaculatus* (F.) and *Coccinella transeversalis* F. occupied ranks 1, 2 and 3, having relative abundance of 24.52 per cent, 18.13 per cent, and 11.99 per cent, respectively. Next 14 species are ranked between 4 and 16 with abundance range of 9.40 per cent to 1.04 per cent and those together accounted 43.02 per cent of total abundance. Remaining 7 species [*Coelophora bisellata* Mulsant, *Pseudaspidimerus* sp., *Cryptolaemus montrouzieri* Mulsant, *Harmonia dimidiata* (F.), *Harmonia sedecimnotata* (F.), *Rodolia* sp., *Cryptogonus quadriguttatus* (Weise)] are ranked between 17 to 21 and cumulatively accounted 2.34 per cent of total abundance (Table 37.2; Figure 37.3).

Table 37.2: List of coccinellid species recorded in agro- and forest ecosystems of Tripura.

Zoological Name	*No. of Specimens Collected in Ecosystems*		*Relative Abundance (per cent)*	*Species Rank*
	Agro	*Forest*		
Anisolemnia dilatata (F.) * +	0	76	4.67	6
Brumoides suturalis (F.) –	45	0	2.77	9
Chilochorous circumdatus Schonherr *	6	17	1.41	14
Chilochorous nigrita (F.)	25	2	1.66	13
Cheilomenes sexmaculatus (F.)	193	102	18.13	2
Coccinella septempunctata L.	80	73	9.40	4
Coccinella transeversalis F.	89	106	11.99	3
Cryptogonus quadriguttatus (Weise)* –	3	0	0.18	21
Coelophora bisellata Mulsant	7	4	0.68	17
Cryptolaemus montrouzieri Mulsant *	3	3	0.37	18
Harmonia dimidiata (F.) * –	5	0	0.31	19
Harmonia octomaculata (F.) *	24	22	2.83	8
Harmonia sedecimnotata (F.) * +	0	4	0.25	20
Harmonia brumoidesaturalis *	4	13	1.04	16
Harmonia eucharis (Mulsant) *	32	2	2.09	11
Jauravia quadrinotata Kapur * +	0	19	1.17	15
Micraspis discolor (F.)	203	196	24.52	1
Oenopia kyrbyi Mulsant * +	0	17	1.04	16
Propylea luteopustulata (Mulsant) * –	28	0	1.72	12
Pseudaspidimerus sp.+	0	6	0.37	18
Rodolia sp.+	0	3	0.18	21
Scymnus (*Pullus*) *latemaculatus* Motschulsky *	66	43	6.70	5
Scymnus sp.	30	38	4.18	7
Synonycha grandis (Thunberg) *	27	11	2.34	10

Symbol following species name denote: * new records for the state, + present only in forest ecosystem,– present only in agro-ecosystem.

Effect of Abiotic Factors on Coccinellids Abundance

Phenological studies were based on the number of individuals caught in each month with relation to environmental factors. Adult predacious coccinellids abundance showed considerable variation throughout the year. Temporal representations of the ladybird populations for the two ecosystems are shown (Figure

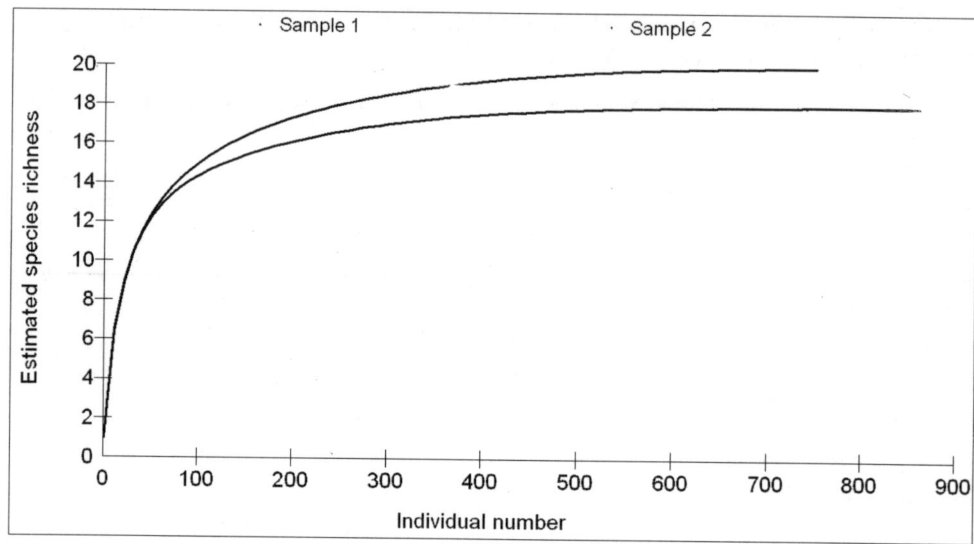

Figure 37.2: Individual based rarefaction curve showing the estimated species richness of predacious coccinellids in agro and forest ecosystem (Sample 1: Agro and Sample 2: Forest).

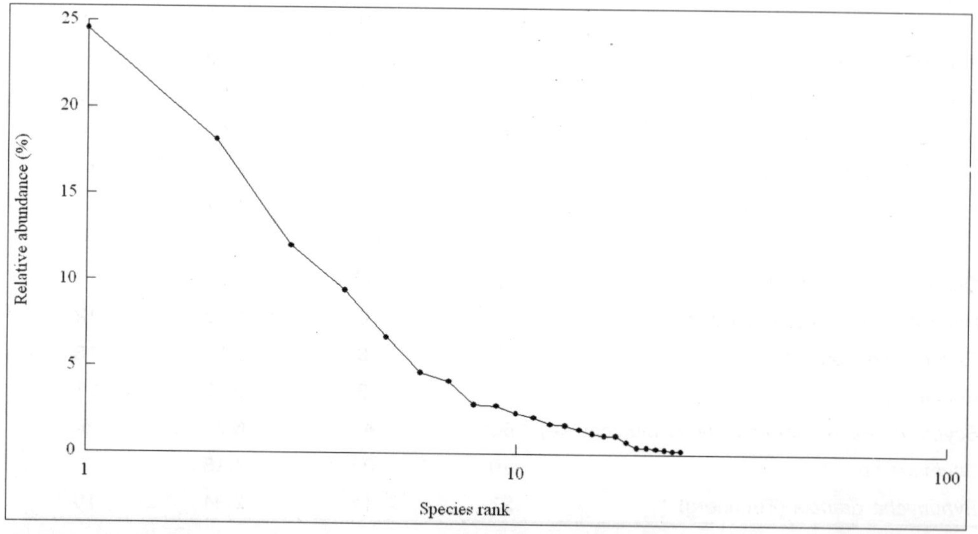

Figure 37.3: Rank abundance diagram showing distribution of predacious lady beetles in Tripura, northeast India.

37.4). The maximum number of individuals recorded in the month of October and November and minimum in the month of March and April (Figure 37.4). Recorded abiotic factors showed significant negative correlations on the coccinellid abundance of agro (temp: $r = -0.29$, RH: $r = -0.35$; $P > 0.05$) and forest (temp: $r = -0.30$, RH: $r = -0.32$; $P > 0.05$) ecosystems.

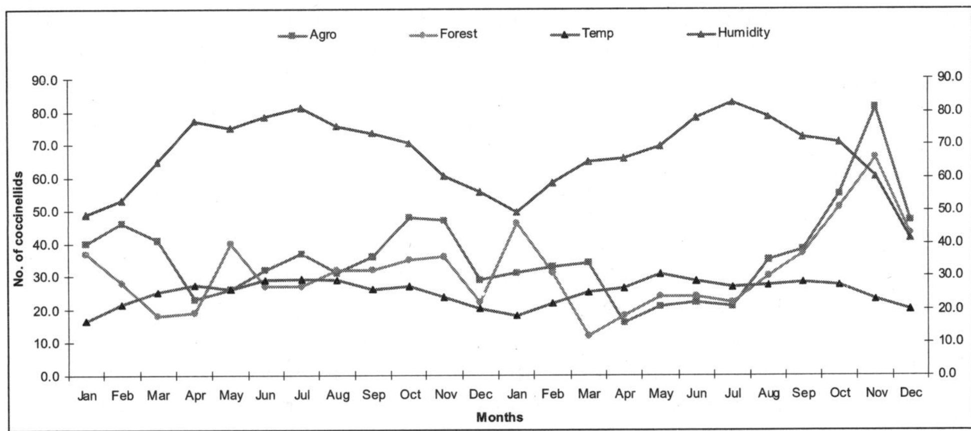

Figure 37.4: Showing the effects of mean temperature and humidity on predacious coccinellids in agro and forest ecosystem during 2008 to 2010.

Table 37.3: Showing number of species, individuals and diversity indices in agro- and forest ecosystems.

Habitat	*No. of Species*	*No. of Individuals*	*Shannon Diversity (Hs)*	*Simpson's Dominance (Dm)*	*Pilou Evenness (J')*
Agro-ecosystem	18	870	2.30	0.14	0.80
Forest ecosystem	20	757	2.33	0.13	0.78
Overall (agro + forest)	24	1627	2.43	0.13	0.77

Discussion

This is the first quantitative study on predacious coccinellids diversity of agro- and forest ecosystems of Tripura state, northeast India. Comparison of abundance, species richness and diversity indices between agro- and forest ecosystems showed that faunal diversity and conservation potential of coccinellids is higher in forest ecosystems. High abundance of *Micraspis discolor* (F.), *Cheilomenes sexmaculatus* (F.) and *Coccinella transeversalis* F. is obvious because these species are generalist in their feeding habit are known to maintain their population throughout the year (Agarwala and Ghosh, 1984; Agarwala and Yasuda 2000; Omkar and Bind 2004; Omkar and James 2004). Ranking of 24 species according to their abundance clearly showed that distribution of different species in the agro- and forest ecosystems is uneven which suggested that the distribution of coccinellids depends on several biotic (prey resource) and abiotic (temperature and humidity) factors of habitats. The biotic properties of the environment have been recognized as strong determinants of the population dynamic of ecosystems (Leonard *et al.*, 1998). Several studies showed that animal numbers are driven by interactions between abiotic and biotic factors, such as competition and predation (Krebs, 2001; Begon *et al.*, 2005). In the present study all measured abiotic parameters showed significant poor negative impact on predator populations. This might be due to the fact that abiotic conditions can modulate predatory impacts on herbivores (Fraser, 1998; Fraser and Grime, 1998) and vice versa. Variation in monthly collected populations was attributed to the ecological conditions of the study locations. The ecological conditions *i.e.* the monsoon season and the rapid growth of plants, habitat quality and climatic factors caused the dispersal of insects within afro- and forest ecosystems.

Conclusion

This is the first quantitative assessment of predacious coccinellids in the state of Tripura. Present study revealed that coccinellid diversity in Tripura state is rich in forests than in agro-ecosystems, confirming the key role of ecological infrastructures within forested ecosystem and our study, improving the knowledge base of the phenology of predacious ladybirds in forested ecosystems of our region. Six species occurred exclusively in forest ecosystems, and these species can be used as bioindicators to evaluate the overall conservation status of forests in Tripura. Further studies were needed of those areas that were not covered in this study to fully explore predatory coccinellids fauna of Tripura state.

Acknowledgements

Authors are thankful to ICAR, New Delhi for financial assistance. We are indebted to Dr. V. V. Ramamurthy for his guidance and suggestions during the study. Thanks are also due to Dr. Jhuma Das, Dr. Parna Bhadra and Dr. P. P. Bhattacharjee for their helps during the study. We appreciate the helping hands of Sri. D. K. Sinha and Sri. Dipten Laskar.

References

Agarwala, B.K. and Ghosh, A.K., 1984. Indian check list. *Record of Zoological Survey of India, Occasional Paper*, 50: 71 pp.

Agarwala, B.K., Bardhanroy, P., Yasuda, H. and Takizawa, T., 2001. Prey consumption and oviposition of the Aphidophagous predator *Menochilus sexmaculatus* (Coleoptera: Coccinellidae) in relation to prey density and adult size. *Environmental Entomology*, 30: 1182–1187.

Agarwala, B.K. and Ghosh, A.K., 1988. Prey records of Aphidophagous Coccinellidae in India. A review and bibliography. *Tropical Pest Management*, 34: 1–14.

Agarwala, B.K. and Yasuda, H., 2000. Competitive ability of ladybird predators of aphids: a review of *Cheilomenes sexmaculata* (Fabr.) (Coleoptera: Coccinellidae) with a worldwide checklist of preys. *Journal of Aphidology* 14: 1–20.

Agarwala, B.K. and Bhowmik, P.J., 2011. Effect of resource gradient on age and size at maturity and their influence on early–life fecundity in the predatory Asian lady beetle, *Harmonia axyridis*. *Entomolia Experimentalis et Applicata*, 141: 97–102.

Begon, M., Townsend, C.R. and Harper, J.L., 2005. *Ecology: From Individuals to Ecosystems*, 4th edn. Blackwell Science, Oxford.

Chakraborty, S.K. and Biswas, S., 2000. Insecta: Coleoptera: Coccinellidae. State Fauna Series 4: Fauna of Meghalaya, *Zoological Survey of India*, 5: 201–233.

Chakraborty, S.K. and Biswas, S., 2002. Insecta: Coleoptera: Coccinellidae. State Fauna Series 7: Fauna of Tripura, *Zoological Survey of India*, 3: 115–137.

Chakaraborty, S.K., 2004. Insecta: Coleoptera: Coccinellidae. State Fauna Seris 10: Fauna of Manipur, *Zoological Survey of India*, 10: 401–419.

Dixon, A.F.G., 1998. *Aphid ecology: an optimization approach.* Chapman and Hall, London.

Dixon, A.F.G., 2000. *Insect Predator–Prey Dynamics: Ladybird Beetles and Biological Control.* Cambridge University Press, London.

Donald, P.F., 2004. Biodiversity impacts of some agricultural commodity production systems. *Conservation Biology*, 18: 17–37.

Fraser, L.H., 1998. Top–down vs. bottom–up control inûuenced by productivity in a North Derbyshire, UK, dale. *Oikos*, 81: 99–108.

Fraser, L.H. and Grime, J.P., 1998. Top–down control and its effect on the biomass and composition of three grasses at high and low soil fertility in outdoor microcosms. *Oecologia*, 113: 239–246.

Fernández–Arhex, V. and Corley, J.C., 2003. The functional response of parasitoids and its implications for biological control. *Biocontrol Science and Technology*, 13: 403–413.

Helden, A.J., Dixon, A.F.G. and Carter, N., 1994. Environmental factors and morphological discrimination between spring and summer migrants of the grain aphid, *Sitobion avenae* (Homoptera: Aphididae). *European Journal of Entomology*, 91: 23–28.

Hodek, I. and Honek, A., 1996. *Ecology of Coccinellidae.* Kluwer Academic Publishers, The Netherlands.

Hammer, O., Harper, D.A.T. and Ryan, P.D., 2001. PAST: paleontological statistical software package for education and data analysis. *Palaeontologia Electronica*, 4: 1.4.

Hemchandra, O., Kalita, J. and Singh, T.K., 2010. Biodiversity of Aphidophagous Coccinellids and Their Role as Bioindicators in Agro–forest Ecosystem. *The Bioscan*, 1: 115–122.

Iperti, G. and Paoletti, M.G., 1999. Biodiversity of Predaceous Coccinellidae in relation to bioindication and economic importance. Special issue: Invertebrate biodiversity as bioindicators of Sustainable landscapes. *Agriculture–Ecosystems–and– Environment*, 74: 323–42.

Krebs, C.J., 2001. *Ecology: The experimental analysis of distribution and abundance.* Harper and Row Pub. N.Y. 800pp.

Kohno, K., 2003. Development and reproduction of *Antilochus coqueberti* (Heteroptera: Pyrrhocoridae), the specific predator of *Dysdercus* spp. (Heteroptera: Pyrrhocoridae). *Applied Entomology and Zoology*, 38: 103–108.

Kshirsagar, R. and Upadhyay, S., 2009. Free radical scavenging activity screening of medicinal plants from Tripura, Northeast India. *Nat Product Radiance*, 8: 117–122.

Lambshead, P.J.D., Paterson, G.L.J. and Gage, J.D., 1997. Biodiversity Professional, version 2.0. *The Natural History Museum and The Scottish Association for Marine Science*

Leonard, G.H., Levine, J.M., Schmidt, P.R. and Bertness, M.D., 1998. Flow driven variation in inter–tidal community structure in a Maine estuary. *Ecology*, 79: 1395–1411.

Murdoch, W.W., 1975. Diversity, Complexity, Stability and Pest Control. *Journal of Applied Ecology*, 12(3): 795–807

Magurran, A.E., 1988. *Ecological Diversity and its Measurement*. Chapman and Hall, London.

Myers, N., Mittelmeier, R.A., Mittelmeier, C.G., da Fonseca, G.A.B. and Kent, J., 2000. Biodiversity hotspots for conservation priorities. *Nature*, 493: 853–858.

McNeely, J.A., 2004. Nature vs. nurture: managing relationships between forests, agroforestry and wild biodiversity. *Agroforestry Systems*, 61: 155–165.

Origin., 1991. Micrococal Software, Inc, Northampton, USA.

Omkar and Pervez, A., 2000. Biodiversity of predacious coccinellids (Coleoptera: Coccinellidae) in India: A review. *Journal of Aphidology*, 14: 41–66.

Omkar and Bind, R.B., 2004. Prey quality dependent growth, development and reproduction of a biocontrol agent, Cheilomenes sexmaculata (Fabricius) (Coleoptera: Coccinellidae). Biocontrol Science and Technology 14: 665–673.

Omkar and James, B.E., 2004. Influence of prey species on immature survival, development, predation and reproduction of Coccinella transversalis Fabricius (Coleoptera: Coccinellidae). *Journal of Applied Entomology*, 128: 150–157.

Putz, F., Blate, *E*.G.M., Redford, K.H., Fimbel, R. and Robinson, J., 2001. Tropical forest management and conservation of biodiversity: an overview. *Conservation Biology*, 15: 7–20.

Poorani, J., 2002. An annotated checklist of the Coccinellidae (Coleoptera) (excluding Epilachninae) of the Indian subregion. *Oriental Insects*, 36: 307–383.

Shantibala, K. and Singh, T.K., 1991. Studies on the vertical distribution of aphidophagous coccinellids (Coleoptera: Coccinellidae) of Manipur and Nagaland. *Journal of Aphidology*, 5: 39–43.

Environment, Biodiversity and Traditional System, 2014 *Pages* 403–409
ISBN 978-81-7622-307-2
Editors: D.R. Khanna, G.S. Solanki and S.K. Pathak
Published by: BIOTECH BOOKS, NEW DELHI

Chapter 38

Concept of Universe in Ancient India

Rahul Verma

Department of Geology,
Pachhunga University College, Aizawl, Mizoram
E-mail: vrahul24@gmail.com

ABSTRACT

The ancient Hindu Rishis' were 'SEER' (the discoverer), Philosophers, Mathematicians and Scientists. They made fundamental discoveries which are very well documented in the great Vedic literature. Most remarkably, all their findings were found to be at par with the results calculated with modern equipments and space technological aids.

Bhuvakosh of the *Padam Puran* gives a clear cut classification of different branches of Earth Science, like- *Bhugol* (Geography), *Khagol Shastra* (Astronony) and *Jyotishya Shastra* (Astrology).

The concept of Universe has well been documented in *Ved, Upanishad, Puran* and *Mahabharat*. Rig Ved mentions the Architectural Origin of Universe (*Rig Ved X 130.1.2; Rig Ved X.72.2*) which mentions the building of universe from the cosmic dust through the architecture of Brahma. Mechanical Origin of universe from the parts of the body of *"Adi Purush"* is depicted in the *Bhagvat Gita* (*Bhag.II.1.26.40*). In the Upanishads, "Philosophical origin" of universe quotes that the earth has originated from water that was produced after the worshipping of death (*Brihadaranyakopanishad.1.2.1-3*). Shiva Puran (*Vidyesvara Samb*.10.3) mentions the "Instrumental Origin of Universe" through the evolution of *"Sarga"* (Creation), *"Pratistha"* (Evolution) and *"Sanhara or Pralaya"* (Deluge).

Matsya Puran (127.73) explains the gaseous origin of universe. Vayu Puran (*Poorvardh 49.151*), *Vishnu Puran* (II.7.22) and *Mahabharat* (Parv.I.58-59) have mentioned the origin of universe through the *"Anda"* (eggs) and also about the elliptical shape of the universe. Vaman Dev discovered the planet Jupiter (*Vrihaspati*) (Rig.IV.50.4) and Vena Bhargava discovered Venus (*Shukra)* in the Vedic period (Rig.123.1).

Aryabhatta's *"Siddhant Shiromani"* (Verse.28) gives an account of Solar System and the orbital circumferences of the planets. The distance from the Earth and Moon has been calculated in *"Surya Siddhanta"* (*Madhyamadhikara*, Verse.25, 26 &27). *"Sathpath Brahman"* (1.3.3.18-19), *Vishnu Puran* (II.12.3) and *Matsya Puran* (127.38.: 149.2) through light upon the origin of moon.

Rig Ved (X.85.2) and *Gita* (X.II) give an account of 27 constellations. *Brahma Puran (II.129.1)* have mentioned about the meteors. *Agni Puran* (129.1) also, gives vivid descriptions of meteorites (N*irghats*).Rig Ved (1.27.11), *"Manu Smriti"* (I.38) give an account of *"Ketus"* and *"Dhumketus"* (Comets). Knowledge of "Pole Star" is reflected in the texts like *"Samhita"* (*Madhya Samh*.XIV.1) and *Bhagwat Gita* (X.II).

Functional mechanism of solar and lunar eclipses is well documented in *"Agni Puran"* (129.1), Rig Ved (IV.16.14; IV.28.2). The Vedic texts also certify the knowledge of the "sun spots (Tamas *and Kilak Ketu*), 23 in number, which may affect the life and destiny of human beings on the earth.

"Dirghatma" had successfully calculated the solar year consisting of 365 ¼ days, that is exactly equal to the modern calculations.

It is evident that the sages of the Vedic period were very well ahead of their times in the field of scientific discoveries. *Rig Ved, Purans, Bhuvankosh, Lokaprajnapti, Samudrakhyayika, Suryaprajnapti, Siddhant Shiromani, Surya Siddhant* etc. provide ample knowledge of all branches of Earth Science such as Climatology, Geology, Geography, Oceanography, Cosmology, Astrology and Astronomy.

The man today, is exploring the Mars and Venus and has now been capable of scaling the solar system and beyond through the satellites and space ships. Surprisingly enough, the mysteries of universe being unfolded in the 21st century, nothing but confirmation of the findings made in Vedic period.

Introduction

The amazing universe has always attracted human since the ancient civilization. All the ancient civilizations, Chinese, Greeks, Egyptians, Persian and Indian have very thoughtfully contributed to solve the mysteries of universe.

The Indian school during the Vedic period, views the universe in a big religious capsule, rather than of Science. The great contribution of our philosophical thinkers of Vedic period is well preserved in the Vedic texts. All Veds and Upanishads have developed their thought under the umbrella of religion and philosophy; however the Mahabharat and Purans view these problems with more scientific approach.

Remarkably, the sages of the Vedic period have made amazing scientific discoveries. Rich texts of *Rig Ved, Purans, Bhuvankosh, Lokaprajnapti, Samudrakhyayika, Suryaprajnapti, Siddhant Shiromani, Surya Siddhant* etc. provide ample knowledge of all branches of Earth Science such as Climatology, Geology, Geography, Oceanography, Cosmology, Astrology and Astronomy. Many of their findings have been confirmed to be of excellent accuracy, at par with those made with modern techniques.

Surprisingly enough, the mysteries of universe being unfolded in the 21st century, nothing but confirmation of the findings made in Vedic period.

Knowledge of Earth and Cosmic Science in Vedic Period

Bhuvakosh of the *Padam Puran* presents a clear cut classification of different branches of Earth Science, like- *Bhugol* (Geography), *Khagol Shastra* (Astronony) and *Jyotishya Shastra* (Astrology). The concept of Universe has well been documented in *Ved, Upanishad, Puran* and *Mahabharat*. Aryabhatta's *"Siddhant Shiromani"* (Verse.28) gives an account of Solar System and the orbital circumferences of the planets.

Agni Puran (129.1) also, gives vivid descriptions of meteorites (*Nirghats*) and Functional mechanism of solar and lunar eclipses. Rig Ved (1.27.11), *"Manu Smriti"* (I.38) give an account of *"Ketus"* and *"Dhumketus"* (Comets). Knowledge of "Pole Star" is reflected in the texts like *"Samhita"* (*Madhya Samh.*XIV.1) and *Bhagwat Gita* (X.II). The Vedic texts also certify the knowledge of the "sun spots (Tamas *and Kilak Ketu*), 23 in number, which may affect the life and destiny of human beings on the earth.

"Dirghatma" had successfully calculated the solar year consisting of 365 ¼ days, that is exactly equal to the modern calculations. Vaman Dev discovered the planet Jupiter (*Vrihaspati*) (Rig.IV.50.4) and Vena Bhargava discovered Venus (*Shukra)* in the Vedic period (Rig.123.1).

Concept of Origin of Universe

The concept of Universe has well been documented in *Ved, Upanishad, Puran* and *Mahabharat*. There were several concepts of the origin of universe namely- (a) Architectural Origin (b) Mechanical Origin (c) "Philosophical Origin" (d) "Instrumental Origin of Universe (e) Gaseous Origin (f) Origin from Egg "Anda".

Vayu Puran (*Poorvardh 49.151*), *Vishnu Puran* (II.7.22) and *Mahabharat* (Parv.I.58-59) have mentioned the origin of universe through the *"Anda"* (eggs) and also about the elliptical shape of the universe.

Architectural Origin

Rig Ved mentions the Architectural Origin of Universe. It explains the building of universe from the cosmic dust through the architecture of Brahma. Gods have woven the universe and deities have given shapes to the universe like skilled blacksmith. (*Rig Ved X 130.1.2; Rig Ved X.72.2*).Cosmic dust is the building material used by the chief architect "Vishwakarma". 'Varuna' and 'Indra' were surveyors. 'Savita' (Sun) holds the Earth through "Yantra" and to check the sky.

Mechanical Origin

"Rig Ved" depicts the "Mechanical Origin" of universe from the parts of the body of *"Adi Purush (Rig.X.90.13, 14*). The moon from mind, Sun from eyes, Indra and Agni from mouth, Wind from breath, Earth from feet, Atmosphere from his navel. Heaven from his head and Quarters have originated from his ears. The same is explained by the following shlok:

Philosophical Origin

The "Philosophical origin" of universe, as quoted in Vishnu Puran (Vishnu I.2.23), mentions that in the beginning, there was neither beginning nor end, no day or night. In the beginning only self born ("Pradhanikam Brahm") existed.

Upanishads deviate to refer that Upanishad, quotes that in the beginning, everything was concealed by death and the Earth has originated from water that was produced after the worshipping of death (*Brihadaranyakopanishad.1.2.1-3*).

Instrumental Origin

Shiva Puran (*Vidyesvara Samb*.10.3) mentions the "Instrumental Origin of Universe" through the evolution of *"Sarga"* (Creation), *"Pratistha"* (Evolution) and *"Sanhara or Pralaya"* (Deluge).

Gaseous Origin

The origin of Universe from gas ("Apah") as proposed in "Matsya Puran" (127.73) and Vayu Puran (Vayu. 41.48). It approves of the origin of universe from the Sun. This particular "Aryan Concept" of Cosmogony explains the size, shape and nature of universe.

Origin from Egg "Anda"

Vayu Puran (*Poorvardh 49.151*), *Vishnu Puran* (II.7.22) and *Mahabharat* (Parv.I.58-59), express the view that the Universe has originated from Egg (Anda). The universe as called "Brahmand" is self explanatory of its elliptical shape. All the galaxies, planet and celestial bodies have elliptical shape (*Andakar*). Sun is considered as one of the million eggs of the universe.

Concept of Celestial Bodies

Solar System

Aryabhatta's *"Siddhant Shiromani"* (Verse.28) gives an account of Solar System and the orbital circumferences of the planets. The distance from the Earth and Moon has been calculated in *"Surya Siddhanta"* (*Madhyamadhikara*, Verse.25, 26 &27). *"Sathpath Brahman"* (1.3.3.18-19), *Vishnu Puran* (II.12.3) and *Matsya Puran* (127.38.: 149.2) through light upon the origin of moon. *"Dirghatma"* had successfully calculated the solar year consisting of 365 ¼ days, that is exactly equal to the modern calculations.

Sun Spots

The Vedic texts also certify the knowledge of the "sun spots (Tamas *and Kilak Ketu*), 23 in number, which may affect the life and destiny of human beings on the earth (Vrihadsanhita" III.7).

Constellations and Stars

The astronomers in the Vedic period were called *"Nakshatradarsas"*. *Rig Ved* (X.85.2) and *Gita* (X.II) give an account of 27 constellations. *Brahma Puran (II.129.1)* have mentioned about the meteors. *"Atharva Ved"* and *"Maitryayani Samhita"* (Atharv.XIX.7), there is mention of 28 nakshatras. The nomenclature of the "nakshatras (constellations') by the Vedic Astronomers, bears legendry significance. For example, "Ardra Nakshatra" refers to the period when the sun sets in the rainier region, and "Kritika nakshatra" signifies that star under this regime of 13°20′ look like the skin of spotted deer (*Mrig Chala*).

Meteors and Comets

Agni Puran (129.1) also, gives vivid descriptions of meteorites (N*irghats*).Rig Ved (1.27.11), "*Manu Smriti*" (I.38) give an account of "*Ketus*" and "*Dhumketus*" (Comets). Knowledge of "Pole Star" is reflected in the texts like "*Samhita*" (*Madhya Samh*.XIV.1) and *Bhagwat Gita* (X.II).

Earth, Sun and Moon Relationship

Functional mechanism of solar and lunar eclipses is well documented in "*Agni Puran*" (129.1), Rig Ved (IV.16.14; IV.28.2). It is also expressed in the Mahabharat (Shanti Parva.201.61) that the moon shines due to reflection of sun light. One half of the moon is always visible to the Earth and the other half always is not visible to the earth.

Spherical Shape of Earth

Rig Ved (I.33.8) explains the spherical shape of the earth as quoting- "*Beautified with golden ornaments, the dutas (vritra), though revolving round the circumference and specially running (about it), have not been capable of being victorious over Indra. Then the Sun has encircled (the dutas)*".

Day and Night on the Earth

In the Rig Ved (4.433), explanation of days and night on various parts of earth given in the following quote-"The Sun spreads is lustrous arms round the earth, helping the world to sleep and awake. Resplendent Savita has filled glow, the region of space, sky and the earth".

Concept of Atmosphere

Bhaskaracharya, in Sidhhanta Shiromani (Goladhya, Madhyagitivasan II), has given a detailed account of structure of atmosphere. He has referred the atmosphere to be of 12 Yojans (96 miles or 153 kilometres). It has been supported by Aryabhatta and Lalla. Remarkably, these calculations made in Vedic period, match with the modern calculations of 100 miles.

References of Universe in Buddhism and Jainism

Buddhist Literature

The Buddhism philosophy advocates refraining from the study of "Astronomy" and devoting on the aspects of human sufferings and purification of life (Tiracchanavijja, Chullavagga).

Jain Literatures

The Jain philosophy believes in the concept of "flat Earth" and two suns shining in the both sides. All the heavenly bodies move parallel to the Earth with centre of their orbits as "*Meru*".

However, the concept of flat earth was not able to justify the alteration of day and night. To overcome this problem, Jain school devised an amazing concept of similar but two opposite sets of suns, two moons and two sets of stars...to solve the day and night problem (Das and Rajan, 1932).

Conclusion

It is self explanatory, that the sages of the Vedic period were very well ahead of their times in the field of scientific discoveries. *Rig Ved, Purans, Bhuvakosh, Lokaprajnapti, Samudrakhyayika, Suryaprajnapti, Siddhant Shiromani, Surya Siddhant* etc. provide ample evidences of knowledge of all branches of Earth Science such as Climatology, Geology, Geography, Oceanography, Cosmology, Astrology and Astronomy.

The man today, is exploring the Mars and Venus and has now been capable of scaling the solar system and beyond through the satellites and space ships. Surprisingly enough, the mysteries of universe being unfolded in the 21st century, are nothing but confirmation of the findings made in Vedic period. We can conclude that the more man explores the universe, the closure they come to the "Veds".

References

Agni Puran (129.1).

Atharv Ved (XIX.7).

Bhagwat Gita *(1.26.40).*

Bhagwat Gita (X.II).

Brahma Puran (II.129.1).

Brihadaranyakopanishad (1.2.1–3).

Das and Rajan., 1932. The Jain School of Astronomy, The Indian Historical Quarterly, Calcutta, Vol.VIII, pp 31.

Mahabharat (Parv.I.58–59).

Mahabharat (Shanti Parva.201.61).

"Manu Smriti" (I.38).

Matsya Puran (127.73).

Matsya Puran (127.38.: 149.2).

Rig Ved (I.33.8).

Rig Ved (IV.16.14; IV.28.2).

Rig Ved (IV.433).

Rig Ved (X.72.2).

Rig Ved (X 149.1.I).

Rig Ved (X.90.13–14).

Rig Ved (I.27.11).

Rig Ved (123.1).

Rig Ved (IV.50.4).

Rig Ved (IV.16.14; IV.28.2).

Rig Ved (X.72.2).

Rig Ved (X.85.2).

Rig Ved (X 130.1.2).

Sathpath Brahman" (1.3.3.18–19).

Samhita" (*Madhya Samh.,* XIV.1).

"Sathpath Brahman" (1.3.3.18–19).

*Shiva Puran (Vidyesvara Samb.*10.3).

"Siddhant Shiromani" (Verse.28).

Sidhhanta Shiromani (*Goladhya, Madhyagitivasan II*).

"Surya Siddhanta" (*Madhyamadhikara,* Verse.25, 26 &27).

Surya Siddhanta" (*Madhyamadhikara,* Verse.25, 26 &27).

Tiracchanavijja, Chullavagga.V.33.3 in Vinaya Text, Sacred Book of the East Vol.XX., p 151.

Vayu Puran (*Poorvardh 49.151*).

Vayu Puran (41.48).

Vishnu Puran (I.2.23).

Vishnu Puran (II.7.22).

Vishnu Puran (II.12.3).

Environment, Biodiversity and Traditional System, 2014 *Pages* 411–418
ISBN 978-81-7622-307-2
Editors: D.R. Khanna, G.S. Solanki and S.K. Pathak
Published by: BIOTECH BOOKS, NEW DELHI

Chapter 39

Epidemiological Survey of Respiratory Dysfunction Caused by Air Pollution and Relief by Nasal Application of Haridra Taila

Surendra Kumar Sharma, Piyush Mehata and Birendra Kumar

Department of Pathology,
National Institute of Ayurveda, Jaipur, Rajasthan

Introduction

Ayurveda has described the natural quality of Air according to Seasons. But in Charak Samhita, a detailed description of unhealthy Vayu (air) is also described. Most off all Pollutant described here are of natural origin.

But, at present era, due to modern civilization, urbanization and industrialization anthropogenic (man made) sources are becoming most responsible for the Air Pollution. Transport, Power Plants, Oil Refining, Burning Crop Waste, Fumes from Paints, Varnish, Aerosol Sprays, nuclear weapons, Toxic Gases etc. are main anthropogenic sources responsible for Air Pollution.

In Ayrveda, so many diseases are described which are produced due to polluted Air (unhealthy air); some diseases of respiratory system are Hikka, Shvasa, Kasa, Pratishyaya. In general one similarity is occurred in these all diseases that these all of are related with Pranavaha Srotasa (respiratory system).

In present research work, extent of Pranvaha Sroto-Dusti was tried to find out with special reference to Vayu Dusti (Air Pollution). Four diseases (Hikka, Shvasa,

Kasa and Pratishyaya) were consider as a symptoms of Pranavaha Srotodushti. After this screening, 60 patients of Pranavaha Srotodusti were registered for Upashayatmak (clinical) study.

Aims and Objectives

The present work was divided into two part first part was Epidemiological Survey and second part was clinical study. The aims of first part were i) To find out the prevalence of Pranvaha Srotodusti in general population of Jaipur. ii) To find out aetiological properties of Vayu Dushti (air contamination) w.s.r. to Respiratory Dysfunction. The aims and objectives of second part were i) To evaluate the efficacy of Haridra taila as Pratimarsh nasya in cases of Respiratory Dysfunction and ii) To evaluate the side effects of medicine, if any occurred.

Materials and Methods

For epidemiological survey a total of 500 subjects were screened from 5 different places of Jaipur, Rajsthan, (100 from each area of Malviyanagar Industrial Area, Chandpola Area, Ajmerigate Area, JhalanaDoongri Area and Govindnagar Area) on subjective parameters of Pranavaha Srotodusti. They were also screened by peak flow meter for the knowledge of the function of their Pranavaha Srotas. The subjects of one place were consider as one group and by this five groups were made and those five groups were specially analyzed for the epidemiological study. For this, The Adults who were aged more than 18 years and less than 70 years and who were used to Live or do their job at a particular places for more then 5 years were surveyed. New borns, children and old persons, Severely diseased person or even with fever or Who were not living in that atmosphere less than 5 years were not included in the survey.

For upasayatmaka study 60 patients, who were clinically found to suffering from Pranvaha Srotodusti were screen and registered. The Patient with severe grade of Respiratory problems or with fever or Who were suffered with other system also as- cardiac, neurological etc. and Pregnant women were also not taking in Upasayatmak study.

Study Drug

Haridra Tail (a Kalpita Yoga) was used as Partimarsh Nasya. The constituents of this Tail were Haridra (*Curcuma longa*) and Til Taila (*Sesamum indicum*). The doses were advised as 2 drops in morning and 2 drops in evening for one month.

Assessments Criteria

Epidemiological study was assessing on the basis of subjective parameters - Lakshan of Pranvaha Srotodusti (Shvasa, Kas, Hikka and Pratishyaya); and also on the basis of PEFR recorded by Peak Flow Meter as Objective parameters. For assessments of Upasayatmak (clinical) study differents morbidity features of Shvasa like Atisrista, Atibadha, Abhikshanam and Sulam, Kas, Pratishyaya were documented on four grading of severity. Hb per cent, TLC, DLC, TEC, FVC, FEV_1, PEFR, and FEV1/ FVC was used as Laboratoric parameters for the assessments.

Observation, Results and Discussion

Epidemiological

This epidemiological survey was conducted from 1 September 2008 to 17 February 2009 in five areas of Jaipur, Rajsthan, which were Malviyanagar Industrial area, Chandpol area, Ajmerigate area, Jhalan Doongari area and Govindnagar area. In this survey total 500 individuals (100 from each area) were observed to know the prevalence of Pranvaha Srotodusti in Jaipur. After the study, total 130 patients of Pranvaha Srotodusti were found (out of 500 individuals). The detailed of which are given below -

The position of Air Pollution in differents area of Jaipur (according to official website of CPCB -Central Pollution Control Board of India; http://cpcbedb.nic.in/) are found as, out of five areas Chandpol residential area was most polluted area with mean of SPM (Suspended Particulate Matter) =404 and for RSPM (Respirable SPM) it was 144, followed by Ajmerigate residential area SPM of which was 276 and RSPM was 112µg for 24 hours. In comparison to these areas the air quality of Malviyanagar Industrial area are quite pollution free as in this area only SPM was slightly increased (208) then its normal limits (200).

It may be explained on the basis this that, Chandpol and Ajmerigate area are one of the oldest and most condensed places with over crowded also. These are the center areas of Jaipur, where traffic load is very high, but the roads are not sufficiently wide to allow all the traffic to move fast and smoothly, results traffic jam, excessive emission of smoke (Dhoom) from these vehicles make atmosphere more polluted. The second causes are looking as, because these areas are most dense and overcrowded area, where many hawkers, and foot-path shopkeepers are sitting in zigzags positions. They also throw a large amount of garbage on the roads. But this garbage is not easily disposed due to excessive scattering of them here and there. Due to these, excessive Raj (dusts) is coming in atmosphere which made these areas most polluted area of Jaipur.

Contrary to this, Malviyanagar Industrial area (MIA) and Jhalana Doongari areas are open areas, so the problem of traffic jam does not occurre generally. So the amounts of exhaust smoke from the vehicles are not accumulated in this are. Even the industries established in this area are not such kinds of industries who emit more smoke and dust as these are pharma companies, pottery, jewelry, mobile, machinery etc. Even the stone processing industries are well organized and so, they also do not emit much dust (Raj) in atmosphere to pollute it. That is why Chandpola and Ajmerygate residential areas are more polluted areas in comperision to Malviyanagar Industrial area.

Jhalana Doongari Area is absolutely pollution free and none parameter was more than its normal limits. The status of Air pollutant for Govindnagar was not published.

Maximum per cent of pts were found from Chandpola area 36.15 per cent fallowed by Ajmerigate area 20 per cent and MIA 16.15 per cent. The probable causes of this may be that, as explain earlier that the air quality of Chandpola area are

Table 39.1: Showing brief presentation of Pranavaha Srotodusti (respiratory dysfunction) according to areas.

Sl.No	*Areas*	*Acute Cases*	*Per cent of Acute Cases*	*Chronic Cases*	*Per cent of Chronic Cases*	*Total*	*Per cent of Total*
1.	MIA	11	16.42 per cent	10	15.87 per cent	21	16.15 per cent
2.	Ajmerigate	12	17.91 per cent	14	22.22 per cent	26	20 per cent
3.	Chandpol	23	34.33 per cent	24	38.10 per cent	47	36.15 per cent
4.	Jhalana Doongari	9	13.43 per cent	7	11.11 per cent	16	12.31 per cent
5.	Govindnagar	12	17.91 per cent	8	12.70 per cent	20	15.38 per cent
	Total	**67**	**100 per cent**	**63**	**100 per cent**	**130**	**100 per cent**

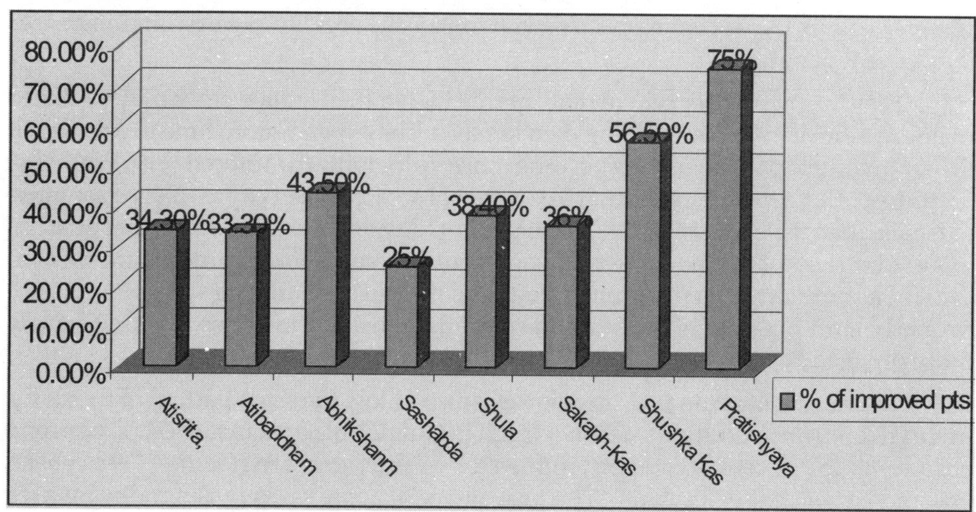

Figure 39.1: Showing per cent of improvement in sign and symptoms of patients.

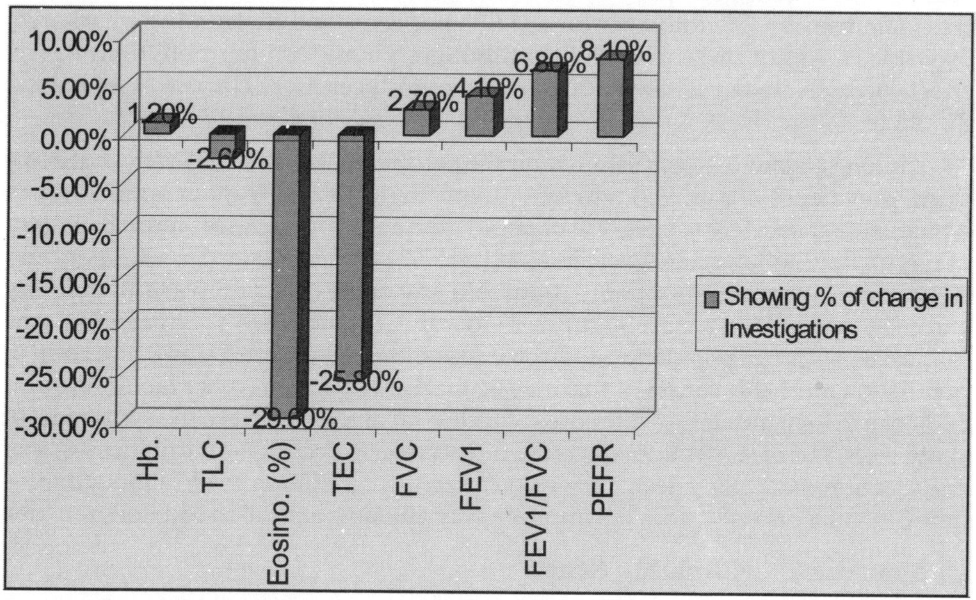

Figure 39.2: Showing per cent of change in investigations.

poorest among all followed by Ajmerigate and then MIA area *i.e.* the prevalence of Pranvaha Srotodusti were occurred according to status of air pollution of that area.

One more factor is also noted that the per cent of smoker are also very high in more polluted areas in comparison to less or no-polluted areas, as per cent of smoker were founds in Chandpole area 22 per cent and in Ajmerigate area 18 per cent. While the per cent of smoker in MIA, Govindnagar and Jhalan Doongri was as 14 per cent,

15 per cent and 13 per cent respectively. It makes the condition worse and increases the per cent of patients.

According to sex, per cent of male pts are higher then female pts, it may be due to male are more in habit of smoking then females. They spend more time in open air; if the air quality is poor then they become more vulnerable to suffered with Pranvaha Srotodusti. But as table, shows that in Chandpol area, per cent of prevalence are more in female than male it may be due to the poor quality of houses located in that area. As in these houses, there is no proper management for ventilation, the rooms are smaller, the smoke emitted from fuel during cooking, the dust fled during sweeping, do not properly emit outside of house and accumulate inside the house, making female more prone to be ill.

The per cent of acute and chronic cases also follow the same pattern of the status of Vayu dusti (air pollution), as it was high in Chandpole area (acute-34.33 per cent, chronic-36.15 per cent of total pts) followed by Ajmerigate area (acute-17.9 per cent, chronic-20 per cent of total pts). The prevalence of acute cases in Govindnagar area was slightly higher (17.91 per cent) then MIA (16.42 per cent) and Jhalana Doongari area (13.43 per cent). The prevalence of chronic cases were more in MIA (16.15 per cent) fallowed by Govindnagar area (15.38 per cent) and Jhalana Doongari area, probable causes of these may be that although, Vayu-dusti (air pollution) has its effects on both acute and chronic cases but it produces more chronic cases in long duration.

Among all four diseases selected for the perceiving the status of Pranvaha Srotodusti, prevalence of Kasa (cough) was quietly high (13.2 per cent in population, in which acute cases were 6.4 per cent of population and chronic cases were 6.8 per cent of population) followed by Pratishyaya (total 7.2 per cent, acute cases 6.2 per cent, chronic cases 1 per cent of population). No any acute case of Shvasa Rog or any chronic cases of Hikka (hiccough) were found. Chronic cases of Shvasa Rog was found 4.8 per cent in population and acute cases, for Hikka were found 0.8 per cent in population, probable causes of this may be that perhaps, some other factors are also responsible for producing acute cases withdrawal of which relieve the symptoms of acute cases, but the air pollution has the direct effect in producing chronic cases, so it does not automatically relieve, till we don't decreases pollution level; (although some times, in some disease it does not improve after attaining a position of deteriorations).

Upasayatmak (Clinical) Study

In present clinical trial Haridra tail (A Kalpit Yog) was found highly significant in relieving Shushka Kas (dry cough) $p<0.01$, followed by Pratishyaya $p<0.05$, Sakaph Kas (productive cough) $p<0.05$, Abhikshanam (increases Respiratory rate) $p<0.05$ and in Atisristam (prolong expiration) $p<0.05$, which are statistically significant. Although, the per cent of improvement was not very high.

The improvement In Atibaddha (obstructed inspiration), Sashabda (additional sounds) and Shula (pain during respiration) was not significant ($p>0.05$).

The improvement in symptoms may be found because most of these symptoms were arises by vitiated Vat and Kaph Dosa, whereas used medicine Haridra Tail (which constituents are Haridra and Til Tail) is Vat Shamak due to Usna, Snigdha and guru gun of Tail and is also Kaph Shamak due to Ruksh, Katu gun and Usna Virya of Haridra and Til Tail.

In laboratory findings decrease in Eosinophils per cent and TEC counts was found highly significant ($p<0.001$), whereas improvement in FEV_1, FEV_1/FVC ratio and PEFR was also found ($p<0.01$), it may be due to properties of Haridra as according to Database Haridra also has Anti-allergic, Anti-pruritic properties and also effective in reliving cough, asthama, bronchitis and hiccough.

References

AMARKOSHA OF AMAR SHIMTHA WITH THE COMMENTORY OF BHANU JI DIKSHITA REVISED BY Pt Vasudeva laxmana panasikara repints 2nd editions 1995.

Bhargava's Standered illustrated Dictionary compiled and edited by Pof. R.C. Pathak, Varanasi.

CHARAKA SAMHITA with "Vidyotini" Hindi commentary by Pt. Kashinath Shastry and Dr. Gorakhnath Chaturvedi, Part 1 and 2, Chaukhambha Bharati Academy, 1996

DATABASE on Medicinal plants used in Ayurveda. CCRAS, Deptt. of ISM and H, Ministry of Health and Family Welfare, Govt. of India, 2002.

DAVIDSON'S Principles and Practice of Medicine, published by Churchil living stone, Robert Stevenson House, 1-3 Baxter's place, leith walk, Edinburgh EH 13 AF; ELBS Edition of 20th addition 2006.

en.wikipedia.org/wiki/**Smoke**

home.iitk.ac.in/~mukesh/ind-aqi/**index**.html - 1k -(basics for Ind-Aqi)

http://cpcbedb.nic.in/ (site for data)

http://rpcb.nic.in/

http://www.cpcb.nic.in/Air_quality_data.php

HUTCHISON'S Clinical Methods by Michael Shvasah. ELBC 20th Ed (1997

MADHAVA NIDAN, with Hindi commentary by Suderson Shastri, published by Chaukhambha Samskrit Sansthana, Varanasi, Ed-28th(1996).

MAHABHARATA KOSH by Ram Kumar Shaha

MAHAJAN B.K.; Methods of Bio-statistics; Published by Jaypee Brothers, Medical Publishers (P) Ltd. New Delhi (India) 6th Edition, 1997

P.J. MEHTA'S Practical Medicine, Published by P.J.Mehta, 13th Ed(1998).

PARK K. Preventive and Social Medicine, 15th Ed M/s Banarasidas Bhanot Publishers (1998).

RAJ NIGHANTU- Ed- by Abtey V.G. Ganthakar-33

Sasakrita-Hindi- English Dictionary by Uma Prakash Panday. Kamal Prakashana, N. Delhi.

SHABDA KALPADRUMA, by Raja Radha Kanta Dev, Published by Chaukhamba Sanskrit Sansthan, Varanasi (1961).

SUSHRUTA SAMHITA with "Ayurveda Tattva Sandipika" commentary by Kaviraj Ambikadutta Shastri part 1 and 2, Chaukhambha Sanskrit Sansthan, Varanasi, 1995.

SUSHRUTA SAMHITA, Sharirsthana with "Ayurveda Rahasyadeepika" commentary by Dr. Bhaskar Govind Ghanekar, Meherchand Lachamandas Publications, New Delhi, 2002.

SVASTHAVRITTAM by Shiva Kumar Gaud 4th Ed (1995).

TABER'S Cyclopedic Medical Dictionary; F.A. Davis Company, Philadelphia, 16th Edi., 1989.

Environment, Biodiversity and Traditional System, 2014 *Pages* 419–426
ISBN 978-81-7622-307-2
Editors: D.R. Khanna, G.S. Solanki and S.K. Pathak
Published by: BIOTECH BOOKS, NEW DELHI

Chapter 40

Ion Pair Formation of BIS-1-Amidino-O-Methylurea Copper (II) Thiocynide and BIS-1-Amidino-O-Methylurea Copper (II) Nitrate in Aqueous Medium at Different Temperatures

N. Mohondas Singh[1], T. Moaienla Ao[2] and Lalrosanga[1]*

[1]Department of Chemistry, Mizoram University, Aizawl, Mizoram
[2]Department of Chemistry, Nagaland university, Lumami, Nagaland

ABSTRACT

Conductometric studies have been made for the solutions of bis-1-amidino-O-methylurea copper (II) thiocyanide and bis-1-amidino-O-methylurea copper (II) nitrate in aqueous solution at different temperatures ranging from 10 – 30^{0}C in the concentration range 1.0 – 0.517 x 10^{-3}. The conductance data in all the cases have been analyzed by Shedlovsky equation to obtain limiting molar conductance (Λ_0) and association constant (K_A) values for the electrolyte. The limiting equivalent conductance values increased linearly with the increase in temperature and association constant values decreased with rise in temperature. The association constant values for bis-1-amidino-o-methylurea copper (II) thiocyanide were found to be higher than those in bis-1-amidino-o-methylurea copper (II) nitrate. The thermodynamic parameters *viz.*, standard free energy (ΔG^0), enthalpy change(ΔH^0) and

* *Corresponding Author:* E-mail: nmdas08@rediffmail.com

entropy change (ΔS^0) for the ion pair formation have been calculated from the values of ion association constant at different temperatures. The chosen metal complexes showed ion-pair association within the experimental different temperature ranges supported by the negative value of ΔG^0. Both the associations are exothermic in nature which was determined by negative value of ΔH^0.

Keywords: ***Conductance, Shedlovsky equation, Bis-1-amidino-O-methylurea copper (II) thiocyanide, Bis-1-amidino-O-methylurea copper (II) nitrate, Thermodynamic parameters, Metal complexes, Ion-association.***

Introduction

The tendency of ions to associate into ion pairs depends on the balanced between the electrostatic force and thermal energy. Ion pair formation refers to the association of cations and anions in solution. An ion-pair is defined to exist when the cation and an anion are close enough in space with a common solvation shell, the energy associated with their electrostatic attraction being larger than the thermal energy (kT) available to separate them. The ions need also to stay associated longer than the time required for Brownian motion to separate non-interacting species[1]. Key to this definition is also the fact that only strict electrostatic interactions should occur between the ions. No interactions other than the Coulombic attraction should formally exist upon the association of the charges- charges that existed prior to the interaction and remaining unchanged in the interaction. In solvents of low dielectric constant, where the electrostatic potential energy of an anion and a cation at contact is large as compared to mean thermal energy kT, it is expected the association of ions as a non-conducting pairs. In any solution of an electrolyte there is always the possibility that the ions of the electrolyte might not be fully dissociated in solution. Ion pairing results when the electrostatic interaction between two oppositely charged ions become sufficiently large for the two ions to move around as one entity, the ion-pair. The extent of association into ion-pairs depends on many factors, with the most important being the nature, charges and sizes of the ions, the characteristics of the solvent and the temperature[2]. As the dielectric constant is increased the extents of association should (and does) decrease, but it should never become zero because there is always a non-zero chance that anion-cation contacts will occur. Conductance of 1:1 electrolytes has been extensively studied[3] by using Shedlovsky equation.

The double layer theory describes the process of ion-exchange and ion-interaction chromatography from the point view of distributed electrostatic field effect on the charged analyte retention. Other approaches have more stoichiometric characters and describe the analyte retention on the basis of the formation of ionic pairs and their subsequent retention on the adsorbent surface. In 1926, Bjerrum used Debye-Huckel theory to describe ion association and took into account the interaction of ions within a short range. He introduced an ion-pair concept, give a definition of ion-pairs as neutral species formed by electrostatic attraction between oppositely charged ions in solution, and showed how ion-pair formation was dependent on the ion size (radius of ions), solvent (dielectric constant), and temperature.

Here, the conductometric technique was applied to study the association of the selected metal ions with the ligands in aqueous solution at 10-30^{0}C. The conductance data were treated using Shedlovsky extrapolation method[3]. We have also investigated the effect of temperature on limiting molar conductance and association constant for the selected compounds which have been supported by the calculated values of standard thermodynamic parameters. These results were discussed in terms of the association constants and also approach of Gibbs energy relationship[4] can be applied to the association process at various temperatures to discuss the thermodynamic features of metal-ion complexes. Also, the thermodynamic parameters ΔG^0, ΔH^0 and ΔS^0 were evaluated by investigating the process over a certain temperature range.

Experimental

The measurements of electrical conductivities of dilute solution of salts or complexes are considered to be one of the important methods for studying the ion pair or multiple-ion association not only in aqueous solutions but also in non-aqueous or mixed ones[5]. The limiting molar conductance and the association constants of the complex ion with anions were calculated by using Shedlovsky method[6].

The electrical conductivities were measured by Systronic-306 digital conductivity bride (accuracy ± 0.1 per cent) with a dip type immersion conductivity cell was used. The solutions of different concentrations (1.0-0.517x 10^{-3}) were carefully prepared by dissolving requisite amount of the sample in conductivity water (*i.e.*, double distilled water) of specific conductance (< 3x10^{-6}). Conductivity measurements were carried out over the temperature range 10-30^{0}C. All the dielectric constants and viscosities were obtained from literature[7]. The temperature control in the ranges 10 -30^{0}C were made by using water bath and circular –Model D8-G of HAAKE Mess-Technik Gmbtlu. Co.Germany. The measurements of weights were done by using a METTER Balance, Model-AE 260, Delta Range. All calculations were done on IBM-PC-AT/386 using a basic programmed.

The experimental data of conductance measurements were analyzed by using Shedlovsky equation. Shedlovsky equation is given by

$$\frac{1}{\Lambda S(Z)}=\frac{1}{\Lambda_0}+\left(\frac{K_A}{\Lambda_0}\right)\left(C\Lambda f_{\pm}^{2}S(Z)\right)\text{.........................}(1)$$

Where is molar conductance at a concentration C (g.mol.dm^{-3}), Λ_0 is the limiting molar conductance and K_A is the observed association constant. The other symbols are given by Z and λ are the valence and conductance of the ions respectively, excluding their signs; D is the dielectric constant of the medium, η the viscosity (c.p.). The degree of dissociation (τ) is related to S(Z) by the equation:

$$S(Z)=\left[\frac{Z}{2}+\sqrt{1}+\left(\frac{Z}{2}\right)^2\right]^2 \qquad Z=\left[\frac{\alpha\Lambda_o+\beta}{\Lambda_o^{\frac{3}{2}}}\right](C\Lambda)^{\frac{1}{2}}$$

$$\alpha = \frac{17.147X10^5 W}{(DT)^{\frac{3}{2}}}$$

$$W = z_+ z_- \frac{2q}{1+q^{\frac{1}{2}}} \qquad\qquad \beta = \frac{151.47}{\eta (DT)^{\frac{1}{2}}}$$

$$q = \frac{z_+ z_-}{z_+ z_-} X \frac{\lambda_+ + \lambda_-}{z_+ \lambda_- + z_- \lambda_+}$$

z and λ are the valence and conductance of the ion respectively, excluding their signs, D is the dielectric constant of the medium, η the viscosity (c.p). The degree of dissociation (τ) is related to S (Z) by the equation: $\tau = \Delta S\,(Z)/\Delta_0$

$f_\pm$ is the activity coefficient of the free ions and was calculated as

$$-\log f_\pm = \frac{Az_+ z_- \mu^{1/2}}{1+BR\mu^{1/2}} \tag{2}$$

Where

$$A = \frac{1.8247X10^6}{(DT)^{3/2}}$$

$$B = \frac{0.5029X10^{10}}{(DT)^{3/2}}$$

$$\mu = \frac{1}{2}\sum_i (C_i \tau_i) Z_i^2$$

R is the maximum centre to centre distance between the ions in the ion-pair. There exists at present no method of determining the value of R precisely. In order to treat the data in our system the R value is assumed to be R = a+d, where a, the sum of crystallographic radii of the ions, is approximately equal to 5A^0 and d (A^0) is given by:

$$d = 1.183(M/\rho)^{1/3} \tag{3}$$

where M is the molecular weight of the solvent and ñ the density of the solution.

An initial value Λ_0 of was obtained by least square method (Λ) and concentration C were introduced as input in computer programme. The mean activity coefficient f was determined by equation (2) for the above chosen complex salts. From the linear plot of $1/\Lambda S(Z)$ versus $C \Lambda f_{\pm}^{2} S(Z)$; and K_A was evaluated from the intercept 1/ and the slope $/\Lambda_0^2$ respectively. The procedure was repeated using these new values of Λ_0 and K_A. All calculations were carried out by IBM-PC.

The Change of free energy for the association process (ΔG^0) was calculated from the equation:

$$\Delta G = -2.303RT\log_{10}K_A \quad (4)$$

The enthalpy change of association (ΔH^0) was obtained from the slope of log K_A vs 1/T. The change of association entropy (ΔS^0) was calculated from the Gibbs-Helmholtz equation:

$$\Delta S^0 = (\Delta H^0 - \Delta G^0)/T \quad (5)$$

Results and Discussion

From Tables 40.1 and 40.2 the values of Λ_0 for the electrolytes increased invariably with increase in temperature in aqueous medium, indicating less solvation or higher mobility of ions (Figure 40.1)[8]. This is due to the fact that the increase thermal energy results in bond breaking and leads to higher frequency and higher mobility of the ions. Values of Λ_0 for $[Cu(AMUH)_2](NO_3)_2$ are always greater than those values of complex $[Cu(AMUH)_2](SCN)_2$ and is most prone to the variation in vibrational, rotational and translational with temperature. The Values of K_As are formed in increasing order$[Cu(AMUH)_2](SCN)_2 > [Cu(AMUH)_2](NO_3)_2$ This is because the size of the anion for thiocyanide is smaller than the nitrite ion. The values of the K_As for these two salts increased with rise in temperature (Tables 40.1 and 40.2 and Figure 40.2). This increase in K_A values is due to the fact that the increased thermal energy results in greater bond breaking and variation in vibrational, rotational and translational energy of molecules leads to dissociation of molecules[9-13]. The values of thermodynamic functions are given in Tables 40.3 and 40.4. The negative value of ΔH^0 indicates that ion association processes are exothermic at all temperatures. Out of these two complexes, negative value of ΔG^0 is more in bis-1-amidino-O-methylurea copper (II) thiocyanide and this complex is more favored in ion–pair formation. A positive entropy change is broken when association takes place leading to an increase in the degree of disorderliness[14].

Table 40.1: The values of limiting molar conductances Λ_0 (Scm³) and Association constants K_A (dm³mol⁻¹) obtained by Shedlovsky technique for bis-1-amidino-O-methylurea copper (II) thiocyanide in aqueous solution at different temperatures.

	283 K	*200 K*	*203 K*	*208K*	*303K*
K_A	50393	591.49	624.56	718.32	864.83
Λ_0	52.51	57.48	62.45	68.41	73.92
$\log K_A$	2.937	2.856	2.796	2.772	2.702

Conclusion

The limiting molar conductance (Λ_0) and the association constant values increased with the increase in temperature. The negative value of "H°indicates that

Table 40.2: The values of limiting molar conductances Λ_0 (Scm^3) and Association constants K-$_A$(dm^{3}mol^{-1}) obtained by Shedlovsky technique for bis-1-amidino-O-methylurea copper (II) nitrate in aqueous solution at different temperatures.

	283 K	*288 K*	*293 K*	*298K*	*303K*
K_A	395.03	495.13	573.96	622.61	717.72
Λ_0	58.52	63.45	69.61	73.37	78.60
$logK_A$	2.856	2.794	2.758	2.695	2.597

Table 40.3: Thermodynamic parameters ΔG° (kJ mol^{-1}), ΔH° (kJ mol^{-1}) and ΔS° (kJK^{-1}mol^{-1}) obtained by Shedlovsky technique for bis-1-amidino-O-methylurea copper (II) thiocyanide in aqueous solution at different temperatures.

	283 K	*288 K*	*293 K*	*298K*	*303K*
ΔG°	-15.91	-15.74	-15.68	-15.82	-15.68
ΔH°	-0.95				
10^3 x ΔS°	0.053	0.051	0.050	0.049	0.048

Table 4: Thermodynamic parameters ΔG° (kJ mol^{-1}), ΔH° (kJ mol^{-1}) and ΔS° (kJK^{-1}mol^{-1}) obtained by Shedlovsky technique for bis-1-amidino-O-methylurea copper (II) nitrate in aqueous solution at different temperatures.

	283 K	*288 K*	*293 K*	*298K*	*303K*
ΔG°	-15.47	-15.41	-15.47	-15.37	-15.06
ΔH°	-1.06				
10^3 x ΔS°	0.051	0.050	0.049	0.048	0.046

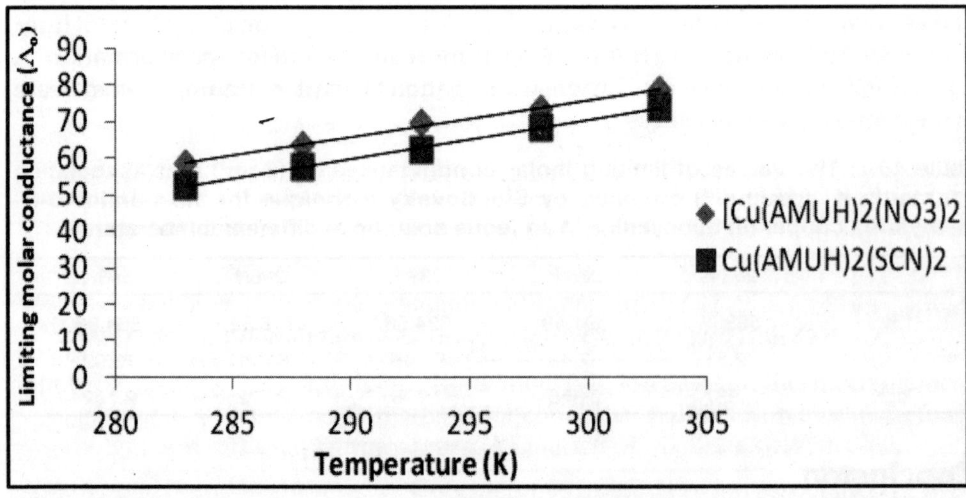

Figure 40.1: Limiting molar conductance for [Cu(AMUH)$_2$(NO$_3$)$_2$] and [Cu(AMUH)$_2$](SCN)$_2$ in aqueous medium.

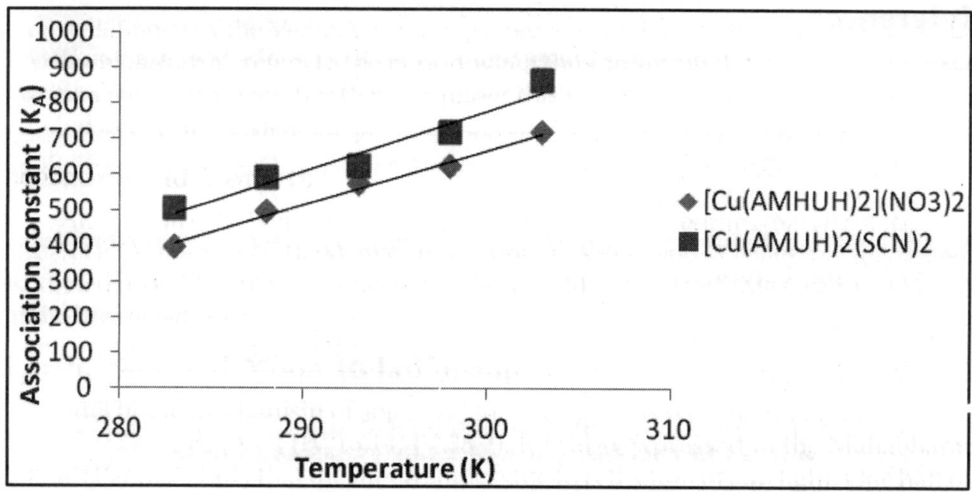

Figure 40.2: Temperature dependence of KA for $[Cu(AMUH)_2(NO_3)_2]$ and $[Cu(AMUH)_2](SCN)_2$.

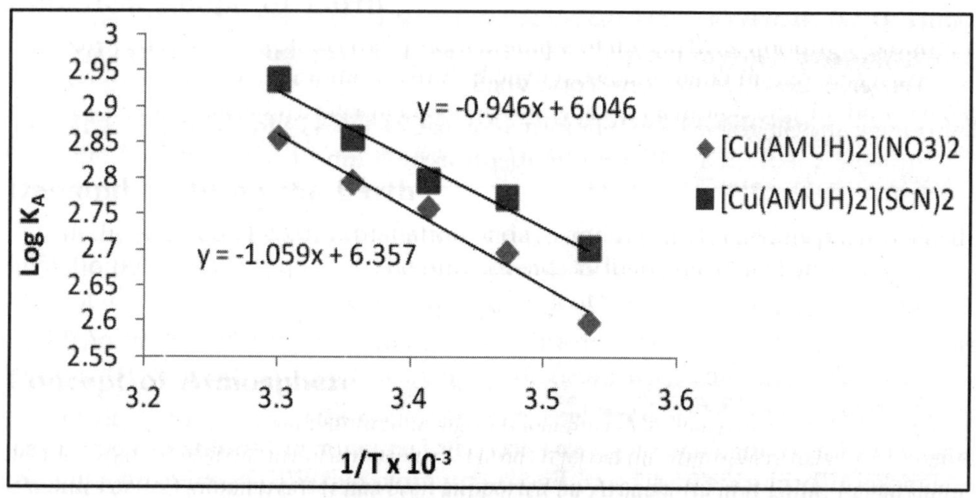

Figure 40.3: Plot of log K_A vs 1/Tx10^{-3} for $[Cu(AMUH)_2(NO_3)_2]$ and $[Cu(AMUH)_2]$ $(SCN)_2$.

both the reaction occur exothermic in nature. The positive value of "S° and negative value of "H° indicates that the ion association processes will occur spontaneously at all temperatures. The chosen metal complexes show ion pair association within experimental obtained by the negative values of ΔG^0 at different temperatures. The K_A values for both complexes are determined and found in order :

$[Cu(AMUH)_2](SCN)_2 > [Cu(AMUH)_2](NO_3)_2$

References

Akhadov,Y. *Dielectric Properties of Binary Solutions*, A Data handbook Pergamon Press, Oxford. 1981; 271p.

Bag, G.C., Mohondas Singh, **N** and Rajmuhon Singh, N. Ionic Association.I. Influence of Ionic Association on the Activity of Ions at a Moderate Degrees of Association. *Indian J Chem.*, 2001; **40A**, 500

Bockris JOM and Reddy AKN. Modern Electrochemistry, Plenum Press, New York, Vol. 1, 2nd Edition, 1998; 251p.

Dash, U.N. and Pasupalak, N.N. Ion Association and Solvation of Octahedral Co(III) Complexes in Water + DMSO mixtures at Different Temperatures, *Indian J. Chem.*, 1997; **36A**, 88.

Debashis Das, Ion Association and Solvation Behavior of Some Lithium Salts in Tetrahydrofuran. A Conductivity and Raman Spectroscopic Studies. *J. Soln. Chem.*, 2008; **37**, 974.

Jenkins IL and Monk CB, The conductivities of some complex cobalt chlorides and sulphates, *J. Am. Chem.* Soc., 1951; **68**: 67-73.

Justice, J.C. *Comprehensive Treatise of Electrochemistry, Thermodynamics and Transport Properties of Aqueous and Molten Electrolytes.*, Conway, B.E., Bockris, J.O'M and Yeager, E. eds.Plenum New York, Vol.**5**

Lalrosanga and Singh NM, Ion pair formation of $CoCl_2.6H_2O$ and $[Co(NH_3)5Cl]Cl_2$ in aqueous medium at different temperatures- A conductance method. Asian J. Chem. 2011; **23(3):** 1120-1122.

Shedlovsky, T and Kay, R.K. The ionization constant of acetic acid in water-methanol mixtures at 25°C from conductance measurements, *J. Phys. Chem.*, 1956; **60**, 151-155.

Singh NM, Lalnunchami and Lalrosanga, Calculation of Ion association constants of $CuCl_2.2H_2O$ and $[Cu(Me\text{-}AMUH)_2]Cl_2$, Indian Journal of Science and Technology, 2013; 6(S3):122-125.

Singh NM, Lalrosanga and Singh KM, Ion pair formation of 1-amidino-o-ethylurea nickel (II) chloride and 1-amidino-o-ethylurea nickel(II) bromide in aqueous medium at different temperatures-A conductance method, *Asian J. Chem*, 2012; **24(5):** 2317-2320.

Singh NM, Singh Th D, Bag G C and Singh NR, Conductometric Studies of the ion association of α-/β-alalnine bis-biguanide cobalt(III) Iodide in methanol + water mixtures at different temperatures, *Asian J. Chem*, 2009; **21(6):** 4215-4223.

Singh NM, Singh Th. D, Yaiphaba N and Singh NR, Ion pair and thermodynamics of glycine bis-1-amidino-O-methylurea cobalt(III) halides in water + methanol solvents at different temperatures, *Asian J. Chem*, 2008; **20(3):**1750-1760.

Yokoyama, H and Nishmura, M. Conductometric study of ion association between tris (oxalate) chromate and alkali metal ions at temperatures from 0 to 500C, *Bull. Chem. Soc. Jpn.*, 1985; **58**, 1094.

Index

B

D

E

N

O

P

Q

R

S

V

W

Y

Z